Stratospheric Ozone and Man

Volume II

Editors

Frank A. Bower, Ph.D.
E. I. du Pont de Nemours & Company
Haskell Laboratory for Toxicology and Industrial Medicine
Newark, Delaware

Richard B. Ward, Ph.D.
Research Associate
E. I. du Pont de Nemours & Company
Petrochemical Department
Wilmington, Delaware

CRC Press, Inc.
Boca Raton, Florida

Library of Congress Cataloging in Publication Data
Main entry under title:

Stratospheric ozone and man.

Includes bibliographies and indexes.
1. Atmospheric ozone. 2. Stratosphere.
3. Man—Influence of climate. 4. Man—Inluence on nature. I. Bower, Frank, A.
II. Ward, Richard B.
QC879.7.S87 363.7′392 80-39562

ISBN 0-8493-5755-1 (v. 2)

This book represents information obtained from authentic and highly regarded sources. Reprinted material is quoted with permission, and sources are indicated. A wide variety of references are listed. Every reasonable effort has been made to give reliable data and information, but the author and the publisher cannot assume responsibility for the validity of all materials or for the consequences of their use.

Direct all inquiries to CRC Press, Inc., 2000 N.W. 24th Street, Boca Raton, Florida 33431.

International Standard Book Number 0-8493-5755-1

Library of Congress Card Number 80-39562
Printed in the United States

PREFACE

Ozone has existed in the upper atmosphere of Earth ever since oxygen gas has been a component of our atmosphere. As our atmosphere developed to its present composition, the ultraviolet absorption properties of ozone helped define the stratospheric temperature profile and the present characteristics of the ozone layer. Qualitatively therefore, the Earth's atmosphere will contain ozone as long as the sun shines and the grass grows.

Quantitatively, we can only speculate on the constancy of the stratospheric ozone concentration profile in view of natural calamities of the ages. Only in about the last dozen years have man's possible perturbing effects on the ozone layer been seriously considered.

Natural perturbations — or more precisely, perturbations unrelated to human technology — are hypothesized to have been caused by periods of high cosmic radiation intensity and energetic electron precipitation, surges in solar radiation and "solar proton events" (one of which is discussed in Chapters 1 and 4), galactically close supernovae, and, closer to home, major volcanic eruptions.

Suggested manmade perturbations include:

- The supersonic transport (water vapor and nitrogen oxides injected into the stratosphere during flight).
- Conventional aircraft (as for supersonic transports but injected into the high troposphere).
- Nuclear weapons (nitrogen oxides formed in the fire ball and convected into the stratosphere).
- Synthetic nitrogenous fertilizers (a source of nitrous oxide which is converted to nitric oxide after transport to the stratosphere).
- The space shuttle (a chlorine-containing oxidizer results in chlorine compounds being injected into the stratosphere).
- Tropospherically stable halocarbons containing chlorine or bromine, (particularly, but not exclusively, the fully halogenated chlorofluoromethanes which photolyse to liberate the halogen atoms after transport to the stratosphere).

The natural perturbations of ozone present some interesting, but rather remote, paleontological questions, but seemingly few concerns during man's development. Most importantly they are, *a priori,* totally beyond our control. Not so the potential manmade perturbations of ozone.

The potential manmade perturbations have stimulated vigorous debate and sophisticated research. To paraphrase Dr. Harvey Brooks, stratospheric ozone grew, in two decades, from an arcane atmospheric science to an environmental concern of the man in the street.

The story of stratospheric ozone and man is the subject matter of this work.

The original impetus for preparing this book came from the publisher who asked whether an account of the current understanding of the factors affecting stratospheric ozone and the consequences of stratospheric change would be useful. Our first reaction was yes, but the timing was wrong. Further reflection led us to the conclusion that in a developing scientific field, there probably is no "right" time, and the decision to move ahead was taken.

The book, as originally conceived, was to be limited to technical considerations, but the scientific course of events has been so interwoven with nonscientific, but nevertheless related, events, we felt it necessary to include an account of this situation. Accordingly, the book is divided into five sections entitled:

- Stratospheric Ozone
- Atmospheric Processes Influencing Stratospheric Ozone
- Does Man Influence Stratospheric Ozone?
- Effects and Research
- Public Policy

In Volume I, the first chapter is devoted to a scientific consideration of ozone measurement techniques and the normal distribution pattern of stratospheric ozone. Chapters 2 through 6 are devoted to the technical basis for mathematical modeling of atmospheric processes, with special consideration of the calculated perturbing effects of added nitrogen oxides, water, or chlorine compounds.

In Volume II, the relationship between the stratospheric issue and human activity is sketched. The historical emergence of the Supersonic Transport (SST) debate is reviewed in Chapter 1, and the debate on chlorocarbons and chlorofluorocarbons is addressed in Chapter 2. The relationship of nitrogen fertilizers to the issue is discussed in Chapter 3.

Biological and climatological effects which might result from changes in stratospheric ozone are discussed in Chapters 4 and 5.

Government and industry research programs related to ozone change are reviewed in Chapters 6 and 7.

Concern over potential ozone depletion is really based on concern over the biological and climatological effects which might result from ozone depletion. Among such effects, the emotive threat of increasing skin cancer incidence has overshadowed other predicted effects. It was this threat which, by attracting much public attention, encouraged a rapid response on the part of federal government agencies. The formation of an *ad hoc* "Committee on Inadvertent Modification of the Stratosphere" (IMOS) was an early step in the formulation of Federal Policy, the subject in Chapter 8.

The continuing public debates have focused attention on the role of the scientist in formulating public policy. Persuasive arguments can be made in favor of the scientist reporting his observations, without comment, in scientific literature. Equally persuasive arguments can be advanced for the scientist making an effort to interpret his findings for the lay public and, in potentially serious situations, attempting to influence public policy. These philosophical points are discussed in Chapter 9.

Repeated calls for regulatory action from environmentalists, concerned scientists and others, led to legislative action and to the early initiation of the rulemaking process by the Environmental Protection Agency. The regulatory process is described in Chapter 10.

Throughout the rather noisy public debate, there were those who counseled that waiting for proof could not be condoned when environmental effects might affect future generations. Others counseled that stratospheric chemistry and circulation were far too complex to permit accurate description by current one-dimensional mathematical models. As a result of a measurement of a previously assumed value, model calculations which indicated ozone depletion from SST exhaust were reversed in 1978 to indicate ozone generation from SST exhaust. At the same time the calculated effect of chlorofluorocarbons was increased. Proponents of both positions used the correction as support.

The fluidity of the scientific situation was further emphasized by events occurring in 1980 and early 1981, between the submission of the chapters and the actual printing of this book. Remeasurement of a number of reactions — especially in the HO_x family of radicals — has led to significant refinement of older data. These rates, when entered into one-dimensional models, produce ozone depletion estimates for chlorofluorocarbon emissions lower than the 16.5% widely reported in 1979. Newer estimates fall in the 5 to 10% range with most values in the range of 5 to 7%. Incorporation of improved transport parameters tends to lengthen the time to maximum depletion, and thus tends to reduce the rate at which ozone depletion is calculated to develop. Uncertainties remain with the latest calculations and further changes in calculated effects must be expected.

A major development which has largely occurred since the writing of this book is the development of two-dimensional modeling capability. Many investigators believe that multidimensional modeling is essential if measurement and theory are to be validated in any meaningful way. Three-dimensional modeling remains the most desirable, but computing capability to handle both the large chemical reaction set and the complex circulation calculations is still well in the future.

The ultimate test of ozone depletion theories remains the detection of a long-term trend in ozone concentration. Recent advances on the original time series analyses described by Hill (Volume I, Chapter 4) have brought the possibility of early detection of trends within our reach.

To the extent that rates of depletion decrease, and to the extent that the sensitivity of trend detection improves, the pressure for early regulation is reduced, and time for more thorough research becomes available.

The ozone depletion issue presents a unique situation: a chain of events — calculated by computers from a theory, which, though unverified quantitatively, is based on sound chemical and physical principles — provides an exquisite predicament which is both created and recognized through technology. Selection of a social response that will withstand the scrutiny of our heirs is imperative. Surely as we seek to balance the pressures between population, quality of life and environmental conservation, our social reaction to the stratospheric ozone depletion issue must become a benchmark.

For this reason we have tried to examine the stratospheric ozone depletion issue from the scientific and social viewpoints. The contributors have given generously of their time and talent in the assembly of this book. We owe them our most sincere and warmest thanks.

The many administrative details involved in correspondence, record keeping, typing, and proofing have been most ably handled by Mrs. Elizabeth M. Lloyd and Mrs. Sally Ann Ward. Many of the figures and illustrations were drafted in final form by Mr. Bruce Esdale. We acknowledge all this help with sincere appreciation.

Frank A. Bower
Richard B. Ward

THE EDITORS

The editors have both been involved in the fluorocarbon-ozone issue since predictions of ozone depletion resulting from emissions of chlorinated organic compounds to the atmosphere were first made.

In 1973, Dr. Frank A. Bower was chosen as the first chairman of the Technical Panel for Fluorocarbon Research, an inter-industry research organization established to explore the environmental effects of chlorofluoromethanes. His involvement in the atmospheric effects of chlorofluoromethanes continued until 1979 when he joined the Haskell Laboratory for Toxicology and Industrial Medicine.

Dr. Richard B. Ward has a technical liaison responsibility broadly covering technical and regulatory aspects of the concern for scientists, managers, technical and trade associations. His work has spanned the preparation of technical summaries and progress reports for the Technical Panel for Fluorocarbon Research to presentation of the industry position to the media.

Dr. Richard B. Ward is a Research Associate in the Environmental Group of the Freon® Products Division, Petrochemicals Department of E. I. du Pont de Nemours and Company, Inc., in Wilmington, Delaware.

Dr. Ward received his B.Sc., Honours Class I in chemistry from Birmingham University, England in 1953, and his Ph.D., also at Birmingham, in 1956. During 1956 to 1958, he held a Research Fellowship at Birmingham from the United Kingdom Atomic Energy Research Establishment, studying the effects of gamma radiation on carbohydrates. He came to the United States in 1958 to take up a postdoctoral fellowship at Ohio State University, and joined the Du Pont Company in 1959.

His work since 1959 has included organic fluorine chemistry, permeation through polymers, dyes, and since 1972, environmental and toxicological aspects of fluorocarbons.

He is a member of the American Chemical Society and of the Royal Society of Chemistry.

Dr. Frank A. Bower is Associate Director of the Haskell Laboratory for Toxicology and Industrial Medicine of the E. I. du Pont de Nemours and Company, Inc., in Wilmington, Delaware.

Dr. Bower received his Bachelor of Science degree in 1949 from Hastings College with majors in Chemistry and Physics. He received his Ph.D. in 1953 from the University of Nebraska with a major in Organic Chemistry and minor in Pharmacology.

He joined Du Pont in 1953 and was engaged in research on polyurethane elastomers and fluorinated organic compounds until induction into the U.S. Army in 1955. During 1955 and 1956, he was engaged in research on organometallic compounds at the U.S. Army Quartermaster Research and Development Command in Natick, Massachusetts. Upon returning to E. I. du Pont in 1957, he was assigned to product development work in the Freon® Products Laboratory, E. I. du Pont de Nemours.

CONTRIBUTORS

James G. Anderson, Ph.D.
Professor of Atmospheric Chemistry
Harvard University
Cambridge, Massachusetts

James K. Angell, Ph.D.
Research Meteorologist
Air Resources Labs, NOAA
Bethesda, Maryland

Carroll Leslie Bastian
Deputy Director (Policy)
Office of International Activities
Environmental Protection Agency
Washington, D.C.

Ferial S. Bishop, M.S.
Biologist
Chair, CFC Interagency Work Group
Environmental Protection Agency
Washington, D.C.

B. Peter Block, Ph.D.
Senior Research Scientist
Pennwalt Corporation
King of Prussia, Pennsylvania

Anthony J. Broderick
Deputy Associate
Administrator for Aviation Standards
Federal Aviation Administration
Washington, D.C.

Harvey Brooks, Ph.D.
Benjamin Peirce Professor of
Technology and Public Policy
Professor of Applied Physics on the
Gordon McKay Endowment
Harvard University
Cambridge, Massachusetts

Perry W. Brunner, M. S.
Team Leader for
Labeling Regulation Chemist
U. S. Environmental Protection Agency
Washington, D.C.

C. C. Delwiche, Ph.D.
Professor of Geobiology
Department of Land, Air, and
Water Resources
University of California
Davis, California

Anna J. Harrison, Ph.D.
Professor of Chemistry
Mount Holyoke College
South Hadley, Massachusetts

William J. Hill, Ph.D.
Manager, Math Sciences
Allied Chemical Corporation
Buffalo, New York

J. Peter Jesson, Ph.D.
Research Supervisor
E. I. du Pont de Nemours & Company
Wilmington, Delaware

Harold S. Johnston, Ph.D.
Professor of Chemistry
University of California
Davis, California

Julius London, Ph.D.
Professor
Department of Astro-Geophysics
University of Colorado
Boulder, Colorado

James E. Lovelock, Ph.D.
Professor of Cybernetics
University of Reading
Reading, England

Michael B. McElroy, Ph.D.
Abbott Lawrence Rotch Professor of
Atmospheric Sciences
Harvard University
Cambridge, Massachusetts

D. Stuart Nachtwey, Ph.D.
Chief, Biomedical Applications Branch
NASA, Johnson Space Center
Houston, Texas

Robert C. Oliver
Associate Director
Science and Technology Division
Institute for Defense Analyses
Arlington, Virginia

R. D. Rundel, Ph.D.
Biomedical Applications Branch
NASA, Johnson Space Center
Houston, Texas

J. R. Soulen, Ph.D.
Associate Manager of
Research and Development
Pennwalt Corporation
King of Prussia, Pennsylvania

Nien Dak Sze, Ph.D.
President
Atmospheric &
Environmental Research, Inc.
Cambridge, Massachusetts

Wei-Chyung Wang, D. Eng. Sc.
Manager
Radiation and Climate Programs
Atmospheric and
Environmental Research, Inc.
Cambridge, Massachusetts

George F. Wirth, M.S.
Physical Science Administrator
U.S. Environmental Protection Agency
Washington, D.C.

DEDICATION

To our environment: physical, biological, and economic.

TABLE OF CONTENTS

Volume I

Volume II

PROLOGUE

Anna J. Harrison

Stratospheric Ozone and Man is an exercise in assessment and communication. In Volume I, a number of specialists address the current state of knowledge concerning the factors which determine the quantity and distribution of ozone in the stratosphere. Two decades ago, ozone in the stratosphere would have been considered an esoteric subject. Today, "the ozone problem" is a topic of public concern. The concern lies, not with ozone itself, but with the response of biological systems on the surface of the Earth to the 290 to 320 nm region of the UV spectrum of sunlight, and with technological practices that modify the intensity of that portion of the spectrum of sunlight reaching the Earth. Ozone is involved in the primary mechanism of controlling that intensity. Ozone absorbs this region of the spectrum, and anything that modifies the quantity of ozone traversed by the sunlight modifies the intensity of the sunlight in that region of the spectrum reaching the Earth. In Volume II, other specialists address the assessment of the impact of current and future technological practices upon the ozone layer. Policy questions relating to the control of technological practices are also addressed.

At the time this prologue is being written, I have not read the manuscripts for either volume. My role is not to act as a critic or as an advocate, but to explore, in a general manner, a few of the many questions related to the assessment of the impact of science and technology upon society and the environment, and the role of the public in delineating technological and societal practices.

In recent years, there have been tremendous advances in the capabilities of the scientific and technological community to expand scientific knowledge and to use that knowledge to achieve specific ends. We now have methodologies and instruments that enable us to make measurements with unprecedented specificity, precision, and speed. As a consequence, we can now monitor very complex systems, both naturally-occurring and synthetic, and undertake the assessment of technological practices, other societal practices, and natural phenomena upon human health and the environment. An essential part of this assessment is the use of modern computer facilities to process large quantities of information and to construct mathematical models. An unprecedented array of questions is being raised. Many of these relate to the quality of life of this and succeeding generations. We are confronted with the necessity of making a large number of far-reaching decisions. To postpone a decision, whether by deliberate intent or by default, is, in itself, a decision.

It is my personal conviction that we are not prepared, either psychologically and/or philosophically, to deal with many of the questions being raised. To make the adjustments to these new capabilities and the new responsibilities that flow from those capabilities requires diligence, cooperation, and candor as we seek to serve the common good.

In a democratic society, decisions relating to the quality of life must ultimately be made by the public, frequently by elected officials or those appointed by elected officials acting as surrogates of the public. Those who are professionally involved with science and technology participate in decisionmaking in exactly the same sense as all other members of the public. Each individual has the right to seek to influence public opinion, and each has the responsibility to vote.

In addition, members of the scientific and technological community have other responsibilities. One is to extend knowledge and develop technologically feasible options.

A second is to assess, and to work with others to assess, the impacts of current practices and proposed innovations upon society. The third is to communicate with the public and its surrogates concerning the nature of technologically feasible options and the nature of the assessments that have been made or could be made concerning the impacts of current practices and proposed innovations. The first is primarily a technical matter. The second has large technical components in terms of the natural sciences, but goes far beyond that to include economic, social, and political impacts. The third is a matter of communication in which the technological community seeks to share what it has learned with the public and/or its surrogates, and also seeks to understand the needs and the concerns of the public.

Assessments are frequently formulated in terms of positive impacts called benefits and negative impacts called costs and/or risks. This approach is admirable in its logic, but misleading in its simplicity. The assessment of benefits, or the assessment of costs, or the assessment of risks, frequently involves the summation of a series of terms, each of which has a degree of uncertainty associated with it. Each of the terms that make up the sum is an extrapolation based upon past experience, or a prediction calculated from a mathematical model, or, in matters dealing with human health, a prediction based upon experimental results with other animals and other organisms. Even so, the most difficult problem is not to make the prediction, but to evaluate how good that prediction is. A further barrier arises in summing terms to obtain a single value for benefits or costs or risks. The various terms may be expressed in different units, and it may be impossible to sum without introducing value judgments in the assignments of conversion factors. How do you assign equivalent values to life, good health, freedom from the necessity of some types of physical labor, availability of clean air, the continuation of a species, economic well-being, political advantage, military security, convenience, etc.?

The diversity and the magnitude of the benefits derived from the goods and services that are a product of science and technology will not be belabored here. In denouncing society, even the most ardent recluse takes with him or her a survival kit of the products of science and technology. Presumably, the ways individuals spend their money indicate the relevant values placed upon the various goods and services perceived by those individuals as viable options; and, in like manner, the ways in which surrogates of the public appropriate public monies indicate the relevant values placed upon the various goods and services perceived by the public as viable options. Presumably, the use of public monies to support research and development is an act of faith that these activities will culminate, either directly or indirectly, in additional viable options that will serve the public good.

The term "cost" is used in at least two senses in the assessment of innovations. One has to do with the premarket cost, and the other has to do with the total cost to society. In making a decision to buy or not to buy, the market cost is the more apparent and is a significant consideration. In making a policy decision, the surrogates of the public have the responsibility to assess the total cost to society. This is much more complex and includes the depletion of nonrenewable resources, the costs associated with the impairment of human health and the environment as a consequence of all processes involved in the manufacture, processing, distribution, use and disposal of all materials, and the dislocation of labor forces. Some of these, such as health care, can be assessed in terms of dollars; others are matters of human trauma and the quality of life of individuals and groups of individuals. I shall use the term in the sense of total cost.

Risk is related to cost, and the two terms are frequently used interchangeably. It is probably more useful to reserve the term "risk" for a potential cost which can be assessed only with a high degree of uncertaintly. The evaluation of confidence limits at a specified confidence level is an essential part of every assessment. In the assessment of a risk, the evaluation of confidence limits is extremely troublesome. Either the nature of the risk being assessed or the current state of knowledge may make it impossible to establish confidence limits that have significance over a relevant time interval. The assessment of confidence limits of a risk associated with natural phenomena can also be very troublesome. The primary differences lie in the body of knowledge that has accumulated in respect to the natural phenomena and the sense of urgency in endeavoring to assess the risk. I suspect that we have become accustomed to (made our peace with) risks that are associated with natural phenomena and other types of risks (such as the risks associated with falling elevators, explosion of gas mains, and the collapse of bridges) that have been around for a long time.

It is probably true that all technological changes, all changes for that matter, have associated with them negative impacts on society regardless of how great the positive impacts inherent in the changes are. The approach to decisionmaking through a consideration of a benefit/cost ratio is entirely realistic in concept, but extremely difficult to deal with in practice. The validity of the ratio can be no greater than the validity of the evaluation of the total benefits and the total costs. Very seldom can the ratio be evaluated as a number without units, and sooner or later value judgments must come into the decisionmaking process.

The "ozone problem" is extremely complex. In principle, at least, a mathematical model describing the quantity and distribution of ozone in the stratosphere must include *all* photochemical processes, *all* chemical mechanisms, and *all* diffusion processes that either directly or indirectly involve the formation or the removal of ozone in the stratosphere. Why the stratosphere? This is a region of the atmosphere in which the absorption of another portion of the spectrum of sunlight by oxygen molecules, O_2, creates the temperature conditions that differentiate the stratosphere from the troposphere and, at the same time, leads to the formation of ozone molecules, O_3. In principle, at least, the model must allow for periodic variation in the intensity of sunlight, and simulate the diurnal, seasonal, and latitudinal effects on substances in the stratosphere. The effects of other changes which, at least superficially, are unrelated chemically, cannot be ignored; for example, the influence of atmospheric carbon dioxide on ozone concentrations. Sources of materials which participate in reactions in the stratosphere may be as diverse as the discharge of volcanoes, lightning in the troposphere, use of nitrogen fertilizers in agriculture, products formed in combustion processes at the surface of the Earth, the exhaust of planes flying in the stratosphere, and the discharge, either accidental or deliberate, of chemicals into the atmosphere. A total assessment of the ozone problem must also deal with the biological effects that flow from changes in the intensity of UV light reaching the surface of the Earth.

Federal regulations are boundary conditions set by surrogates of the public. To be effective, regulations concerning the use of science and technology must have creditability with the public, and also with the subset of the public known as the scientific and technological community. To have creditability with the public in general, the goals of the regulations must be consistent with the mores of the society. To have creditability with the scientific and technological community, the mechanisms of the regulations must be consistent with scientific knowledge and realistic with respect to technological capabilities.

Federal regulations are one mechanism through which the resolution of a problem can be approached, but not the only mechanism. An informed public can exert effective economic pressure in the marketplace, and corporate leadership can influence, even determine, the array of options available to the public.

A problem which pivots through reactions in the stratosphere of substances introduced into the atmosphere at or near the surface of the Earth is a global problem, and ideally would be approached as such by all parties and/or all nations significantly involved in the initial introduction of the materials into the atmosphere. The "ozone problem" will undoubtedly become a significant case study in the resolution of an international problem. *Stratospheric Ozone and Man* is a progress report.

Section 1: Does Man Influence Stratospheric Ozone?

Chapter 1

THE SUPERSONIC TRANSPORT

Anthony J. Broderick and Robert C. Oliver

TABLE OF CONTENTS

I. INTRODUCTION

By the beginning of the 1970s, the supersonic transport (SST) designs of the U.S., Great Britain and France, and the Soviet Union had been in the process of continual refinement for some 10 years. Prototypes of the Anglo-French Concorde and Soviet TU-144 had made their first flights. Simultaneously, a general environmental awareness had been building in the U.S. The year 1970 saw the initial implementation of the National Environmental Policy Act[1] (NEPA), with its clear declaration of U.S. policy on environmental preservation and enhancement; and the Clean Air Act Amendments of 1970[2] (CAAA), which charged the infant U.S. Environmental Protection Agency with establishing regulations which would reduce automobile emissions by a factor of 10 within about 5 years. NEPA and the Clean Air Act can, in retrospect, be viewed as major turning points on a road toward growing national concern for environmental protection and enhancement, and 1970 is the year that the SST collided with environmental interests.

The U.S., for the first time in many decades, was providing direct federal financial support for the development of commercial aircraft—the supersonic transport. Under contract to the government, the Boeing and the General Electric companies were to design, build, and test a commercially viable SST which was to carry nearly 300 passengers at 2.7 times the speed of sound between points as distant as New York and Paris. A key provision in this government-industry agreement was that development costs were to be paid back to the government by means of a royalty which would be part of the sales price for each aircraft subsequently sold by the Boeing/G.E. team. The government projected a substantial future profit from these sales, a contention viewed as overoptimistic by critics.

The direct-subsidy aspect of this arrangement rankled many. The long history of indirect federal support of commercial aircraft development—via research and development of military aircraft—was not something people seemed to be generally familiar with or, if they were, it seemed to present less of a philosophical problem. If the development of the world's largest commercial aircraft, the Boeing® 747, had been materially enhanced by Boeing's early work on the Air Force C-5A program; if the development of the modern high-bypass ratio turbofan engines for the Boeing® 747, the McDonnell-Douglas® DC-10, and the Lockheed® L-1011 widebody jets had been based on similar efforts, so be it. Apparently, such indirect subsidies could be rationalized on the basis of original defense needs driving the early development work, and commercial ingenuity subsequently making a profit. But the SST program was viewed by many as, purely and simply, a distasteful case of government subsidy to big business.

Concern had been expressed in the 1960s about the environmental effects of the SST, but that discontent centered almost exclusively on two issues: extreme noise on takeoff, compared to even the noisy early jets, the result of the enormous exhaust jet velocity of its four afterburning turbojet engines; and the prospect of frequent sonic booms. With regard to the sonic boom issue, it was finally agreed after many years of study and testing that sonic booms would not be tolerated by the average citizen. As a result, the U.S. Government considered that commercial supersonic flights over populated areas could not be permitted. Opponents of the SST then raised the issue of possible harm to wildlife in uninhabitated areas and to sealife from sonic booms. So in 1970 the sonic boom issue cannot really be said to have been resolved. Takeoff noise, on the other hand, was a distinctly more difficult problem. This environmental issue remains today as one of two key roadblocks, the other being energy consumption to the development of a commercially successful SST.

But there was yet another set of environmental concerns to be raised widely in 1970:

what would be the effect of exhaust *gases* from the SST fleet? Would the large amounts of exhaust water vapor form persistent contrails or clouds? Would the amount of stratospheric ozone be reduced by the exhaust water vapor? Would nitric oxide emissions of the SST reduce stratospheric ozone? Would global climate changes result from SST operations?

II. SST EXHAUST COMPOSITION

The basic combustion chemistry of hydrocarbon fuels results in exhaust emissions of carbon dioxide (CO_2) and water vapor (H_2O). Owing to the heating of inlet air to very high temperatures, oxides of nitrogen (NO_x) are also found in the exhaust. Incomplete combustion—though combustion efficiencies of jet engines are remarkably high—results in emission of small amounts of carbon monoxide (CO) and unburned or only partially burned hydrocarbons (HC). Trace elements in the fuel itself also result in oxidized forms of emissions, with oxides of sulfur (SO_x) being of primary interest.

It is conventional to speak of the "emission index" of a jet engine when discussing exhaust emissions of a substance, citing the number of grams of the substance emitted per kilogram of fuel burned. Thus, Table 1 shows the assumed emission indexes calculated from data in the Study of Critical Environmental Problems[3] (SCEP). These data were based on calculations supplied by General Electric Company, who was designing the Boeing® SST powerplant. These estimates were assumed valid for cruise conditions in the stratosphere. The tentative nature of the data available then is a result of the fact that these data had never been requested before and no actual measurements were available. Thus, the emission rate of nitrogen oxides was estimated to be higher than one would now estimate, as discussed in the appendix.

These data, or similar data for other aircraft, form the starting point for analyses of atmospheric effects from an SST fleet. For each aircraft, one first estimates the fuel burned per year in the stratosphere, and then derives an annual stratospheric emission rate by each aircraft of the substance of interest. Multiplication by the number of aircraft expected to be in the fleet, finally leads to an estimate of the fleet's annual stratospheric emissions.

Considerable confusion reigned over the appropriate value for the oxides of nitrogen emission index and other details of annual emission rate estimates. The appendix reviews the aircraft NO_x emission estimates made by several groups, and explains their differences. As shown there, though fleet emission values as high as 4×10^{12} g/year NO_x were estimated, a thorough consideration of all the data leads to an estimate of about 3.6×10^{11} g/year NO_x for a hypothetical fleet of 500 SST aircraft of the 1970 U.S. design, with each aircraft emitting 10.8g of NO_x per kilogram of fuel burned at cruise altitudes. For the Concorde, a number of similar calculations were made, supplemented by measurements. Without unnecessary detail, we can cite a reasonable estimate for the Concorde's NO_x emission index of 18, which is also representative[4] for the Soviet TU-144.

A final point about engine NO_x emissions is worth making. In the course of the engine emission measurements described by English,[4] two different systems were sometimes employed for NO_x measurements. The data reported above are based on results of techniques used which closely followed standards established by the Environmental Protection Agency (EPA) for such measurements.[5] An alternative approach yielded considerably higher values for certain engine operating conditions, giving rise to concern (and sometimes heated discussion) that established procedures tend to greatly underestimate—by a factor ranging from 2 to 6—NO_x emissions. A discussion of this point is given by Oliver et al.,[6] and a multi-agency study aimed at resolution of this discrepancy is expected to be completed shortly by the Federal Aviation Administration (FAA).

Table 1
1970 — ASSUMED EMISSIONS OF SST[a]

Exhaust constituent	Emission index (g emitted/Kg fuel burned)
CO_2	3100
H_2O	1300
CO	42
NO_x (as NO_2)	65
SO_2	1
HC	0.5

[a] Calculated from data in SCEP[3].

III. THE GENESIS OF THE SST-OZONE CONTROVERSY

Contrary to popular impression, the controversy over ozone destruction by SST's did not play a major role in the eventual cancellation of the U.S. SST development program. The publicized "SST debates" of 1969 to 1970 were essentially political discussions in the U.S. Congress which centered on whether or not to provide funds for the SST development program, as were requested in the President's Fiscal Year 1970 budget. Opponents of the SST program seized upon environmental grounds to base some of their arguments against continuation of the SST program.

Their environmental arguments can be broken down into two broad areas: noise (excluding sonic boom) and upper atmospheric effects from SST exhaust. The strongest environmental arguments were made on the basis of airport takeoff noise and sonic boom, which we do not discuss here. If environmental considerations played any real role in the eventual decision, it was these. Upper atmospheric pollution arguments were, unlike the more readily grasped noise issues, much more technical and difficult to understand. Thus, though issues of potential climatic change were from time to time brought to public attention, for instance by the National Academy of Sciences (NAS),[7] possible climatic effects of the SST retained a status of only secondary environmental arguments. They provided more justification to oppose the SST, but with little real substantiation.

It is fair to say, in retrospect, that the slow popularization of the concept of potential climatic effects of the SST was in large measure the result of broad-based skepticism on the part of some members of the scientific establishment. The inertia so displayed by them was, in the specific case of *direct* climatic effects from SST exhaust, well-founded. It was not possible to make a good case for significant direct climatic effects simply because the magnitudes of the calculated concentration perturbations of climatically significant trace species (carbon dioxide, water, aerosols) were small or insignificant. It is unfortunate, however, that this reluctance spilled over into the ozone reduction issue, since the ozone reduction arguments turned out to be, scientifically, a horse of a different color. These had more credibility or, at least, were more long-lived.

In 1934, Chapman[8] discussed means of improving astronomical observations and noted that creating a "hole in the ozone layer" would permit otherwise unobtainable data to be gathered in the UV region of the spectrum. He noted that, to do this, one might use "some catalyst which, without itself undergoing permanent change, could promote the reduction of large numbers of ozone molecules..." Despite this prescience, he is best known for his basic studies[9] of the fundamental role of oxygen chemistry in maintaining the ozone balance. For years, though, observers had recognized that Chapman's oxygen chemistry alone was not adequate in accurately describing the ozone balance of the natural stratosphere, since models which employed only these reactions consistently predicted the presence of more ozone than was observed. In an attempt to rationalize this situation, Hampson[10] discussed the role that hydrogen species might play; Hunt[11] showed that with certain *assumptions* about rates of reaction involving HO and HO_2 chemistry, the inclusion of an odd hydrogen catalytic cycle could provide a balance in the atmospheric modeling of ozone. Hampson[12] noted that if this *postulate* were true, artificial stratospheric water vapor increases caused by an SST fleet would act to reduce atmospheric ozone, since water was the principal source of stratospheric odd hydrogen species. Leovy,[13] in 1969, built on Hunt's analysis to develop an ozone model; Leovy's model was then used and extended in 1970 by Harrison[14] to estimate that water from the exhausts of a fleet of 500 U.S. and 350 non-U.S. SST aircraft might diminish the atmospheric ozone amount by a maximum of about 4%.

Up until this time, it was thought that the consequences of such an ozone reduction were limited to climatic effects. But in 1970, accepting these data as valid analyses of potential atmospheric response, McDonald[15] provided a landmark analysis of another ramification of ozone reduction, one which had far greater public impact. He argued that ozone reduction might be expected to be shortly followed by increased skin cancer incidence, resulting from the increased flux of UV radiation which could then be transmitted to the surface of the Earth. This information proved useful to SST opponents, and was the full extent of the SST ozone-skin cancer link when the decisive Congressional votes to cancel the U.S. SST project were taken in March 1971. Oxides of nitrogen effects? These had been examined and rejected as negligible by the *Report of the Study of Critical Environmental Problems*[3] in 1970.

The March 1971 votes of the U.S. Congress were decisive, but supporters of the SST continued to mount vigorous attempts to reverse that decision over the succeeding several months. Consequently, there remained an environment conducive to public airing of anti-SST environmental arguments. On the day prior to the critical House of Representatives vote of March 17, 1971, a meeting of the chemistry subgroup of the Department of Commerce Panel on Supersonic Transport Environmental Research had begun. Presentations made to that panel suggested no substantial change in the above-mentioned estimates of ozone destruction by SST water vapor. But Professor Harold S. Johnston (University of California, Berkeley) and Professor Arthur A. Westenberg (Johns Hopkins University) disagreed with this conventional wisdom, and pointed out that it was clear to them that a serious technical error had been made in estimating the degree to which oxides of nitrogen would reduce atmospheric ozone inasmuch as their role in a catalytic cycle had not been considered. On March 18, Johnston provided the panel with an analysis showing that his estimate of ozone reduction which might be attributed to oxides of nitrogen present in the SST exhaust exceeded, perhaps by a factor of 10, that which might be attributed to water vapor.

The nitric oxide-ozone issue would probably not have received so much attention had it not been for the subsequent abortive attempts of SST supporters to reverse the SST program-termination decision. But the Congressional debate dragged on, and Johnston[16] was convinced of the fundamental soundness of his analysis which, when coupled with McDonald's linking of ozone reduction and increased skin cancer,[15] proved to be worthy of national news coverage (see, for example, Sullivan[17]), some of it rather sensational. In August 1971, with the ozone issue still being widely debated in scientific circles (though the movement to revive the SST project had by then faltered), the Congress made available funding for the Climatic Impact Assessment Program (CIAP)[18] of the U.S. Department of Transportation (DOT). The funds had originally been requested in 1970 as part of the SST development monies. Though the SST program was now clearly finished, it was believed that the proposed CIAP investigations were deserving of support, since other SST aircraft were still being developed in Europe and the Soviet Union and the environmental questions had not yet been answered.

By the beginning of CIAP in late 1971, there was additional important evidence available bearing on the SST-ozone matter, not all of which had gotten the extensive news coverage that Johnston's and McDonald's analyses had experienced. With regard to odd-hydrogen catalysis of ozone reduction, Nicolet[19] had suggested, and Langley and McGrath[20] had experimentally confirmed, that the rate of ozone destruction computed in Hunt's 1966 analysis[11] was excessive, and SST water vapor was *not* likely to be a problem. In addition, support for Johnston's NO_x thesis had come from Crutzen's independent analysis[21] of the role of oxides of nitrogen in maintaining the stratospheric ozone balance. Interestingly, Crutzen's analysis took a similar approach to that documented by Hunt.[11] However, instead of finding, as Hunt did, what rate constants would satisfy a proposed ozone controlling mechanism, Crutzen used a known mechanism (the NO_x catalytic cycle) with estimates for rate constants, and calculated the concentrations of NO_x needed in the stratosphere for the cycle to be largely controlling. The concentrations looked reasonable in light of what was known.

IV. CIAP — NAS — COVOS*—COMESA**: THE SST-OZONE STUDIES

The CIAP program had, as its objective, "to assess, by a report in 1974, the impact of climatic changes which may occur from the operation of a worldwide stratospheric fleet in 1990, in order to determine the regulatory constraints on flight in the stratosphere required to prevent adverse environmental effects".[22] The objective was broad, and interpretation of its bounds was flexible. Consequently, this major ($23 million) DOT study evolved in six major program areas: aircraft engine emissions, atmospheric modeling, atmospheric monitoring and experiments, biological monitoring and experiments, atmospheric chemistry, and analysis, integration, and assessment. The major documentation of the results of the study, in addition to the chronology of Mormino,[22] may be found in six monographs and the proceedings of four international conferences, which together total about 9000 pages.[18a-18f] In order to maintain objectivity, the DOT-CIAP management elected to commission the U.S. National Academy of Sciences (NAS) to conduct a parallel study of the SST climatic impact problem—a type of continuing review of the CIAP itself. The NAS Climatic Impact Committee was established for this purpose in 1972 and completed its work in 1975.[23] The NAS Committee reviewed all aspects of the CIAP work program, and was given access to all of CIAP's data.

In the period 1972 to 1976, two other major studies of SST climatic impact are worthy of special note. In Great Britain, COMESA was established to direct a study which covered only the physical and chemical aspects of the question. COMESA organizers were of the belief that devoting scarce resources to investigation of biologic, social, and economic consequences of an as yet unproven stratospheric effect would be counterproductive, and limited their study to atmospheric science. In France, COVOS went a step further than COMESA, and included biological effects studies. The COMESA results were made available in 1975, and those of COVOS in 1976.[24]

The CIAP, NAS, COVOS, and COMESA efforts actually followed, in large part, the work of the Australian Academy of Science (AAS)[25] carried out between September

* COVOS — Committee on the Consequences of Stratospheric Flight (France).

** COMESA — Committee on Meteorological Effects of Stratospheric Aviation (U.K.).

1971 and February 1972. This effort, however, was concentrated on the climatic effects of SST's and on the photochemical effects of water vapor, pleading that "the photochemical problem involving nitrogen oxides is difficult to assess...on the one hand the implication is of no noticeable effect, on the other hand there is a possible cause for concern..."

It is of interest to highlight a few of the points made by the AAS in their report,[25] as these give a reasonably good indication of the early 1972 positions taken by many experts on the issues:

- Water vapor and clouds: "Neither case [increased global average water vapor amounts, or buildups which might be expected in flight corridors] would be expected to cause a significant increase of cloud...Stratospheric contrail formation should remain transient...No noticeable climatic effects would be expected...from these causes..."
- Particulates: "It is estimated that particles created by carbon, unburnt hydrocarbons, or oxidation of sulphur content of the fuel would accumulate...even with...unknowns it seems unlikely that increases caused by aircraft using the stratosphere will present a real hazard..."
- Oxides of nitrogen: "Although there has already been a considerable amount of lower stratospheric flying over the last decade, the ozone concentrations have not decreased...We therefore believe that the effect of supersonic aircraft on the ozone layer is not likely to be serious..."
- Atmospheric motions: "...transport processes in the stratosphere...will ameliorate the impact of supersonic aircraft on the ozone layer..."

CIAP was a multidisciplinary impact assessment which, by choice of its management, did not focus on producing "best estimates" of probable effects of a 1990 stratospheric fleet. Instead, the CIAP was driven to provide a highly conservative (from the environmental viewpoint) assessment of atmospheric effects that might result from optimistic expansion of airline travel in general, and SST travel in particular, and what could be the biological, economic, and social consequences of these changes. CIAP did not actually focus on the original question—what would happen *if* 500 "Boeing® SST's" were flown? Nevertheless, it produced a wealth of valuable atmospheric data, and provided a welcome stimulus to the research funding for many in the atmospheric sciences community. Its approach was more or less parametric in nature, with its studies all proceeding in parallel, to be drawn together in the final analysis and assessment. The CIAP study soon became organized around the development of the six monographs, which were intended to be independent, self-contained analyses whose parametrically expressed conclusions could be used in the final CIAP assessments.[18a-18l] As given by Mormino,[22] the contents of each monograph may be summarized as follows:

- Monograph 1, *The Natural Stratosphere of 1974,* describes what is known of the natural stratosphere, including the new knowledge contributed by CIAP.[18b]
- Monograph 2, *Propulsion Effluents in the Stratosphere,* describes what is known about aircraft engines, both present and future, including CIAP-sponsored measurements, and experiments. It postulates the routes and frequencies of travel for future aircraft (foreign and domestic), and the engine emission products introduced into the stratosphere.[18c]
- Monograph 3, *The Stratosphere Perturbed by Propulsion Effluents,* estimates possible perturbations of the stratosphere resulting from pollution by engine effluents.[18d]

- Monograph 4, *The Natural and Radiatively Perturbed Troposphere,* describes possible resulting effects in the troposphere involving climatic change.[18e]
- Monograph 5, *The Impacts of Climatic Change on the Biosphere,* describes effects on plants and animals of climatic and radiation changes.[18f,18g]
- Monograph 6, *Economic and Social Measures of Biologic and Climatic Change,* illustrates some economic and social costs of potential climatic and biologic change, and compares them with the costs of feasible abatement measures.[18h]

The CIAP conclusions evolved in an interesting manner. The thesis presented by Johnston was, naturally, the focus of the ozone studies. (As mentioned earlier, the work of Langley and McGrath[20] had already served to discount the water vapor-ozone destruction hypothesis.) A major portion of the CIAP effort went into stratospheric measurements of trace species. Atmospheric models were developed, some in one dimension (1-D, a "global" average, variable altitude only), some in two dimensions (2-D, variable altitude and longitude, a "zonal" average), and one in three dimensions. A relatively small portion of the effort (about $1 million or less) went into measurement of chemical reaction kinetics; in large measure the chemistry employed was thus "pre-CIAP" but with refinements (as, e.g., to include HNO_3 in the modeling) which reduced effects some twofold. An attempt was made to estimate uncertainties, but the true uncertainties (now evident) remained unappreciated.

More specifically, the chemistry set employed in CIAP was that of the O_x-HO_x-NO_x system. In this set, with the rate constants employed, the NO_x catalytic cycle provides the most important sink for ozone in the natural atmosphere (the NO_x coming from N_2O via $O[^1D]$), so that NO_x additions necessarily led to a decrease in total ozone. The CH_4 oxidation reactions, which were shown by Johnston[26] to be capable of producing ozone with NO_x additions (by "smog" reactions), and which he found to lead to a crossover in ozone effect at about 13 km (depletion above, enhancement below), were not included in the models,* even when the models were used to evaluate effects of subsonic aircraft at 9 to 12 km.** With this possible exception the chemistry set was intended to be the best available, having been examined and reexamined by a committee of experts and the U.S. National Bureau of Standards. Modelers were in general asked to use the data set recommended by this group in order to provide comparability: this approach, of course, tended to provide uniform answers, varying primarily with model representations of transport processes, and quite effectively hid the real uncertainties.

The CIAP conclusions[18a] (as well as the NAS[23] results) were based principally on results from 1-D models, with some adjustments based on 2-D results.*** In 1-D models, vertical transport is treated as a diffusion process, with transport proportional to a concentration gradient times an "eddy diffusivity" (K_z) which varies with altitude. K_z profiles for the stratosphere are generated from data on a tracer, such as CH_4 or N_2O, which falls off in concentration with height and is assumed to be at steady state, and for which chemical or photochemical loss processes are thought to be known. In principle a great deal of such data is needed (at various latitudes and times, etc.; see Schmeltekopf et al.[29]); in practice only limited data were available and used in CIAP.

* Some authors argued that the CH_4 "smog" reactions are too poorly known to be included.

** The COMESA group included these reactions in the early phases of their study (and showed some enhancements for NO_x injections from subsonics) but the reactions were deleted in a later model to reduce computer running time.

***The 3-D model results were not considered reliable in terms of longitude, so they were reduced to two dimensions (Cunnold et al.[27,28]). In fact, the CIAP "results" represented an arithmetic average of the "CIAP model" results and the "NAS model" results — a sort of compromise.

Changes in understanding of loss processes change K_z profiles so derived. K_z profiles also differ with solar zenith angle and other modeling details.

The lower-altitude portion of the K_z curves must be developed by different procedures; residence time data from injections of gaseous or particulate (usually radioactive) tracers have been used. The lower portion of the curves, which involves great uncertainty, and some conceptual difficulties,* is of considerable significance to the aircraft problem; the model tropopause height and minimum K_z value are particularly important. The minimum K_z value acts as a diffusion barrier upwards if flight is below the model tropopause, and as a barrier downward for flight above the tropopause. The model values chosen for the tropopause height and minimum K_z are somewhat arbitrary, but are obviously important in computing effects of subsonic aircraft. (An aircraft at a given altitude might be above a model tropopause but below the true tropopause.) The various aspects of the treatment were extensively debated[30] during CIAP.

The choice of K_z profile in CIAP was of considerable significance in a quantitative, and thus potential federal regulatory, sense. With the chemistry employed, NO_x injections at all altitudes led to ozone depletion; these results implied that some depletion of ozone due to aircraft was inevitable. It was suggested,[18a] based on a detectability argument, that 0.5% depletion of the ozone column might be acceptable. The corresponding permissible number of aircraft then depended on the operating hours at altitude per year per aircraft, the operating altitude, the fuel flow, and the NO_x emission index. Water vapor was of no concern in the CIAP ozone calculations; in fact, Crutzen,[31] using CIAP chemistry, showed that increases in water vapor alone, by reducing the effectiveness of the NO_x cycle, led to a slight increase in total ozone (a doubling of water vapor led to 1.5% increase in ozone). With given fleet projections, the 0.5% depletion figure would thus be reached at different times by different models, but inevitably would be reached as aircraft (particularly SST) traffic grew, unless the NO_x emission index, by presumably costly means, could be reduced. The NO_x emission index is a function of combustor operating conditions for conventional combustors, as noted in an earlier section, but techniques are known (lean premixed prevaporized homogeneous or catalytic combustor systems) by which the NO_x emissions index can be greatly reduced.[4] Problems were evident with these techniques, but a committee set up under NASA sponsorship (at CIAP request) suggested some possible goals. These were interpreted by the CIAP office as being 6-fold or 60-fold reduction goals, to be introduced prior to the time which the total ozone depletion would otherwise have been 0.5%.[18a] It might be added that these goals now seem much less easily attainable.

While CIAP models all showed ozone depletions with NO_x injections, a nagging uncertainty was introduced by Foley and Ruderman.[32,33] These authors pointed out that very large injections of NO_x into the stratosphere had occurred during the period of intensive nuclear testing in 1958 to 1962 which should, by then-current models, have led to substantial ozone depletion in the Northern Hemisphere. However, ozone records for this period showed no unambiguous evidence of ozone depletion. Extended arguments followed thereafter about "detectability" and the validity of the ozone data results. Perhaps the best argument for evidence of depletion was given by NAS,[23] in which a "missing peak" in a cyclic ozone record was noted as occurring about the time of the tests and thereafter.

As CIAP approached an end, with Concorde SST sales lagging, the Concorde pro-

* The question is whether the so-parameterized transport processes relate in the proper way to the transport processes (meridional or zonal) which bring reactive species together in the real atmosphere. This problem is believed to be more severe in the lower stratosphere, where chemical lifetimes are long, than in the upper stratosphere where lifetimes are short.

duction line approaching shutdown, and the Soviet SST development moving slowly, the SST threat to ozone began to seem more remote. However, the subsonic question began to be of greater concern. Thus the NAS report[23] showed that 100 "advanced subsonics" (such as the 747SP) cruising at 12- to 15-km altitude would reduce Northern Hemisphere ozone by 0.16% per 100 aircraft. Only 300 such aircraft would thus give 0.5% ozone depletion, and some projections[4] suggested that this number might be reached within a few years, certainly before a radically new combustor technology could be available. It was clear that subsonics had to be given more detailed attention, and this is a principal reason why the FAA continued these studies after completion of the CIAP effort.

Summarizing, the CIAP-period ozone studies yielded results which supported, using much the same chemistry, Johnston's contentions[16] as to SST ozone depletion due to NO_x additions. The models were developed for SST's, where other reactions (e.g., CH_4 oxidation) were believed to be insignificant, but were applied much more broadly. Water vapor additions by SST's, at least in one model run reported, by tying up NO_x as HNO_3, tended with this chemistry to slightly increase model ozone rather than to deplete it, as had been postulated earlier.

V. THE YEARS 1975 TO 1976: CH_4 OXIDATION EFFECTS, UNCERTAINTIES, AND HALOCARBON EFFECTS

The 2-year period following delivery of the CIAP Report of Findings[18a] in December 1974 involved three developments which were particularly significant to the calculation of aircraft effects. The first, in terms of its implication to the aircraft industry, was a further study by the FAA's High Altitude Pollution Program of subsonic aircraft effects, including the methane oxidation reactions which had not been included in CIAP. The results, showed small ozone enhancements rather than significant ozone depletions and were thus of obvious policy significance; certainly the nature of the problem was totally altered. The second of these was a study by Duewer et al.,[34,35] which focused on uncertainties in the modeling of the O_x–HO_x–NO_x atmosphere and in the effects of perturbations to it; the results showed that NO_x from SST's at 17 to 20 km *could* increase the ozone column within the uncertainties of the chemical kinetics or could cause much more severe effects than previously thought. A third development, one of obviously more significance in an environmental and economic sense, was the recognition of the impact of chlorofluorocarbons, and especially the influence of chlorine chemistry, on stratospheric NO_x chemistry.

A. Uncertainties in the O_x–HO_x–NO_x Atmosphere

Each reaction rate in a photochemical model has associated with it an estimated uncertainty. These uncertainties have been found at times (as shown here; see also Smith[36]) to be seriously underestimated; also no guarantee exists that all pertinent reactions have been included. (Thus the uncertainty analysis is itself highly uncertain.) Various procedures to estimate uncertainty have been used, but the one which appears to have given the most insight is the one adopted by Lawrence Livermore Laboratory in examining uncertainties in the O_x–HO_x–NO_x model atmosphere. The procedure used was to simultaneously vary a number of key reaction rates, within the stated uncertainties, towards or away from ozone sensitivity to NO_x (see for instance, Duewer et al.[34,35]).

Five reactions were found by Duewer et al.,[34,35] to be of particular significance. Without going into extensive detail, it was found, as noted earlier, that ozone changes with NO_x injections at SST altitudes could vary widely, ranging from relatively large negative values to slight positive values; a change to one uncertainty limit in as few as four reactions could lead to ozone enhancements. They also showed that water injection effects on ozone go oppositely to NO_x injections. If NO_x increases O_3, water decreases it, and conversely. (This is now a commonly observed result.) The water effect proved to be small but clearly was in need of more study.

B. Chlorofluorocarbons

The chlorine cycle at first seemed to be of minor interest in terms of the natural atmosphere, primarily a matter of concern to the National Aeronautical and Space Administration (NASA) and its space shuttle. The picture changed dramatically with publication by Rowland and Molina,[37] pointing out the potential effects of chlorofluorocarbons, particularly $CFCl_3$ and CF_2Cl_2 usage (and release) of which had grown dramatically through widespread use as propellants in aerosol spray cans and refrigerants, etc. The only significant sink for these materials appeared (and still appears to be) photolysis in the stratosphere; the materials had been detected and measured on a worldwide basis by Lovelock.[38] Congressional hearings were thus held in December 1974.[39] The issue obviously had the potential for large economic impact, as a total of some 800 million pounds of these two compounds had been produced in the U.S. alone in 1973. A study by the NAS[40] resulted, with conclusions released in late 1976, and discussed[41] in a conference in September 1976. Large depletions of ozone (2 to 20%) were forecast at 1973 release rates (if continued) requiring however perhaps 50 years to equilibrate.

The halocarbon work stimulated much new research, most of which is outside the topic of this chapter. However, the aircraft NO_x problem was affected, as it became clear with further stratospheric measurements that naturally-occurring and manmade chlorine species are and will be important in the stratosphere, and that these interact with previously known NO_x (e.g., $ClO + NO_2 + M \rightarrow ClONO_2 + M$); and HO_x (possibly as HOCl) cycles.

During this work, various reaction rates were examined. One reaction

$$OH + HO_2 \rightarrow H_2O + O_2 \tag{1}$$

was particularly significant to both aircraft effects and to the halocarbon question. In the NO_x cycle, OH combines with NO_2,

$$OH + NO_2 + M \rightarrow HNO_3 + M \tag{2}$$

whereas in the Cl_x cycle, OH regenerates Cl otherwise held in a reservoir of HCl,

$$OH + HCl \rightarrow H_2O + Cl \tag{3}$$

The rate assigned in CIAP was very fast (2×10^{-10} cm^3/sec) which tended to maximize NO_x effects. In the halocarbon work, this reaction was reexamined and a much slower rate (2×10^{-11}) assigned. This change, when applied to the NO_x problem, reduced calculated ozone depletions by nearly a factor of two.[42]

Aircraft effects in a chlorine-containing atmosphere were later shown to be sensitive to the amount of chlorine present; NO_x had less negative effect on ozone as the total chlorine content was increased. As an example in calculations performed at Lawrence Livermore Laboratory in 1976[43] (see Figure 3.26 in reference) a large (2.46×10^9 kg/

year) injection at 17 km of NO_x led to a 1.2% depletion of ozone at zero total chlorine but a +0.3% increase at 4 ppb total chlorine. Water injections in a chlorine-containing atmosphere were found to decrease ozone; thermal feedback effects, however, appeared to largely negate the water effect.

Summarizing, at the end of 1976, the subsonic NO_x effect on ozone had been reversed from that found in CIAP; calculated effects of SST's were smaller, uncertainties in SST effects were greater, and interaction with more current threats was evident.

VI. THE KINETICS REVOLUTION, 1977 to 1978

The reactions discussed by Duewer et al.,[34,35] in their uncertainty analysis included several involving the HO_2 radical. The effects of a rate change in one of these ($HO_2 + \rightarrow H_2O + O_2$) have already been noted. In general, during the CIAP period, rates for these reactions had been derived indirectly, as direct means for measuring HO_2 concentration had not been available. In 1977, however, Howard and Evenson,[44] utilizing a laser-magnetic resonance procedure to follow the decay of HO_2 in the presence of NO, found that, at 296K and 140 to 220 Pascals, the reaction rate constant for

$$HO_2 + NO \rightarrow HO + NO_2 \tag{4}$$

is $(8.1 \pm 1.5) \times 10^{-12}$ cm³/mol-sec. This reaction, during CIAP, had been given[45] a preferred value of 2×10^{-13} cm³/sec (with an uncertainty factor of 3). The measured rate was some 40 times the prior preferred value. This unexpectedly high rate "short-circuits" the NO catalytic cycle by providing a second path, which does not destroy ozone, to NO_2.

$$NO + O_3 \rightarrow NO_2 + O_2 \tag{5}$$

followed by Reaction 4.

NO_2 of course continues to react with O atoms in the catalytic cycle to form NO, but an increased NO_2 concentration (relative to that in a model using a slow rate) also provides, by photolysis, a pathway which forms ozone. The high rate for Reaction 4, coupled with the reduced rate for Reaction 1 (and other less significant changes), tended initially to increase the model natural atmospheric ozone column above that measured,[6] creating doubts about model validity. It also implied that SST NO_x injections, except at very high injection rates, would *increase* the ozone column through enhancements of ozone in the lower stratosphere.

Crutzen and Howard[46] pointed out that computed stratospheric ozone enhancements due to NO_x injection required a fast rate for the reaction

$$HO_2 + O_3 \rightarrow OH + 2O_2 \tag{6}$$

The argument is subtle inasmuch as a fast rate for this reaction clearly destroys ozone. The issue however relates to NO_x injections into a model atmosphere. The fast rate decreases the model stratospheric ozone (a needed effect as noted above); NO_x injections, however, *reduce the reduction,* leading to an ozone enhancement relative to the model atmosphere, by reducing HO_2 concentrations through reactions such as Reaction 4. If Reaction 6 is slow, this mode of O_3 destruction is less important, and SST NO_x injections may deplete ozone (depending on choice of reaction rates).

This issue was apparently resolved in spring 1978 when the rate for $HO_2 + O_3$ was measured[47] again using a laser magnetic resonance technique, and found to be much *faster* than expected, being $(1.4 \pm 0.4) \times 10^{-14}$ e − $(580 \pm 100)/T$. At 230K, the central value from this expression is 1.1×10^{-15} cm^3/sec. This is to be compared with the rate from NASA[48] [7.3×10^{-14} e − 1275/T or 2.9×10^{-16} at 230K] or the value from CIAP [1×10^{-13} e − 1250/T or 4.4×10^{-16} at 230K]. A slower rate had been recommended in a 1977 NASA workshop[46] [1×10^{-13} e − 1525/T or 1.3×10^{-16} at 230K].

These developments in reaction rates have had a profound effect on the understanding of ozone, both stratospheric and tropospheric, raising a number of questions in the process. NO_x injections in the various models, with this chemistry, except at very high rates, whether from SST's or subsonics, increase the ozone column. Calculated ozone depletion due to chlorine has been substantially increased by these rate changes, as has the slight ozone reduction attributed to water.

VII. 1979 STATUS OF COMPUTED SST EFFECTS ON OZONE

As understanding of the many species and processes controlling ozone has grown, and models have become more complex, it has become necessary to be more precise in describing a model result and the model reference atmosphere when reporting the computed effect of a perturbation to that atmosphere. The time-dependent aspects for example can be troublesome, as stratospheric chlorine compounds, CO_2, and NO_x (from N_2O) and other species all will change with time; a model computation may (withough necessarily so stating) be of a year 2000 perturbation in a 1978 atmosphere, or a 1978 perturbation in a halocarbon-equilibrated 2050 atmosphere. Groves and Tuck[49] for example, have shown that an increase to 600 ppmv of CO_2, which might occur by about 2020, reduces the chlorofluoromethane effect on ozone by a factor of about 25%. H_2O and NO_2 effects are sensitive to chlorine content. Many other details also have a strong influence on results; inclusion or noninclusion of thermal feedback effects for example has strong effect on computed water effects on ozone. Similarly, inclusion or noninclusion of the troposphere in a model may affect the sign of a computed change in an ozone column. Any given model results thus must be examined and carefully evaluated in some detail.

Available model results (not all of which provide enough detail to be examined in terms of the foregoing) include the model predictions from a number of groups provided at the June 1979 NASA Harper's Ferry Workshop (unpublished at the time of preparation), and a recent (but unfortunately already dated) published 1-D assessment from NASA-Ames[50] and later 1-D results from the same group.[51]

One part of the Harper's Ferry Workshop involved a comparison of models, using in the case of interest here a single NO_x perturbation. The results are reasonably consistent, presumably reflecting each modeler's use of the same chemistry (specified as a "package" identified as Jet Propulsion Laboratory-79 or "JPL-79"), and all models show ozone column enhancement for the specified (large) injection rates of NO_x. This column enhancement is made up of an enhancement in the lower stratosphere and a depletion in the upper stratosphere. The enhancement and depletion regions will vary with K_z profile, feedback effects, chemistry used, with whether or not a troposphere is included in the model, etc.

VIII. DISCUSSION

A. Attempts to Establish Predictive Validity

It has been shown that dramatic changes in computed effects of perturbations to the stratosphere have resulted as new reactions, new rates, and new species have been introduced into models. Yet at each stage in this process (with minor exception) the models have appeared to duplicate reasonably well the always scattered measurements of trace species. These observations, coupled together, demonstrate that while a model may (in fact, must) duplicate reasonably a natural atmosphere in various particulars, history shows that it may be totally incorrect when used as a predictive tool. For this reason, observations of effects following known perturbations are invaluable and must always be critically examined; unfortunately, as was the case with the nuclear test injections of NO_x[23,32] such observations tend to involve many ambiguities.

It should be noted that various categories of validation can be described. Thus a prediction of a change in the concentration of a trace species with altitude, later supported by measurement, provides an obvious, if partial, form of model validation. Ratios of reactive species in a region of photochemical equilibrium, if enough species can be measured simultaneously, can also provide extremely useful information and, should anomalies appear in concentration, ratios, etc., provide even more useful information as the model undergoes critical examination for possible missing reactions, improper treatment of transport processes, etc. (Some examples of these anomalies will be given in the next section.) These partial validation efforts must proceed but "global" tests of the models, if such can be found, would obviously be of more interest.

Several "global" (or at least quasiglobal) tests of the models can be noted. These come about from the nuclear weapons tests of the 1958 to 1962 period, the solar proton event of August 1972, solar variability (of UV output) over the 11-year cycle, and from Umkehr measurements of ozone at altitudes where theory says ClX species should be causing a decrease in ozone with time. A proper discussion of these effects would be lengthy and is beyond the scope of this chapter. It should be noted, however, that difficulties have generally been encountered when calculations and observations of these events have been compared in detail.

B. Current Uncertainty of Predictions

With all said and done, and assuming (quite reasonably) that no attempt to establish predictive validity has been fully successful and in spite of prior failures in estimating uncertainties (as discussed earlier), users and other interested parties are still frequently asked to estimate the "error" of their current model's prediction. While we recognize the validity of such a request, we contend that, at present, an accurate estimate of error is simply not possible. In order to estimate the accuracy of a prediction, one must know what factors affect that prediction, what the interactions are among those factors, and the amount of uncertainty associated with each factor and interaction mechanism. If a model is generally complete from the viewpoint of an acceptably accurate physical representation of interaction mechanisms (i.e., the model is a good system representation), uncertainties in these mechanisms will be well behaved. This is the case (at least to first order) in most current 1-D stratospheric models, and thus the sensitivity of a predicted ozone change to uncertainty in, say, reaction products is calculable in a straightforward manner.

Unfortunately, the situation is not the same for either our current understanding of what factors affect the model predictions (i.e., what species or chemical reactions are important and how do atmospheric motions affect them) or our current understanding of the uncertainty associated with these factors (e.g., the accuracy of a given reaction rate estimate). One has only to look at the lack of appreciation of the role of naturally occurring chlorine compounds in the 1974 to 1975 period (see, e.g., NAS[23]) to see how valid this statement has been in the past with respect to completeness of the chemical reaction set. With regard to the accuracy of "expert" judgments of the uncertainty in various reaction rate estimates, Smith[36] has amply demonstrated the case: the estimates of uncertainty are frequently so far in error that it is difficult to understate their validity.

What one is presently faced with, in estimating the uncertainty of predictions of aircraft effects on ozone, is a problem in inductive reasoning. One does not now know if he has considered all the factors which bear on the answer. It is not now known if our estimates, of the magnitudes of those factors we believe to be important, are accurate. Thus, one cannot now deduce a valid estimate of probable error in prediction. Such estimates, no matter how complex and sophisticated they may be, should be viewed as only educated guesses. The problem of developing better means to address uncertainties merits considerably more thought than it is presently receiving. With anomalies and possible "missing" reactions still identifiable, as discussed below, it is wise to bear these factors in mind when interpreting model results.

C. Anomalies

A number of current modeling questions exist, which merit continued investigations:

1. Stratospheric OH uncertainty, the HNO_3/NO_2 ratio, and the NO/NO_2 ratio: Turco[52] cites various measurements on OH and on the HNO_3/NO_2 ratio; Evans et al.[53] discuss the NO/NO_2 ratio. The various data imply at least a factor of 2 uncertainty in OH concentrations and imply there is less OH in the real stratosphere than is currently computed by the models. The point is important as reduced OH would increase ozone destruction by NO_x (SST's) and reduce ozone destruction by halogens. Various questions of the measurements and models are involved.
2. Stratospheric free chlorine: Anderson et al.[54] have shown that the measured ClO value in the stratosphere varies widely, which fact alone provides evidence of a poor horizontal mixing of the stratosphere, a fundamental postulate of 1-D models. A more important problem, however, is that measured values have *exceeded* maximum predicted model values, based on known sources and sinks of chlorine, in one case by a factor of 7 or so.[55] And finally, where the ClO was measured to be very high, the ozone concentration seemed not abnormal. A number of possible explanations have been offered; one possibility, however, aside from the obvious one of measurement error, is that an odd oxygen production mechanism exists which is not presently in the chlorine model chemistry.[55]

D. Missing Reactions

Reactions about which nothing is known are necessarily left out of models or used as adjustable parameters to fit a postulate, as implied earlier. For example, the $HO_2 + NO$ reaction was noted by Crutzen[21] but left out because no information on it existed. Similarly, the $HO_2 + O_3$ rate used in the early (1966 to 1970) water effects work was one found to be necessary to adjust the ozone distribution to observations. These facts were carefully reported, but the caveats tended to be lost when interpreting model results. A current question relates to the significance of the reaction between OH and ClO which may go according to $OH + ClO \rightarrow HCl + O_2$ or $OH + ClO \rightarrow Cl + HO_2$ or to other products. The first path being a four-center reaction, was originally considered to be slow and unimportant. Neither was included in the original halocarbon modeling. The overall reaction has recently been shown to be fast, with the majority (a minimum of 65%) of the reaction going according to the second path.[56] If 35% (which is a maximum, the range being 0 to 35%) goes by the first path, in which two reactive radicals are destroyed, a twofold reduction in halocarbon effect results. Aircraft effects would be altered, but no computations have been reported.

Bromine and to a much less extent fluorine reactions may also affect the sensitivity of ozone to chlorine and NO_x species.[52]

E. Multiple Solutions: A Modeling Problem

Prather et al.[57] have shown that current models may yield more than one set of solutions and suggest that such multiple solutions may have physical validity. Widely varying species measurements with latitude may be the result of such phenomena. This observation, if confirmed, offers an entirely new class of model validation and interpretation problems.

F. Concluding Comments

Predicted effects of NO_x and H_2O on stratospheric ozone have gone through wide changes in the last decade as understanding has grown. Currently, NO_x additions by aircraft at all plausible altitudes tend to increase the model ozone column modestly; water, if thermal feedback effects are ignored, tends to decrease ozone. The net effect for realistic NO_x emission rates is one of a small increase in the ozone column. The predicted net combined effect of aircraft with halocarbons and CO_2 changes on a time-dependent basis is not available, but interactions are unavoidable. Halocarbon ozone depletion effects will probably dominate at least over the near term.

Aircraft effects on the ozone column now seem to be of relatively little concern. However, questions and uncertainties persist whose resolution could be essential for future SST proponents. Effects on the distribution of ozone with height, a much more subtle question, may be of more significance; these changes would have climatic consequences and need study in light of other potential aircraft climatic effects, related to water vapor and sulfur dioxide emissions.

The review of these changing computed effects suggests that future changes may yet occur. The predictive capability of models, as opposed to the descriptive capability, is not yet demonstrated. No alternative yet exists, however, to the use of such models in formulating policy; their development, along with measurements of various kinds which provide the necessary supporting data, must be continued. And results must be interpreted with great care.

IX. APPENDIX

Oxides of Nitrogen Emission Estimates

SCEP Estimates

Statistics presented in SCEP[3] included an estimate of 1400 lb NO (nitric oxide) emissions per hour at cruise, with a fuel flow of 33,000 lb/hr/engine. This equates to the NO_x emission index of 65 given in Table 1. It has become conventional to quote the "oxides of nitrogen" (NO_x) emission index, since both NO and NO_2 are present in the exhaust. Thus, the volume fractions of NO and NO_2 estimated (or measured) for the exhaust are summed, and the mass emission rate of NO_x is reported as if all the NO were in the form of NO_2. These figures were qualified by a footnote which read, "The General Electric Company advises (Thompson, 1970) that the true NO output is likely to be no more than one-half to one third of the calculated values. A few past comparisons suggest that measured values will be values that are 10 to 15% of the calculated numbers (Thompson, 1970)."*

SCEP cited an FAA report[58] as the basis for the following fleet estimate: 500 SST aircraft to be flying during the period 1985 to 1990. Each of these jets would fly in the stratosphere 2500 hr/year. Of the 500 jets, 334 would be U.S. SST's and would be equipped with four engines, and 166 would be of non-U.S. fabrication and would have the equivalent of two U.S. SST engines. The fleet was presumed to cruise at 65,000 ft (20 km).

* The reference to "Thompson, 1970" in this quotation cites a personal communication from Thompson, not published.

Table 2
COMPARISON OF VARIOUS "BOEING SST" ESTIMATES

Estimate	Number of aircraft	Cruise (hr/day)	Fuel flow per aircraft (kg/hr)	Emission index (g/kg)	NO_2 injection (g/y)	Cruise altitude (km)
SCEP[3]	336 (U.S.)	6.85[a]	60,000	65.1	3.28×10^{12}	20
	164 (Non-U.S.)	6.85	30,000	65.1	8.01×10^{11}	20
	500 (Mix)	6.85	N/A	65.1	4.08×10^{12}	20
Johnston[16]	336 (U.S.)	6.85	60,000	22.8	1.15×10^{12}	20
	164 (Non-U.S.)	6.85	30,000	22.8	2.80×10^{11}	20
	500 (Mix)	6.85	N/A	22.8	1.43×10^{12}	20
"Canonical"	N/A	N/A	N/A	N/A	1.8×10^{12}[b]	N/A
Broderick/Oliver	336 (U.S.)	4.5	51,877	10.8[c]	3.09×10^{11}	19.5[d]
	164 (Non-U.S.)	4.5	25,939	10.8	7.55×10^{10}	19.5
	500 (Mix)	4.5	N/A	10.8	3.55×10^{11}	19.5

[a] 2500 hr/year.
[b] Based on assumed stratosphere fuel burnup of 1×10^{14} g/year and assumed emission index of 18.
[c] Based on core engines emission index of 16.3 multiplied by fraction of fuel burned in core engine (66%) to derive an "effective" engine emission index.
[d] Midpoint of 18 to 21 km range of cruise.

Table 2 shows the results of using the SCEP assumption to calculate stratospheric NO_x emissions of this hypothetical fleet.

Johnston's Estimates

Johnston,[16] citing his lack of expertise in the field of turbine engines, references several literature sources and a conversation with Professor R. Sawyer (University of California, Berkeley) as the basis for his reducing the original SCEP estimate of the NO_x emission rate by 65%. Though not using the words, Johnston effectively estimated, therefore, the Boeing NO_x-emission index at 23, instead of the 65 estimated by SCEP.

"Canonical"

Many atmospheric modelers have used an NO_2 injection of 1.8×10^{12} g/year as an "SST Perturbation". This number was originally based on an order-of-magnitude approximation to the amount of fuel which might be consumed "at cruise" for a large, end-of-century fleet of SST aircraft. The assumed fuel consumption was 1×10^{14} g/year; at an emission index of 18, this yields an NO_2 injection of 1.8×10^{12} g/year. There is no basis for assigning a number of aircraft to this injection.

In order to document an estimate of what the SCEP-projected emissions of a fleet of U.S. SST aircraft *should* have been, we examine each of the key variables, as described below. This estimate is also included in Table 2.

Number of Cruise Hours in the Stratosphere

The SCEP estimate cites an FAA projection (see above) as its basis for estimating 7 hr/day of flight in the stratosphere, at a height of 20 km. In order to make such an estimate, it is necessary to construct a worldwide route network, estimate the passenger traffic, take account of maintenance lay-ups, turnaround time between flights, etc. Two groups had published such studies by 1974: English and co-workers at the University of California at Los Angeles[59] and Leach and co-workers[60] at the British Aircraft Corporation. The upper-bound traffic estimate of English[59] implies (i.e., the value is *not assumed,* but may be *derived from* the data) an average annual SST cruise of 4.83 hr in the 15 to 21 km band. (This value, and others discussed herein, was calculated by working backwards from total NO_2 emissions as a baseline; using other bases, such as fuel flow, will give comparable results differing only by the effect of round-off error accumulation.)

Independently, Leach[60] presents data on a different fleet estimate. From his data for, say, a 1980 Concorde/TU-144 fleet of 150 SST's, one can derive an estimate of 4.65 hr per average day which is spent cruising in the 15 to 18 km band. Unfortunately, Leach is not specific on his definition of per-plane fuel flow for his 1990 mix of 300 Concorde/TU-144 plus 75 "second generation" SST's. However, if we assume that the second generation SST has a fuel flow equal to that of Concorde (20,000 kg/hr), we can derive a 1990 estimate of 4.36 hr/day. If, on the other hand, we assume that the "second generation" SST has a fuel flow twice that of Concorde, the estimated average daily cruise time in the 15 to 18 km band is 4.14 hr.

In many cases, these estimates agree well and we adopt a value of 4.5 hr/day for SST cruise above 15 km, as shown in Table 2.

As corroborative information, consider the following: a sample (Boeing ® #1[61]) mission summary calculation for the Boeing® 747 shows a maximum-range (5245 miles) international airborne flight time of 12.0 hr, of which 92% of this time is spent at cruise. However, 1971 Civil Aeronautics Board (CAB) operating cost and performance figures show the average trip (stage) length of the 747 in calendar year 1971 to be 2136 miles. Removing the difference (5245 to 2136) from the cruise portion of the flight allows us to approximate the average airborne flight time at 4.6 hr, of which only 3.6 hr is spent at cruise (77%). The CAB data show an average of 9.14 hr/day airborne for all Boeing® 747s. (This does not include maintenance.) Thus, the CAB data imply an average figure of 7.04 hr/day at cruise for current Boeing® 747 aircraft, not including maintenance. This value, incidentally, is in good agreement with data which can be derived from estimates of subsonic aircraft utilization.[59]

Consider now the January 21, 1969 validation data (Boeing® #2[62]) for the Boeing® SST. Maximum range is 3537 miles; total airborne time, 2.74 hr; total cruise time (2946 miles), 1.9 hr (69%). Assume an average trip length of 2136 miles (same as the Boeing® 747): The total cruise time (1545 miles) is 1.0 hr; the total airborne time (2136 miles) is 1.84 hr. In this case, total cruise time is only 54% of the total airborne time. Thus, cruise times of 4.5 hr/day for the Boeing® SST imply aircraft utilization on the order of 8.33 hr/day, not including maintenance. In addition, taxi time (which, where added to average daily airborne time, yields so-called "block-to-block" utilization) adds another fraction of time (say, 10% for a Boeing® 747, possibly more for the SST) to aircraft use. Certainly, the SCEP value of 7 hr/day at cruise is a highly optimistic estimate and can only be regarded as an "upper limit" which is difficult to justify.

Fuel Consumption

SCEP quotes an estimate of the "federal SST office" (not formally cited by SCEP) of 33,000 lb/hr for the fuel flow of each Boeing® SST engine at cruise conditions. In fact, this is a valid estimate for the *beginning* of the cruise phase, but the fuel flow varies during the course of a flight.[4] The total fuel consumed during the cruise portion of the SST's maximum-range, maximum-payload flight was designed[62] to be 216,834 lb, from which we can derive an estimated average per-engine fuel flow of 28,532 lb/hr. (On shorter flights, this value would be *lower.*) We see no justification for not using this value in calculations.

Engine-Emission Index for NO_x

Since the GE-4 SST engine was never fully developed, no measurements of its NO_x

emission index were performed. In 1970, there were no valid data upon which to base a good estimate of the emission index. For these reasons, a great deal of time was spend in 1972 to 1973 in performing measurements, under simulated flight conditions, on a close relative of the SST engine, the GE J-93 engine which powered the XB-70 supersonic bomber. These measurements are well-documented.[4] It may be assumed that the GE-4, being a sort of second-generation J-93, would have behaved in the same manner as the J-93.

Since the GE-4 engine was to use varying amounts of afterburning during cruise for thrust augmentation, it is necessary to separate afterburning NO_x emissions from the core-engine emissions. The modest temperatures and pressures of the afterburner are not high enough to form significant amounts of NO_x. During the cruise portion of the SST mission, the core engine was operated at "military" (maximum) power. Performing a least-squares exponential fit to the J-93 data described[4] (assuming a square-root dependence of NO_x formation on pressure), we arrive at the following relationships: $EI\ (NO_2) = 5.07 \times 10^{-2}\ P^{0.5}\ \exp.\ (T/435)$ where P is in psia and T is in °R. The relationship has a correlation coefficient of 0.96. The "average" cruise point for the GE-4 is 65,000 ft, Mach 2.7, where $P = 79.2$ and $T = 1559$. Thus, the average core engine emission index at cruise for the GE-4 is estimated to be 16.3 g NO_2 per kilogram of fuel burnt. But during the cruise portion of the flight, approximately 33% of the fuel is burned in the afterburner, which produces no NO_x. Thus, the "effective" emission index for the GE-4 engine is 10.8 g of NO_2 per kilogram of fuel (main engine *plus* afterburner) burnt.

REFERENCES

1. National Environmental Policy Act of 1969, Public Law 91-190, 42USC4321, 1969.
2. Clean Air Act Amendments, Public Law 91-604, December 31, 1970.
3. Man's Impact on the Global Environment, Rep. Study of Critical Environmental Problems, MIT Press, Cambridge, 1970.
4. English, J. M., Oliver, R. C., and Forney, A. K., Propulsion effluents in the stratosphere, Monograph 2, Climatic Impact Assessment Program, U.S. Department of Transportation, DOT-TST-75-52, September 1975.
5. Environmental Protection Agency, *Fed. Regist.*, 38FR19088, July 17, 1973.
6. Oliver, R. C., Bauer, E., and Wasylkiwskyj, W., Recent Developments in the Estimation of Potential Effects of High Altitude Aircraft Emissions on Ozone and Climate, Institute of Defense Analysis Paper P-1343, October 1978.
7. National Academy of Sciences, *Weather and Climate Modification, Problems, and Prospects*, Vol. 2, Research and Development, National Academy of Sciences, Washington, D.C., 1966.
8. Chapman, S., The gases of the atmosphere, *Q. J. R. Meteorol. Soc.*, 60, 127, 1934.
9. Chapman, S., A theory of upper atmospheric Ozone, *Q. J. R. Meteorol. Soc.*, 3, 103, 1930.
10. Hampson, J., Photo Chemical Behavior of the Ozone Layer, Canadian Armament Research Division Establishment Technical Note #1627, Val Cartier, Quebec, 1964.
11. Hunt, B. G., Photochemistry of ozone in a moist atmosphere, *J. Geophys. Res.*, 71, 1385, 1966.
12. Hampson, J., Atmosphere, Canadian Armament Research Division Establishment Technical Note #1738, Val Cartier, Quebec, 1966.
13. Leovy, C. B., Atmospheric ozone: an analytic model for photochemistry in the presence of water vapor, *J. Geophys. Res.*, 74(2), 417, 1969.
14. Harrison, H., Stratospheric ozone with added water vapor: influence of high-altitude aircraft, *Science*, 170, 734, 1970.
15. McDonald, J. E., Hearings before a Subcommittee of the Committee on Appropriations, House of Representatives, Civil Supersonic Aircraft Development (SST), 1971.
16. Johnston, H. S., Reduction of stratospheric ozone by nitrogen oxide catalysts from supersonic transport exhaust, *Science*, 173, 517, 1971.

17. Sullivan, W., Sorry but there's still more to say about the SST, *New York Times*, Section 4, May 30, 1971.
18a.* Grobecker, A. J., Coroniti, S. C., and Cannon, R. H., Jr., Report of Findings: The Effects of Stratospheric Pollution by Aircraft, U.S. Department of Transportation Rep. No. DOT-TST-75-50, 1975. (Available from NTIS as AD-A005458.)
18b. Reiter, E. R., Bauer, E., and Coroniti, S. C., Eds., The Natural Stratosphere of 1974, CIAP Monograph No. 1, U.S. Department of Transportation Rep. No. DOT-TST-75-51, 1975. (Available from NTIS as PB246318.)
18c. English, J. M., Oliver, R. C., and Forney, A. K., Eds., Propulsion Effluents in the Stratosphere, CIAP Monograph No. 2, U.S. Department of Transportation Rep. No. DOT-TST-75-52. (Available from NTIS as PB246319.)
18d. Robinson, G. D., Hidalgo, H., and Greenstone, R., Eds., The Stratosphere Perturbed by Propulsion Effluents, CIAP Monograph No. 3, U.S. Department of Transportation Rep. No. DOT-TST-75-53, 1975. (Available from NTIS as PB249684.)
18e. Leith, C. E., Hidalgo, H., and Sundararaman, N., Eds., The Natural Radiatively Perturbed Troposphere, CIAP Monograph No. 4, U.S. Department of Transportation Rep. No. DOT-TST-75-54, 1975. (Available from NTIS as PB249316.)
18f. Nachtwey, D. S., Caldwell, M. M., Biggs, R. H., Cutchis, P., and Gwiazdowski, J., Eds., Impacts of Climatic Change on the Biosphere, CIAP Monograph No. 5, Part 1, in Ultraviolet Radiation Effects, U.S. Department of Transportation Rep. No. DOT-TST-75-55, 1975. (Chapters 1 to 3 available from NTIS as PB247724; Chapters 4 to 7 available as PB247725.)
18g. Bartholic, J., Jensen, R. E., Caldwell, M. M., Cutchis, P., and Gwiazdowski, J., Eds., Impacts of Climatic Change on the Biosphere, in Climatic Effects, CIAP Monograph No. 5, Part 2, U.S. Department of Transportation Rep. No. DOT-TST-75-55, 1975. (Available from NTIS as PB247726.)
18h. d'Arge, R. C., Daly, G., and Patten, C. W., Eds., Economic and Social Measures of Biologic and Climatic Change, CIAP Monograph No. 6, U.S. Department of Transportation Rep. No. DOT-TST-75-56. (Available from NTIS as PB247727.)
18i. Barrington, A. E., Ed., Proceedings of the Survey Conference, February 15 to 16, 1972, Climatic Impact Assessment Program, U.S. Department of Transportation Rep. No. DOT-TSC-OST-72-13, 1972. (Available from NTIS as PB212427.)
18j. Broderick, A. J., Ed., Proceedings of the Second Conference on the Climatic Impact Assessment Program, November 14 to 17, 1972, U.S. Department of Transportation Rep. No. DOT-TSC-OST-73-4, 1973. (Available from NTIS as PB221166.)
18k. Broderick, A. J. and Hard, T. M., Eds., Proceedings of the Third Conference on the Climatic Impact Assessment Program, February 26 to March 1, 1974, U.S. Department of Transportation Rep. No. DOT-TSC-OST-74-15, 1974. (Available from NTIS as ADA003846.)
18l. Hard, T. M. and Broderick, A. J., Eds., Proceedings of the Fourth Conference on the Climatic Impact Assessment Program, February 4 to 7, 1975, U.S. Department of Transportation Rep. No. DOT-TSC-OST-75-38, 1976. (Available from NTIS as ADA068982.)
19. Nicolet, M., Ozone and hydrogen reactions, *Ann. Geophys.*, 26, 531, 1970.
20. Langley, K. F. and McGrath, W. D., The ultraviolet photolysis of ozone in the presence of water vapor, *Planet. Space Sci.*, 19, 413, 1971.
21. Crutzen, P. J., Ozone production rates in an oxygen-hydrogen-nitrogen oxide atmosphere, *J. Geophys. Res.*, 76(30), 7311, 1971.
22. Mormino, J., Sola, D., and Patten, C., Climatic Impact Assessment Report, prepared for the U.S. Department of Transportation, Contract No. DOT-TST-76-41, Washington, D.C., December 1975.
23. National Academy of Sciences, *Environmental Impact of Stratospheric Flight: Biological and Climatic Effects of Aircraft Emissions in the Stratosphere*, National Academy of Sciences, Washington, D.C., 1975.
24. Comite d'Etudes sur les Consequences des Vols Stratospheriques Activites 1972-1976, COVOS, Societe Meteorologique de France, Boulogne, France, 1976.
25. Australian Academy of Science, Atmospheric Effects of Supersonic Aircraft, Report of the Australian Academy of Science, No. 15, February 1972.
26. Johnston, H. S. and Quitevas, E., The Oxides of Nitrogen with Respect to Urban Smog, Supersonic Transports, and Global Methane, paper presented at the Int. Congr. Radiation Research, Seattle, July 14 to 20, 1974.
27. Cunnold, D., Alyea, F., Phillips, N., and Prinn, R., A three dimensional dynamical-chemical model of atmospheric ozone, *J. Atmos. Sci.*, 32(1), 170, 1975.
28. Cunnold, D. M., Alyea, F., and Prinn, R., Relative effects on atmospheric ozone of latitude and altitude of supersonic aircraft, *AIAA J.*, 15(3), 337, 1977.

* References 18a-18l are Climatic Impact Assessment Program (CIAP) publications.

29. Schmeltekopf, A. L., Albritton, D. L., Crutzen, P. J., Goldan, P. D., Harrop, W. J., Henderson, W. R., McAfee, J. R., McFarland, M., Schiff, H. I., Thompson, T. L., Hofmann, D. J., and Kjome, N. T., Stratospheric nitrous oxide altitude profiles at various latitudes, *J. Atmos. Sci.*, 34, 729, 1977.
30. Chang, J. S., Uncertainties in the validation of parameterized transport in 1-D models of the stratosphere, in Proc. 4th Conf. on CIAP, Hard, T. M. and Broderick, A. J., Eds., U.S. Department of Transportation, Cambridge, 1975.
31. Crutzen, P. J., A discussion of the chemistry of some minor constituents in the stratosphere and troposphere, *Pure Appl. Geophys.*, 106, 1973.
32. Foley, H. M. and Ruderman, M. A., Stratospheric Nitric Oxide Production from Past Nuclear Explosions and its Relevance to Projected SST Pollution, Institute of Defense Analysis Paper P-894, August 1972.
33. Foley, H. M. and Ruderman, M. A., Stratospheric NO production from past nuclear explosions, *J. Geophys. Res.*, 78, 4441, 1973.
34. Duewer, W. H., Wuebbles, D. J., Ellsaesser, H. W., and Chang, J. S., NO_x catalytic ozone destruction: sensitivity to rate coefficients, *J. Geophys. Res.*, 82(6), 935, 1977.
35. Duewer, W. H., Wuebbles, D. J., Ellsaesser, H. W., and Chang, J. S., Reply, *J. Geophys. Res.*, 82(18), 2599, 1977.
36. Smith, W. S., Jr., Uncertainties in Evaluated Atmospheric Rate Constants, paper presented at World Meteorological Organization Symp. on the Geophysical Aspects and Consequences of Changes in the Composition of the Stratosphere, Toronto, June 26 to 30, 1978, 36.
37. Molina, M. J., and Rowland, F. S., Stratospheric sink for chlorofluoro-methanes: chlorine atom-catalyzed destruction of ozone, *Nature (London)*, 249, 810, 1974.
38. Lovelock, J. E., Maggs, R. J., and Wade, R. J., Halogenated hydrocarbons in and over the Atlantic, *Nature (London)*, 241, 194, 1973.
39. Fluorocarbons — Impact on Health and Environment, Hearings before the Subcommittee on Public Health and Environment of the Committee on Interstate and Foreign Commerce, H. R. 17577 and 17545, Ser. No. 93-110, 93rd Congress, 1974.
40. National Academy of Sciences, Halocarbons: Environmental Effects of Chlorofluoromethane Release, Committee on Impacts of Stratospheric Change, National Research Council, National Academy of Sciences, Washington, D.C., 1976.
41. Kaufman, F., Halocarbons: Effects of Stratospheric Ozone, Summary of Findings, paper presented to the Panel of the National Research Council, Int. Conf. on Problems Related to the Stratosphere, Utah State University, September 15 to 17, 1976, National Academy of Sciences, Washington, D.C., 1976.
42. Broderick, A. J., Stratospheric effects from aviation, *J. Aircraft*, 15(10), 643, 1978.
43. Oliver, R. C., Bauer, E., Hidalgo, H., Gardner, K. A., and Wasylkiwskyj, W., Aircraft Emissions: Potential Effects on Ozone and Climate, prepared for High Altitude Pollution Program, FAA-77-3, Federal Aviation Administration, U.S. Department of Transportation, Washington, D.C., March 1977.
44. Howard, C. J. and Evenson, K. M., Kinetics of the reaction of HO_2 with NO, *Geophys. Res. Lett.*, 4(10), 437, 1977.
45. Reiter, E. R., Bauer, E., and Coroniti, S. C., The Natural Stratosphere of 1974, CIAP Monograph No. 1, Climatic Impact Assessment Program, U.S. Department of Transportation Rep. No. DOT-TST-75-51, 1975.
46. Crutzen, P. J. and Howard, C. J., The effect of the HO_2 + NO reaction rate constant on one-dimensional model calculations of stratospheric ozone perturbations, *Pure Appl. Geophys.*, 116, 1978.
47. Zahniser, M. S. and Howard, C. J., A Direct Measurement of the Temperature Dependence of the Rate Constant for the Reaction $HO_2 + O_3 \rightarrow OH + 2O_2$, paper presented at the World Meteorological Organization Symp. on the Geophysical Aspects and Consequences of Changes in the Composition of the Stratosphere, Toronto, June 26 to 30, 1978.
48. Hudson, R. D., Ed., Chlorofluoromethanes and the Stratosphere, NASA Publication 1010, 1977.
49. Groves, K. S. and Tuck, A. F., Simultaneous Effects of CO_2 and Chlorofluoromethanes on Stratospheric Ozone, *Nature (London)*, 280, 127, 1979.
50. Poppoff, I. G., Whitten, R. C., Turco, R. P., and Capone, L. A., An Assessment of the Effect of Supersonic Aircraft Operations on the Stratospheric Ozone Content, NASA Reference Publication 1026, 1978.
51. Turco, R. P., Whitten, R. C., Poppoff, I. G., and Capone, L. A., SST's, nitrogen fertilizer and stratospheric ozone, *Nature (London)*, 276, 805, 1978.
52. Turco, R. P., Atmospheric halogens: the status of current model predictions, position paper prepared for NASA Stratospheric Workshop, Harper's Ferry, W. Va., June 4 to 8, 1979.

53. **Evans, W. F. J., Fast, H., Kerr, J. B., McElroy, C. T., O'Brien, R. S., and Wardle, D. I.,** Stratospheric Constituent Measurements from Project Stratoprobe, paper presented at the World Meteorological Organization Symp. on the Geophysical Aspects and Consequences of Changes in the Composition of the Stratosphere, Toronto, June 26 to 30, 1978.
54. **Anderson, J. G., Grassl, H. J., Shetter, R. E., and Margitan, J. J.,** Stratospheric free chlorine measured by balloon-borne in situ resonance fluorescence, position paper prepared for NASA Stratospheric Workshop, Harper's Ferry, W. Va., June 4 to 8, 1979.
55. **Anderson, J. G., Margitan, J. J., and Stedman, D. H.,** Atomic chlorine and the chlorine monoxide radical in the stratosphere: three in situ observations, *Science,* 198, 501, 1978.
56. **Leu, M. T. and Lin, C. L.,** Rate constants for the reactions of OH with ClO, Cl_2, and Cl_2O at 298K, *Geophys. Res. Lett.,* 6(6), 425, 1979.
57. **Prather, M. J., McElroy, M. B., Wofsy, S. C., and Logan, J. A.,** Stratospheric chemistry: multiple solutions, *Geophys. Res. Lett.,* 6(3), 163, 1979.
58. The U.S. SST, A Report on Economic Feasibility, Federal Aviation Administration, Washington, D.C., 1967.
59. **English, J. M.,** Air traffic predictions for 1990-2000, in *Proc. 3rd Conf. on the Climatic Impact Assessment Program,* Broderick, A. J. and Hard, T. M., Eds., U.S. Department of Transportation Rep. No. DOT-TSC-OST-74-15, Cambridge, 1974.
60. **Leach, J. F., Morris, M. A., and Wardman, P.,** Distribution of emissions in the stratosphere from present and future high altitude aircraft, in *Proc. 3rd Conf. on CIAP,* Broderick, A. J. and Hard, T. M., Eds., U.S. Department of Transportation Rep. No. DOT-TSC-OST-74-15, Cambridge, 1974.
61. Document No. D6-30187, Boeing Commercial Airplane Company, Seattle, Washington.
62. Document No. D6A10634-1, Boeing Commercial Airplane Company, Seattle, Washington.

Chapter 2

HALOCARBONS

J. Peter Jesson

TABLE OF CONTENTS

I. INTRODUCTION

This chapter begins with a brief description of the halocarbon industry in terms of production of halogen-containing compounds and their rate of release to the atmosphere together with currently identified natural sources of chlorine compounds. Subsequent sections attempt a systematic survey of the present understanding of the impact of the compounds released on the stratospheric ozone layer.

II. HALOCARBON SOURCES

A. The Halocarbon Industry — Production and Release

A primary prerequisite for understanding the possible impact of commercially-produced halocarbons on atmospheric chemistry is a detailed knowledge of the types of compounds produced, the quantities in which they are produced, and the rate at which they are released to the atmosphere.

The quantities of the major halocarbons produced in 1973 are given in Table 1.[1-3] Figures for production are relatively accurate ($<\pm5\%$) for the U.S. since they are available through U.S. Tariff Commission reports. World production figures are, in general, less accurate. For the particular case of CCl_3F and CCl_2F_2 (FC-11 and FC-12) a concerted effort by the major manufacturers has produced accurate ($<\pm5\%$) worldwide production figures.[2,4]

Atmospheric release is much more difficult to determine and must be assessed for each material on the basis of use. Assessments of this type have been carried out for the halocarbon industry.[1,2,5-7] The results for 1973 are given in Table 2. It can be seen that during 1973, for most compounds, about half the release was in the U.S. Note also that in the cases of CCl_4 and CH_3CCl_3 there are several estimates which differ signficantly.

The compounds in Table 2 may be divided into two categories — (1) those that are produced in large quantities and at the same time are expected to have a relatively long lifetime in the troposphere (>5 years half-life), and (2) those which are produced in relatively small amounts and/or are expected to have relatively short lifetimes. This distinction is important since only if a compound is produced in large quantities and is relatively long-lived in the troposphere can significant quantities survive to reach the

Table 1
QUANTITIES OF PRINCIPAL HALOCARBONS PRODUCED IN 1973 (10^3 METRIC TONS)

Compound	U.S. production	World production	Ref.
CCl_3F	150	354	2
CCl_2F_2	220	447	2
$CHClF_2$	60	77	3
CCl_4	470	950	1
CH_3CCl_3	250	420	1
$CHCl_3$	115	225	1
C_2H_5Cl	300	550	1
CH_2ClCH_2Cl	4224	12,000	1
CH_3Cl	247	400	1
CH_2Cl_2	236	425	1
$CCl_2:CCl_2$	321	750	1
$CCl_2:CClH$	205	700	1
CH_2Cl	2432	7100	1

Table 2
ESTIMATED RELEASE OF PRINCIPAL HALOCARBONS IN 1973 (10^3 METRIC TONS)

Compound	U.S. release[1]	World release[1-3,5-7]
CCl_3F	127	290
CCl_2F_2	178	386
$CHClF_2$	28	36
CCl_4	21	42
		61
		90
CH_3CCl_3	195	324
		387
		423
$CHCl_3$	6.2	12.4
C_2H_5Cl	8.0	14.6
CH_2ClCH_2Cl	199	565
CH_3Cl	4.8	7.9
CH_2Cl_2	197	346
$CCl_2:CCl_2$	272	609
$CCl_2:CHCl$	194	648
$CH_2:CHCl$	89	352

stratosphere. Production and release of compounds in category (1) will be considered in more detail in Section II B of this chapter.

It should be noted that the natural product CH_3Cl is a partial exception to this classification: the biosphere produces an enormous flux of this compound and, even though it has a comparatively short tropospheric lifetime, CH_3Cl represents a significant source of stratospheric chlorine.

We conclude the present section by examining the factors which allow us to divide the industrial halocarbons into the two categories. One factor which will assure a relatively short lifetime in the lower atmosphere is high water solubility. Large amounts of HCl, inorganic halides, and molecular chlorine are produced commercially. Although they are, in general, intermediates in manufacturing processes and are largely converted to other chemicals instead of being released to the atmosphere, some undoubtedly escape. They have not been included in Tables 1 and 2 because they are highly water soluble and, under normal circumstances, are prevented from reaching the stratosphere by washout in rain water. In the normal troposphere, the concentration of water soluble halogen compounds drops off very rapidly with altitude.

It might be noted that bulk processes (volcanic eruption, violent storms) may transport soluble material such as HCl and NaCl to the stratosphere directly. Such activity must be considered in establishing a total chlorine budget for the stratosphere.

Organic halocarbons such as those listed in Tables 1 and 2 have very low water solubility and are not subject to efficient rainout. Some of them can, however, react with other trace species in the troposphere. The end products of these interactions are usually water soluble so that, if the conversion of the halocarbon to the soluble form is sufficiently rapid, only a small fraction of the chlorine content of the compound will reach the stratosphere. Examples are the reactions of compounds containing C−H or C=C functions with hydroxyl radicals which are present in the troposphere: $CHCl_3 + OH \rightarrow H_2O + CCl_3 \cdot \rightarrow$ water soluble products and $CCl_2{=}CCl_2 + OH \rightarrow HOCl_2C - CCl_2 \cdot \rightarrow$ water soluble products.

Fully halogenated compounds such as CCl_3F or CCl_4 do not undergo rapid reaction with trace species such as OH. They are, therefore, relatively long-lived in the troposphere and appreciable fractions survive to reach the stratosphere.

We may now divide the compounds in Table 2 into the two categories. In harmony with the above discussion CH_2Cl_2, $CHCl_3$, $CCl_2{=}CCl_2$, and $CHCl{=}CCl_2$, are found in the troposphere in concentrations <30 parts per 10^{12} by volume (ppt) relative to >300 ppt for CCl_3F and CCl_2F_2 combined. The only halocarbons found in amounts >50 ppt are CCl_3F, CCl_2F_2, CCl_4, CH_3CCl_3, and CH_3Cl. These are considered in more detail in the following sections of this chapter.

Detection of CH_2ClCH_2Cl in the atmosphere has not been reported in the literature. Singh[8] indicates that it is difficult to detect by gas chromatography since it elutes close to the CH_3CCl_3 peak, and has much lower intrinsic sensitivity in electron capture detection. Reaction with OH appears to be an order of magnitude faster[9] than for CH_3CCl_3. This, combined with the release rates given in Table 2 suggests that CH_2ClCH_2Cl will not be an important source of stratospheric chlorine.

B. Long-Lived Halocarbons — Production and Release

1. CCl_3F and CCl_2F_2

Accurate data for the production and release of CCl_3F and CCl_2F_2 have been obtained through a survey of the manufacturers carried out by independent examiners. The totals to the end of 1975 have been published.[4] Totals to the end of 1979 are available[2] and annual updates are planned. Release to the atmosphere is estimated as being close to 85% of the amount produced at any given time; this value was obtained by breaking down sales in terms of end use (refrigeration, aerosols, closed cell foam, etc.) and combining it with a release pattern for each application.[2-4] For instance, the CCl_2F_2 in domestic refrigerators is released slowly, whereas aerosol propellants are released quickly.

Total world release to the atmosphere by the end of 1979 was 4.01×10^9 kg of CCl_3F and 5.82×10^9 kg of CCl_2F_2.[2] Overall estimates of release to the atmosphere are probably accurate to ±5%. World production and release of CCl_3F and CCl_2F_2 on a year-by-year basis to the end of 1979 are given in Table 3.[2,4] The exponential growth in CCl_3F release to the end of 1974 is illustrated in Figure 1. Since 1973 the release has been relatively constant.

Table 3
CCl_3F AND CCl_2F_2 PRODUCTION AND RELEASE SUMMARY — WORLD TOTAL[2,4] (10^3 METRIC TONS)

	CCl_3F				CCl_2F_2			
	Annual		Cumulative		Annual		Cumulative	
Year	Production	Released	Production	Released	Production	Released	Production	Released
1931	0.0	0.0	0.0	0.0	0.5	0.0	0.5	0.0
1932	0.0	0.0	0.0	0.0	0.1	0.0	0.7	0.1
1933	0.0	0.0	0.0	0.0	0.3	0.1	1.0	0.2
1934	0.0	0.0	0.0	0.0	0.7	0.2	1.7	0.3
1935	0.0	0.0	0.1	0.0	1.0	0.2	2.7	0.6
1936	0.1	0.0	0.2	0.0	1.7	0.4	4.4	0.9
1937	0.1	0.0	0.3	0.1	3.1	0.6	7.5	1.5
1938	0.1	0.0	0.4	0.1	2.8	0.9	10.3	2.4
1939	0.1	0.1	0.5	0.2	3.9	1.3	14.2	3.7
1940	0.2	0.1	0.7	0.3	4.5	1.7	18.8	5.5
1941	0.3	0.1	1.0	0.4	6.3	2.3	25.0	7.8
1942	0.3	0.1	1.3	0.5	5.9	2.9	31.0	10.6
1943	0.4	0.2	1.7	0.7	8.2	3.5	39.2	14.2
1944	0.4	0.2	2.0	0.9	16.7	4.7	55.9	18.9
1945	0.4	0.3	2.4	1.2	20.1	6.1	76.0	25.0
1946	0.7	0.6	3.1	1.8	16.6	11.9	92.7	36.9
1947	1.3	1.2	4.4	3.0	20.1	19.0	112.8	56.0
1948	3.0	2.3	7.4	5.3	24.8	22.2	137.6	78.2
1949	4.5	3.7	11.9	9.1	26.1	24.2	163.7	102.4
1950	6.6	5.4	18.6	14.4	34.6	27.1	198.3	129.5
1951	9.1	7.5	27.6	21.9	36.2	30.2	234.5	159.6
1952	13.6	10.8	41.2	32.7	37.2	31.5	271.7	191.1
1953	17.3	14.7	58.5	47.4	46.5	35.5	318.2	226.6
1954	20.9	18.3	79.4	65.7	49.1	40.3	367.4	266.9
1955	26.3	22.6	105.6	88.3	57.6	45.2	425.0	312.1
1956	32.5	28.2	138.1	116.5	68.7	52.6	493.6	364.7
1957	33.9	31.6	172.0	148.1	74.2	59.8	567.8	424.5

Table 3 (continued)
CCl_3F AND CCl_2F_2 PRODUCTION AND RELEASE SUMMARY — WORLD TOTAL[2,4]
(10^3 METRIC TONS)

	CCl_3F				CCl_2F_2			
	Annual		Cumulative		Annual		Cumulative	
Year	Production	Released	Production	Released	Production	Released	Production	Released
1958	29.5	29.7	201.6	177.8	73.4	62.6	641.2	487.1
1959	35.6	30.3	237.1	208.1	87.6	69.6	728.8	556.7
1960	49.7	39.7	286.9	247.9	99.4	83.2	828.3	639.9
1961	60.5	51.2	347.3	299.0	108.5	93.2	936.8	733.2
1962	78.1	64.1	425.4	363.1	128.1	107.1	1064.9	840.3
1963	93.3	78.5	518.7	441.7	146.4	125.8	1211.3	966.1
1964	111.1	93.2	629.8	534.9	170.1	146.6	1381.4	1112.7
1965	122.8	106.3	752.6	641.2	190.1	165.6	1571.4	1278.4
1966	141.0	119.0	893.7	760.2	216.2	184.3	1787.6	1462.7
1967	159.8	135.1	1053.4	895.3	242.8	208.3	2030.4	1670.9
1968	184.8	154.6	1238.2	1049.9	278.8	238.0	2309.2	1909.0
1969	219.8	180.2	1458.0	1230.1	311.4	270.0	2620.6	2179.0
1970	241.1	205.1	1699.1	1435.2	336.9	296.2	2957.5	2475.2
1971	266.6	225.4	1965.7	1660.6	360.5	319.1	3318.1	2794.3
1972	310.5	253.6	2276.2	1914.5	401.7	348.3	3719.7	3142.6
1973	354.3	290.6	2630.5	2205.0	447.5	386.2	4167.2	3528.8
1974	377.6	320.9	3008.1	2526.0	473.6	420.3	4640.8	3949.1
1975	323.5	312.9	3331.6	2838.8	418.6	412.0	5059.4	4361.1
1976	349.9	304.1	3681.5	3142.9	449.8	395.7	5509.3	4756.8
1977	332.2	306.4	4013.7	3449.3	424.4	376.5	5933.6	5133.4
1978	321.2	291.0	4334.9	3740.3	414.1	347.6	6347.7	5481.0
1979	302.0	272.0	4636.9	4012.3	400.3	338.0	6748.0	5818.9

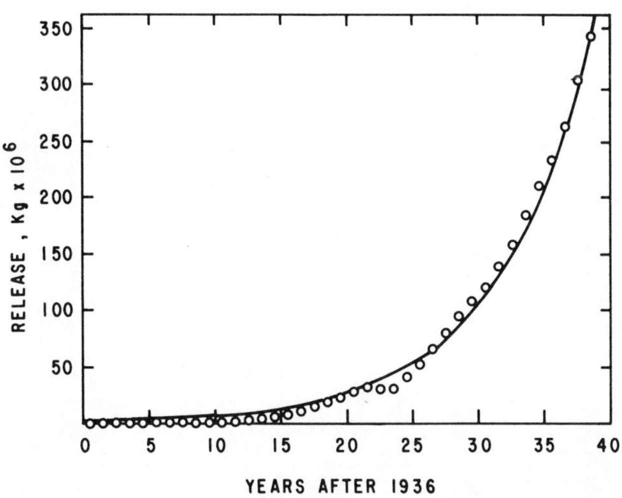

FIGURE 1. Plot of release vs time for FC-11 from 1936 to 1974 together with an exponential fit (solid curve). (Reprinted with permission from Meakin, P., et al., *Atmos. Environ.*, 12, 1271, 1978. Copyright Pergamon Press, Ltd.)

2. $CHClF_2$

Other chlorofluorocarbons (CFC) $CHClF_2$ (FC-22), CCl_2FCClF_2 (FC-113), and $CClF_2CClF_2$ (FC-114) are produced in much smaller amounts. Of these, CHF_2Cl is manufactured in the largest volume, world production through the end of 1977 for uses other than as an intermediate in the production of other chemicals being $\sim$1000 metric kilotons[3] ($\sim$1/10th the combined CCl_3F and CCl_2F_2 production). Additionally as discussed in Section III, $CHCl_2F$ is subject to attack by hydroxyl radicals in the troposphere giving it a comparatively short tropospheric lifetime. World production and release of FC-22 on a year-by-year basis to the end of 1975[3,4] is given in Table 4.

3. CCl_4

CCl_4 is currently manufactured largely as an intermediate for the production of CCl_3F and CCl_2F_2. In this application it is not released to the atmosphere in significant amounts. The use of CCl_4 for other purposes (solvent applications, fire extinguishing applications, use as a fumigating agent, etc.) has declined significantly since the 1950s.

Values for production and atmospheric emissions through 1976 are given in Table 5. The estimates are from several sources[6,10-12] and show considerable scatter (release in recent years is only a small fraction of total production and could, therefore, be subject to large errors.) Total worldwide production is also somewhat uncertain.

The estimates of Galbally[6] assume that the only significant production between 1914 and 1955 was in the U.S., U.K., France, and Germany. Estimates for later dates include Japan and other European countries. No data are available for the USSR and China. The emission estimates[6] were obtained by subtracting CCl_4 production used as a feedstock and assuming 90% of the remainder is lost to the atmosphere. Linear interpolation was used[6] to obtain estimates for those years where no production and usage figures were available. An uncertainty figure of ±30% was given for the final estimates.

4. CH_3CCl_3

Production and release figures for methyl chloroform[13,14] through 1978 are given in Table 6.

Table 4
$CHClF_2$ PRODUCTION[a] AND RELEASE — WORLD TOTAL (10^3 metric tons)

Year	Annual production	Released[b]	Cumulative production	Released[b]
1943	0.0	0.0	0.0	0.0
1944	0.0	0.0	0.1	0.0
1945	0.1	0.0	0.2	0.0
1946	0.0	0.0	0.2	0.0
1947	0.1	0.0	0.3	0.0
1948	0.2	0.0	0.5	0.1
1949	0.3	0.1	0.8	0.1
1950	0.8	0.1	1.6	0.2
1951	1.0	0.2	2.6	0.4
1952	1.6	0.3	4.2	0.7
1953	2.2	0.4	6.4	1.1
1954	2.9	0.6	9.3	1.6
1955	3.7	0.8	13.0	2.5
1956	6.1	1.2	19.1	3.7
1957	6.3	1.1	25.4	4.8
1958	7.6	1.4	32.9	6.2
1959	11.2	2.7	44.2	8.9
1960	12.2	3.3	56.3	12.2
1961	12.2	3.6	68.5	15.9
1962	15.4	4.6	83.9	20.4
1963	17.6	5.6	101.5	26.1
1964	22.4	6.8	123.8	32.9
1965	25.1	8.1	148.9	40.9
1966	31.5	9.8	180.4	50.7
1967	37.3	12.3	217.8	63.0
1968	45.6	16.3	263.4	79.3
1969	55.9	20.3	319.3	99.6
1970	58.6	23.2	377.9	122.7
1971	64.5	26.2	442.4	149.0
1972	70.0	30.9	512.4	179.9
1973	76.5	35.7	588.9	215.6
1974	87.7	41.8	676.6	257.4
1975	73.6	47.1	750.2	304.5

[a] The amounts exclude production as an intermediate for chemical or plastic production.

[b] Release obtained using scaling factors from Table 7 of Reference 4.

C. Natural Sources of Halocarbons

The presence of natural sources of halocarbons may be inferred by the detection of concentrations of halocarbons in the troposphere significantly in excess of the concentration which would be expected from human activities.

Methyl chloride is a clear example of a halocarbon with a natural source. It is released commercially in comparatively small amounts (the emissions are about 1/100th the combined emissions of CCl_3F and CCl_2F_2 on an equal weight basis [Table 2]) and yet it is the most abundant halocarbon in the troposphere (concentrations are 700 to 1000 ppt).

It is produced at least in part by a metathesis reaction of methyl iodide with the chloride ion in seawater.[15] Methyl iodide itself is found in high concentrations over kelp beds.[16] Suggestions have also been made that CH_3Cl is a product formed in the partial combustion of vegetation.[15,17] Convection columns from large fires reaching to

Table 5
PRODUCTION AND RELEASE RATES OF CARBON TETRACHLORIDE THROUGH 1976 (10^3 METRIC TONS)

	U.S. production		U.S. production not used for FC-11 & FC-12 manufacture		World				
Year	Annual	Cumulative from 1923[a]	Annual	Cumulative from 1948[b]	Production (Annual[c])	Production not used for FC-11 & FC-12 manufacture (Annual[c])	Emissions (Annual[d])		
1934	21	145			30	30	27	31	36
1939	41	316			51	51	46	54	61
1941	55	417			69	69	62	63	80
1943	79	562			99	99	89	75	117
1945	87	744			108	108	97	76	105
1946	67	812			82	82	74	75	78
1947	90	902			113	113	102	73	109
1948	97	999	63	63					
1950	98	1187	49	163	109	98	88	62	83
1955	130	1752	49	402	145	67	60	52	83
1956	137	1889	48	450	156	67	60	—	81
1957	144	2033	47	497	170	66	59	—	80
1958	142	2175	40	537	171	60	54	—	69
1959	167	2341	45	582	206	64	58	—	77
1960	169	2510	36	618	216	61	55	50	62
1961	174	2684	28	646	229	55	50	—	47
1962	219	2903	36	682	297	59	53	—	61
1963	235	3139	39	721	—	—	—	—	66
1964	243	3382	36	757	—	—	—	—	62
1965	269	3651	26	783	408	69	62	62	45
1966	294	3944	42	825	—	—	—	—	72
1967	323	4268	52	878	—	—	—	—	89
1968	346	4614	54	932	—	—	—	—	93

Table 5 (continued)
PRODUCTION AND RELEASE RATES OF CARBON TETRACHLORIDE THROUGH 1976 (10^3 METRIC TONS)

	U.S. production		U.S. production not used for FC-11 & FC-12 manufacture		World				
Year	Annual	Cumulative from 1923[a]	Annual	Cumulative from 1948[b]	Production (Annual[c])	Production not used for FC-11 & FC-12 manufacture (Annual[c])	Emissions (Annual[d])		
1969	400	5014	67	999	—	—	—	—	114
1970	458	5473	118	1117	—	—	—	78	123
1971	457	5930	102	1219	—	—	—	—	96
1972	451	6382	46	1265	904	87	78	—	79
1973	475	6858			998	100	90	—	53
1974	459	7317			1014		—	94	—
1975					907		—	—	—
1976					857		—	—	—

[a] Cumulative production prior to 1923 is estimated to be ∿50 metric kilotons.[12]
[b] Cumulative production not including use in fluorocarbon manufacture, prior to 1948 is estimated to be ∿900 metric kilotons.[12]
[c] Reference 6.
[d] References 6, 10, and 11, respectively.

Table 6
PRODUCTION AND RELEASE RATES OF METHYL CHLOROFORM THROUGH 1978[14]
(10^3 METRIC TONS)

	Annual		Cumulative	
Year	Production	Release	Production	Release
1951	0.14	0.14	0.14	0.14
1952	0.26	0.23	0.40	0.36
1953	0.98	0.95	1.38	1.32
1954	2.9	2.7	4.28	4.04
1955	8.4	8.0	12.7	12.0
1956	13.8	12.4	26.5	24.4
1957	20.6	19.6	47.1	44.0
1958	21.8	20.7	68.9	65.8
1959	31.9	30.3	100	95
1960	38	36.1	139	131
1961	40.0	38.0	179	169
1962	59.4	56.2	238	225
1963	58	50.8	296	276
1964	64.4	56.7	361	333
1965	82.5	73.0	443	406
1966	116	109	559	515
1967	146	131	705	645
1968	158	145	863	790
1969	168	148	1031	939
1970	181	155	1212	1093
1971	191	167	1403	1260
1972	267	230	1670	1490
1973	350	340	2020	1831
1974	389	363	2409	2194
1975	374	365	2783	2558
1976	438	416	3221	2974
1977	444	422	3665	3396
1978	498	473	4163	3869

the tropopause[18] provide a more direct stratospheric injection process than eddy diffusion from the surface.

Other compounds for which a natural component has been suggested are $CHCl_3$, CH_2Cl_2, and CCl_4. In all these cases the evidence is far less clear than for CH_3Cl.

D. Processes Which Could Lead to Direct Injection of Chlorine into the Stratosphere

In previous sections we have considered potential sources of stratospheric chlorine which could reach the stratosphere by relatively slow mixing from the earth's surface. This automatically leads to the establishment of criteria based on large sources and long tropospheric lifetimes. If there are mechanisms which result in direct injection of chlorine compounds into the stratosphere, clearly these criteria do not apply. The most obvious example is direct injection of inorganic chlorine from volcanoes.

A listing of the major volcanic eruptions[5] from 1955 to 1976 is given in Table 7. Table 8 gives two analyses[19,20] of the amounts of material reaching the stratosphere from the Agung eruption in 1963. Table 7 also contains a set of scaling factors giving estimates for the various major eruptions relative to the size of the Mt. Agung event.[5]

The chlorine atom injection rates to the stratosphere estimated by various groups cover a range from 1×10^6 to 300×10^6 atoms/cm^2-sec.[21,22] This gives estimates of stratospheric mixing ratios for odd chlorine of volcanic origin from 0.1 to 30 ppb.

Table 7
STRATOSPHERIC INJECTIONS DUE TO MAJOR VOLCANIC ERUPTIONS (1955-1975)[5] SCALED IN TERMS OF THE MT. AGUNG INJECTION[a] (TABLE 8)

Date of eruption	Volcano	Latitude	Magnitude of injection relative to Agung	Maximum height of injection (km)	Local tropopause height (km)
30 Mar 1956	Bezymianny	56° N	1.4	47	9.5
21-24 May 1960	Puyehue	40° S	0.2	18	11.5
17 Mar, 5 May 1963	Agung	8.5° S	1.0	23	16.3
12 Nov 1964	Sheveluch	56° N	(0.3)	(25)	9.8
12 Aug 1966	Awu	3.5° N	0.1	(20)	16.5
11-12 Jun 1968	Fernandina	0.5° S	0.1	22	16.3
10-23 Oct 1974	Fuego	14.5° N	0.3	19	16

[a] Preliminary estimate. Material injected should be distributed uniformly between the local tropopause height and the maximum height of injection. Numbers in parentheses are especially uncertain.[5]

Reproduced from Bauer, E. J., *J. Geophys. Res.*, 84, 6929, 1979 with permission. Copyright American Geophysical Union, 1979.

Alternatively the data given for Mt. Agung in Table 8 correspond to an injection of 1.5×10^{34} chlorine atoms. Globally averaged this gives 1.2 ppb ClX in the stratosphere. The lifetime of stratospheric chlorine is ∿1.5 years. An eruption the size of Agung every 5 years would give an average stratospheric loading of 2 ppb ClX from volcanoes alone.

Other possible direct injection mechanisms for stratospheric chlorine are sea salt from major storms and methyl chloride from burning vegetation. These factors must be considered in establishing the atmospheric chlorine budget. There is, however, little information on which to base a quantitative estimate.

III. TROPOSPHERIC CONCENTRATIONS

After release, halocarbons enter the troposphere, are transported upwards slowly, cross the tropopause (∿15 km) and enter the stratosphere (∿15 to 50 km). Any processes which degrade the halocarbons in the troposphere will diminish the fraction which survives to reach the stratosphere. The lifetime for any particular removal process is defined as the time required for that process to reduce a given halocarbon concentration to 1/e (0.37) times its original value.

The measured tropospheric concentrations for a variety of halocarbons[23-25] are given in Tables 9 and 10, respectively. As anticipated from the discussion in Section II, the compounds can be divided into two categories, those with average tropospheric concentrations >90 ppt (CCl_3F, CCl_2F_2, CCl_4, CH_3CCl_3, CH_3Cl) and those with average concentrations <40 ppt. In the following subsections we consider each of the compounds in the first category individually with the exception of the natural product CH_3Cl.

A. CCl_3F

It is still not known whether there are processes in the troposphere which destroy CCl_3F at a rate sufficiently fast to compete with photolysis in the stratosphere as a removal process. One way to check this point is to compare the amount of CCl_3F and

Table 8
STRATOSPHERIC INJECTIONS FROM THE MT. AGUNG ERUPTION[5]

	Mass, kg	
	Estimate of Cadle et al.[20]	Estimate of S. Self[21]
Fine ash	1×10^{10}	9×10^{9}
Total gas	—	2.9×10^{10}
Water vapor	1.2×10^{11}	2.2×10^{10}
Total S as H_2SO_4	1.8×10^{10}	2.1×10^{9}
HCl	1.2×10^{9}	7.3×10^{8}

Reproduced from Bauer, E. J., *J. Geophys. Res.*, 84, 6929, 1979 with permission. Copyright American Geophysical Union, 1979.

Table 9
BACKGROUND CONCENTRATIONS OF HALOCARBONS AT THE END OF 1977[23,24]

	Concentration[a] (ppt)[b]		
Compound	N. Hemisphere average	S. Hemisphere average	Global average
CCl_2F_2 (FC-12)	230	210	220
CCl_3F (FC-11)	133	119	126
CCl_2FCClF_2 (FC-113)	19	18	18
$CClF_2CClF_2$ (FC-114)	12	10	11
$CHCl_2F$ (FC-21)	5	4	4
CCl_4	122	119	120
CH_3CCl_3	113	75	94
CH_3Cl	611	615	613
CH_3I	2	2	2
$CHCl_3$	14	<3	8
CH_2Cl_2	44	20	32
C_2HCl_3	16	<3	8
C_2Cl_4	40	12	26

[a] For those species where significant variations within each hemisphere were observed, the average concentration within each hemisphere is the concentration that, when uniformly mixed in the hemisphere, represents the total burden of the species in that hemisphere.

[b] ppt = 10^{-12} v/v.

Reprinted from Singh, H. B., et al., *Science*, 203, 899, 1979 with permission. Copyright American Association for the Advancement of Science.

CCl_2F_2 known to have been released to the atmosphere with the amount found to be present by direct measurements.

1. Estimation of the CCl_3F Tropospheric Lifetime from a Determination of the Atmospheric Burden

There have been extensive measurements of CCl_3F over a period of 8 years. Most of these measurements have been carried out at mid-latitude locations in the Northern

Table 10
ATMOSPHERIC HALOCARBON CONCENTRATIONS (ppt) (1976-1978)

		1976	1977	1978
CCl_2F_2	North	230	258	271
	South	217	246	258
	Average	223	252	264
CCl_3F	North	134	154	164
	South	127	148	151
	Average	130	151	157
CH_3CCl_3	North	108	108	119
	South	84	91	94
	Average	96	99	106
CCl_4	North	131	137	137
	South	124	131	135
	Average	127	134	136

Hemisphere. Whatever the latitude, several problems must be resolved before a meaningful comparison can be made between measured values and the amount released,[26]

1. The measurements must represent *background* concentrations at the latitude of measurement. This is particularly difficult in the Northern Hemisphere because of proximity to the sources of release.
2. The measurements must be accurately calibrated on an *absolute* scale.
3. Since the background concentration varies as a function of latitude, any given set of measurements must be converted to a global average, using *latitude concentration adjustment factors,* before it is possible to compare the measurements with the results of a one dimensional (1-D) modeling calculation.
4. Calculated concentrations for a particular release scenario are sensitive to the *eddy diffusion scheme* chosen.
5. The accuracy of the calculated concentrations also depends on the accuracy of the *release* figures.
6. A two dimensional (2-D) model is essential for proper interpretation of the data. We consider each of these factors in turn.

a. Background Concentrations

Measurements by Singh,[23,24] Rasmussen,[25] and by Lovelock[27] show that local variability is small at remote points in the Southern Hemisphere (i.e., points well-removed from sources of release). These measurements may reasonably be judged to represent background concentrations at their respective locations. Measurements by Singh[23,24] and Rasmussen[25] at relatively remote locations in the Northern Hemisphere suggest that Northern Hemisphere background concentrations are less than 20% greater than the Southern Hemisphere background. The South/North ratio of $\sim$0.5 found by Lovelock between Capetown (35°S) and Adrigole (51°N) suggest that a significant contribution to the Adrigole measurements is arising from local sources (*vide infra*).

It appears that with careful selection of a number of locations and repeated measurement over a significant period of time, reliable samplings of background air could be obtained.

b. Absolute Calibration

In determining tropospheric lifetimes for chlorofluoromethanes (CFM) from a com-

Table 11
HALOCARBON RESULTS RETURNED FROM PARTICIPATING LABORATORIES[a]

Sample no.	FC-12 (ppt v/v)	FC-11 (ppt v/v)	CH_3CCl_3 (ppt v/v)	CCl_4 (ppt v/v)
Nominal	250	150	100	150
1	243 −2.8%	145 −3.3%		
2	246 −1.6%	136 −9.3%	97 −3%	167 +11%
3	223 −10.8%	120 −20.0%		
4	250 0%	148 −1.3%	100 0%	144 −4%
5		145 −3.3%		
6	227 −9.2%	149 −0.7%		130 −13%
7	246 −1.6%	150 0%	112 +12%	105 −30%
8	322 +28.8%	176 +17.3%		
9	250 0%	120 −20.0%	80 −20%	120 −20%
10	362 +44.8%	150 0%	103 +3%	108 −28%
11	249 0%	141 −6.0%	130 +30%	71 −53%
12		156 +4.0%	37 −63%	126 −16%
13	224 −10.4%	124 −17.3%	106 +6%	122 −19%
14		145 −3.3%		
15	243 −2.8%	123 −18.0%		
16	204 −18.4%	120 −20.0%	30 −70%	321 +114%

[a] The % figures are percent different from the submitted value.

Reprinted with permission from Rasmussen, R. A., *Atmos. Environ.*, 12, 2502, 1978. Copyright Pergamon Press, Ltd.

parison of the burden calculated from release and that found from measurement the greatest difficulties are presented by absolute calibration problems. The results of an interlaboratory calibration exercise for CCl_3F[28] involving 16 blind determinations are given in Table 11. The extremes of the range, for a single standard sample, were 120 and 176 ppt, and, although eight of the measurements were within 6 ppt of the nominal value (150 ppt), five of the measurements were at the 120 ppt extreme (20% below the nominal value). Corresponding data for CCl_2F_2, CH_3CCl_3 and CCl_4 intercalibrations are also given in Table 11. Here the results were even less satisfactory. The highest CCl_2F_2 value was 45% above the standard, the highest CCl_4 value was 114% above the standard, and the highest CH_3CCl_3 value was 30% above the standard. The overall scatter was worse for these three cases than for CCl_3F. Additionally, comparison of the data in Tables 9 and 10 shows that Rasmussen[25] finds a global CCl_3F concentration of 151 ppt for 1977 while Singh[24] finds a value of 126 for the end of 1977. Clearly both numbers cannot be correct. The difference is 20% and is equal to the calculated difference between an infinite tropospheric lifetime and a lifetime of less than 30 years.[29]

A further comparison, which is critical to the discussion of interhemispheric gradients which follows, is between the Rasmussen,[25] and Lovelock[26,27] Southern Hemisphere data for 1976 — Lovelock found 85 to 95 ppt CCl_3F at Capetown in mid 1976 (consistent with many earlier measurements taken at this location over a period of 4 years) while Rasmussen's Southern Hemisphere value for 1976 was 127 ppt. The difference is 40% and corresponds to the calculated difference between an infinite tropospheric lifetime and one of less than 20 years.[29] Since there is comparatively little difficulty in sampling background air in the Southern Hemisphere, the differences noted must be due to problems of absolute calibration. It is this problem, and not the question of interhemispheric gradients, which leads to large differences in tropospheric lifetimes calculated from the data of Rasmussen, Lovelock, and Singh.

c. Latitude Distribution

There is a higher concentration of chlorofluoromethanes (CFM) in the Northern Hemisphere (where they are released) than in the Southern Hemisphere. Measurements of Singh[23,24] and Rasmussen[25] show that currently, for background air, the difference between the two hemispheres is less than 20%. These results appeared to be in conflict with the earlier data obtained by Lovelock[30] on a voyage of the research ship Shackleton from north to south and back in 1971 to 1972 (Figure 2). A least squares fit to these data has been used to generate empirical latitude conversion factors.[29]

A set of theoretical curves for CCl_3F concentrations, calculated as a function of tropospheric lifetime, is compared with the data of Lovelock[27] in Figure 3. The data were adjusted for latitude using the empirical conversion factors from the Shackleton voyage.[29] The data, collected in both the Northern and Southern Hemispheres over a period of 5 years, lie between the 15- and 20-year lifetime curves. The fact that the Northern and Southern data, after adjustment for latitude, lie close together, is an indication of the internal consistency of the procedure. In all cases the relatively slow Hunten diffusion profile was chosen. Faster diffusion would have lead to longer estimated tropospheric lifetimes.

The Rasmussen data lead to a much longer tropospheric lifetime (close to infinite) based on a similar calculation.

We now try to identify the source of this discrepancy. The National Academy of Sciences[31] suggested that the Lovelock data for 1971 to 1972 might be less reliable than his more recent data. This explanation is not viable. Table 1 of Reference 29 summarizes the Lovelock Northern and Southern Hemisphere data from 1970 to the end of 1975. From this, one can estimate a South/North ratio for 35°S to 51°N for each of the years as shown below:

1971	0.48
1972	0.53
1974	0.51
1975	0.54

The actual value used[29] from the full data set of the 1971 to 1972 cruise corresponds to a ratio for 35°S to 51°N of 0.54, exactly the 1975 spot value.

Another explanation which has been suggested is that the decrease in the rate of annual increase of CFM production in recent years accounts in part for the discrepancy.[32] The Lovelock data (for 1976) shows a continuation in the Southern Hemisphere trend with a value of 85 to 95 ppt for mid 1976 and little change in the South/North

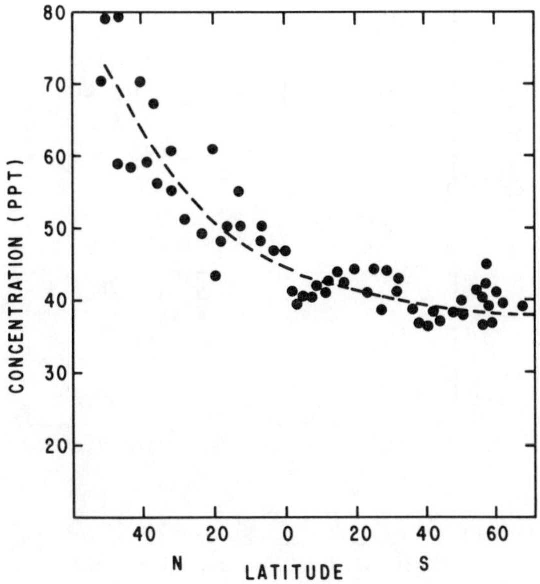

FIGURE 2. FC-11 concentrations as a function of latitude measured on a cruise of the Shackleton in 1971 to 1972. The dotted curve is a polynomial fit. (Reprinted with permission from Jesson, J. P., et al., *Atmos. Environ.*, 11, 449, 1977. Copyright Pergamon Press, Ltd.)

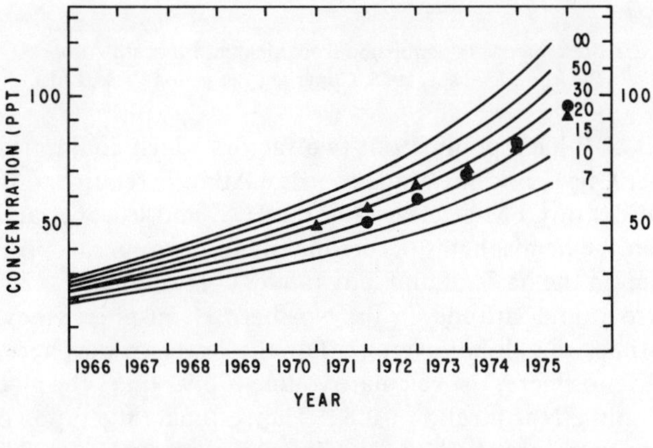

FIGURE 3. A comparison of FC-11 mixing ratios in the troposphere calculated for tropospheric lifetimes of ∞, 50, 30, 20, 15, 10, and 7 years with the measurements of Lovelock.
▲ Data from exponential least squares fit to all Bowerchalke and Adrigole data (51°N) adjusted for latitude. • Southern Hemisphere data adjusted for latitude. (Reprinted with permission from Jesson, J. P., et al., *Atmos. Environ.*, 11, 449, 1977. Copyright Pergamon Press, Ltd.)

ratio. This suggests that changes in production levels were not of major importance up to mid 1976. Lovelock's Southern Hemisphere data from the end of 1971 to mid 1976, plotted in Figure 4, show a smooth trend.

We are left with the conclusion that the North/South ratios found by Lovelock and by Rasmussen and Singh are real (apart possibly from an absolute calibration prob-

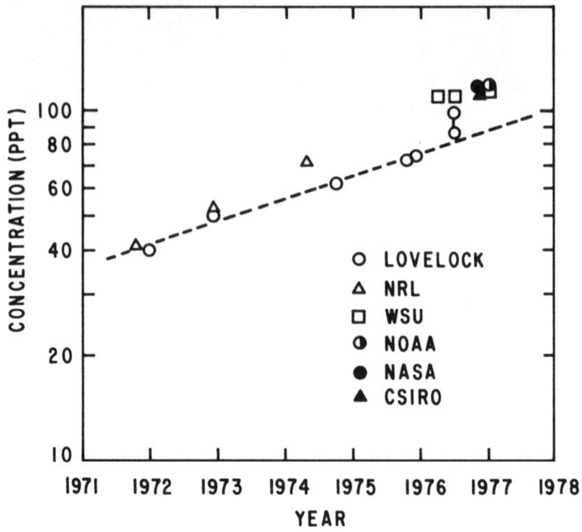

FIGURE 4. Southern Hemisphere FC-11 concentrations vs. time. The straight line is the data of Lovelock (1971 to 1976).

NRL - Naval Research Laboratory
WSU - Washington State University
NOAA - National Oceanic and Atmospheric Administration
NASA - National Aeronautics and Space Administration
CSIRO - Commonwealth Scientific and Industrial Research Organization

(Reprinted with permission from Meakin, P., et al., *Atmos. Environ.*, 12, 1271, 1978. Copyright Pergamon Press, Ltd.)

lem). Cunnold et al.[32] have pointed out two factors which could make the data compatible: (1) higher CFM concentrations over the Atlantic relative to the Pacific due to proximity to sources of CFM's released in the U.S. and Europe, and (2) the fact that exchange between the hemispheres occurs mainly in the upper troposphere. They analyzed the problem on the basis of an 8 box model (4 latitudinally × 2 vertically). In the lower troposphere at mid-latitudes in the Northern Hemisphere, they calculate a concentration 1.14 times the global average. For the lower troposphere at mid-latitudes in the Southern Hemisphere, the calculated value is 0.93 times the global average leading to a surface South/North ratio of 0.82. They estimate the ratio could be as low as 0.50 when proximity to sources in the North is taken into account. The corresponding value in the upper troposphere is 0.89 with a mid-latitude Southern Hemisphere value in the upper troposphere 0.94 times the global average.

This analysis provides the most reasonable explanation for the observed distributions. If it is correct, it leads to the conclusion that the Lovelock Southern data (lower troposphere) at any given time should correspond to the global mean concentration multiplied by a factor of the order 0.93, while the Rasmussen Southern data (upper troposphere) should correspond to the global mean concentration multiplied by a factor of the order 0.94. In other words, at any given time the two sets of data should agree in the Southern Hemisphere, with the Lovelock data being higher than the Rasmussen data in the Northern Hemisphere due both to proximity to sources of release and to the tropospheric circulation patterns. From Figure 4 we get a Southern value of 85 to 95 ppt for Lovelock in mid 1976. This translates into an approximate global mean of $90/0.93 = 97$ ppt which corresponds to a tropospheric lifetime of between

15 and 20 years, assuming the Hunten profile. In other words, if the Lovelock Northern data are high for the reasons outlined by Cunnold et al.,[32] the reanalysis leads to a shorter tropospheric lifetime.

On the other hand, the Rasmussen Southern Hemisphere data (also shown in Figure 4) give values of ∿120 ppt for 1976 leading to a global estimate of 128 ppt and a lifetime of ∿50 years for the Hunten profile.

On the basis of the analysis given above, the Lovelock and Rasmussen data should agree in the Southern Hemisphere, while Lovelock's data should be much higher than Rasmussen's in the Northern Hemisphere. What has been consistently found is approximate agreement between the two groups in the North and much higher values for Rasmussen than Lovelock in the South.

Assuming the Cunnold analysis[32] is correct, there is a systematic absolute calibration difference between the Lovelock and Rasmussen groups with the Rasmussen data being uniformly higher. The Singh data lie between the other two sets (see Tables 9 and 10 and Figure 4).

d. Eddy Diffusion

In trying to establish tropospheric lifetimes from global CFM burdens, the stratospheric photolytic sink must be properly taken into account. The effectiveness of this sink is quite sensitive to the rate of transport to the stratosphere.[26]

For example, if the hypothetical global average concentration of CCl_3F in mid 1975 was 80 ppt, the tropospheric lifetime, assuming a Hunten diffusion profile would be ∿12 years. The corresponding lifetimes for a range of other diffusion profiles[26] would be Crutzen III, 15 y; Chang I, 16 y; Chang II, 16 y; McElroy, 18 y; Crutzen II, 22 y. Thus, within a range of commonly treated diffusion profiles, the estimated tropospheric lifetime can vary by a factor of close to two.

The accuracy with which transport of CFM's to the stratosphere must be specified places a significant limitation on the accuracy with which tropospheric lifetimes can be calculated from global burdens.

e. Release

Figures for CFM release are now believed to be accurate to ±5%. This is one of the smaller uncertainties in the problem but must be allowed for in comparing measurements and theory.

f. 2-D Modeling

With the current emphasis on the development of 2-D models, it should be possible to empirically describe the transport in a much more realistic fashion. Interhemispheric gradients, as well as seasonal effects, would be expected to be correctly reproduced. A 2-D model with a convincing description of transport and proper inclusion of the photolytic sink would appear to be necessary for a sound evaluation of tropospheric lifetimes from global burdens. Such models will soon be available.

The combination of problems listed above, relative to estimation of tropospheric lifetimes from global burdens, make it unlikely that the approach will prove satisfactory in the immediate future.

2. Estimation of the CCl_3F Tropospheric Lifetime from the Rate of Increase of Atmospheric Concentration with Time

The difficulties associated with the determination of tropospheric lifetimes from absolute atmospheric burdens, have led to the examination of alternative approaches. The most promising of these is to determine lifetimes from trends in concentration as a function of time.[32] Although CFM growth rate differences for different tropospheric lifetimes are small (<3% y^{-1} between an infinite and a 10 y lifetime), the approach has the advantage that it is largely independent of absolute calibration, detailed release history, and eddy diffusion scheme. Moreover, the year to year trend at a single station is likely to be not much different from the global average trend.

In assessing the quantitative aspects of the trend approach, Cunnold et al.[32] estimated that the yearly CFM production figures are subject to a random error of ±5% and a systematic error of ±5%. An additional ±5% systematic error was assumed in the yearly release figures.

Instrumental uncertainties were estimated to give 2σ errors of ±10% for systematic effects and ±5% for random effects.

Variability arising from single station observations was assessed as ±10% for systematic and ±20% for random errors.

While the global burden approach is subject in general to both the systematic and random errors, the trend analysis approach is primarily sensitive to random errors.

It was assumed that a sufficient number of observations would be made to eliminate instrument precision problems and that repeated calibrations would allow for any long-term drift. Additionally, it was assumed that seasonal variations were either repeatable or small. (Seasonal variations appear to be very small in the Southern Hemisphere.)

Using the above criteria, it was calculated that, for a network of four stations spaced around the globe, it would be possible to establish a 10-year tropospheric lifetime from observations over a period of 3-year and a 20-year lifetime from observations over a period of 5-year (2σ confidence limits) assuming relatively constant production and release.

A network of five measuring stations has been established under the auspices of the Chemical Manufacturers Association (Chapter 7) to acquire long-term CFM growth trend data with the specific objective of establishing tropospheric lifetimes. The network has been operational since mid 1978.

B. CCl_2F_2

Measured atmospheric concentrations for CCl_2F_2 are given in Tables 9 and 10. Most of the detailed discussion of concentrations and lifetimes has been focused on CCl_3F since there are many more CCl_3F measurements over a longer period of time. All the considerations in the previous section apply, essentially unchanged, to CCl_2F_2.

C. Other Chlorofluorocarbons

Table 9 gives global average concentrations for CCl_2FCClF_2 at 18 ppt, $CClF_2CClF_2$ at 11 ppt, and$CHCl_2F$ (FC-21) at 4 ppt. Based on production, $CHClF_2$ might be expected to be the most abundant CFC after CCl_3F and CCl_2F_2. However, it is subject to attack by OH and, additionally, it is difficult to detect by electron capture gas chromatography. No reliable estimates of the atmospheric concentration of $CHClF_2$ are available.

Of the three other CFC's which were detected, CCl_2FCClF_2 and $CClF_2CClF_2$ are produced industrially. Presumably their concentrations can be roughly accounted for on the basis of historical release. $CHCl_2F$, on the other hand, is not produced commercially and total release from industrial sources could account for only 0.03 ppt.[33]

$CHCl_2F$ has been detected by a number of groups[34-37] and has been unambiguously identified using gas chromatography mass spectrometry.[37]

It has been suggested that $CHCl_2F$ may arise from Teflon®, poly (tetrafluoroethylene), off-gassing in the measuring equipment. This has been disproved by the following studies: $CHCl_2F$ is still detected in Teflon®-free equipment;[36] no increase in $CHCl_2F$ is observed when the off-gas from large amounts of finely divided Teflon® are sampled;[36] there is no indication in the manufacture, chemistry, or industrial experience with Teflon® that $CHCl_2F$ is a contaminant.[33]

Ausloos et al.[38] have observed the formation of $CHCl_2F$ following irradiation of CCl_3F with near UV light ($300<\lambda<400$ nm) in the presence of sand. Although the size of this effect in terms of a possible tropospheric sink has not been determined, it does provide direct evidence for conversion of CCl_3F into $CHCl_2F$. Thus, if $CHCl_2F$ is a real atmospheric constituent, its most likely source is as a degradation product of CCl_3F.

An alternative explanation is that $CHCl_2F$ arises from decomposition of CCl_3F on the chromatographic columns, or from degradation of stored CCl_3F standards. Either of these explanations would raise questions about the reliability of current CCl_3F measurements.

D. CCl_4

Tables 9 and 10 give measured atmospheric concentrations for CCl_4. Unlike the CFC's, CCl_4 is very uniformly distributed, presumably because the growth of emissions over the last decade has been relatively small. It is difficult to determine, because of uncertainties in the release figures for CCl_4, whether all of the current atmospheric burden is of industrial origin. The observation of elevated CCl_4 concentrations in urban areas is evidence for continued industrial emission[39] as is the observation of a continued growth rate of $\sim$2%/y in the atmospheric background concentration.[23] The ocean appears to be a somewhat more efficient sink for CCl_4 than for CCl_3F (*vide infra*).

E. CH_3CCl_3

CH_3CCl_3 is the only abundant industrial halocarbon which is subject to attack by tropospheric OH at a significant rate: $CH_3CCl_3 + OH \rightarrow H_2O + \cdot CH_2CCl_3$. For this reason most of the studies of the tropospheric concentration of CH_3CCl_3 have been used as an indirect route to determining the average tropospheric OH concentration.

Singh has noted that a fairly sharp change in the latitude gradient for CH_3CCl_3 between 20°N and 20°S is probably due to the fact that this is the region of highest tropospheric OH concentration and of highest tropospheric temperature. He finds[23,24] the average concentration in the Northern Hemisphere to be 113 ppt and in the Southern Hemisphere 75 ppt for late 1977 (Table 9). The large interhemispheric difference (50%) is clearly suggestive of a latitude dependent removal mechanism. In this regard it must be noted, however, that the hemisperic difference found by Rasmussen[25] for CH_3CCl_3 is only 24% (Table 10).

The tropospheric lifetimes and average tropospheric OH concentrations estimated in a variety of studies are given in Table 12. While there is a grouping of many of the estimates for the tropospheric lifetime around 8 y and while the estimated average tropospheric OH concentration ($\sim 5 \times 10^5$ molecules/cc) suggests a lower tropospheric OH concentration than is given by most 1-D models, it is clear that the quantitative aspects of the situation are even less satisfactory than for CCl_3F. We briefly examine the same factors for CH_3CCl_3 that we considered earlier for CCl_3F Section III A 1.

1. Background Concentrations

As for CCl_3F it should be possible to obtain background readings if sufficient samplings are taken in well-chosen areas. Comparison of Tables 9 and 10, however, suggests that this may not be possible with current measurements. The results in Table 9 show an interhemispheric difference of 50% while those in Table 10 show an interhemispheric difference of 18% for the same year.

2. Absolute Calibration

The results in Table 11 show this to be a bigger problem for CH_3CCl_3 than for CCl_3F, as the values span the range +30% to −70% from the submitted value of 100.

Table 12
ESTIMATIONS OF THE TROPOSPHERIC LIFETIME, τ, OF CH_3CCl_3 AND OF THE AVERAGE TROPOSPHERIC OH CONCENTRATION, C_{OH} BY A VARIETY OF AUTHORS

τCH_3CCl_3(y)	C_{OH}(radicals/cc)	Ref.
7.2	$4 (\pm 2) \times 10^5$	40
5—10	$<7 \times 10^5$	42
8—11	4.1×10^5	41
8	—	7
12	—	43
3.3	1×10^6	13
1.4	—	31
10	—	44

3. Latitude Distribution

As noted under (1) above there are large differences in the measured latitude distribution.

4. 2-D Modeling

To model properly the CH_3CCl_3 situation, a 2-D model is required which generates realistic relative OH distributions as a function of latitude, properly simulates inter-hemispheric mixing, and latitudinal temperature gradients, and properly includes the stratospheric sink.

5. Release

The results in Table 2 show that, unlike the CCl_3F situation, estimates of CH_3CCl_3 release differ appreciably (the highest value is 30% greater than the lowest value).

6. Rate of Reaction with OH

There is some disagreement between currently available measurements of the rate of reaction of CH_3CCl_3 with OH.

The cumulative effect of the uncertainties listed above leads to large uncertainties in both the estimated tropospheric lifetime for CH_3CCl_3 and the estimated tropospheric OH concentration.

F. Removal Mechanisms for Chlorocarbons in the Troposphere

Tropospheric removal processes become significant when their rate approaches or exceeds the overall rate of removal via stratospheric photolysis. That removal however, is a slow process, and therefore tropospheric processes need not be very fast to be important. In the case of CH_3CCl_3, removal by OH is clearly a process which competes effectively with stratospheric photolysis *(vide supra).*

In the case of fully halogenated hydrocarbons, processes of comparable efficiency have not been established to date. The National Academy of Sciences[31] identified three removal processes for CCl_3F and CCl_2F_2 which they believed warranted more study. These processes are solution in surface waters (estimated lower limit on removal time 100 y) ion molecule reactions (1000 y) and photodissociation in the troposphere (5000 y). As they stand, none of these is likely to present significant competition to stratospheric photolysis.

Singh has investigated the possibility that the ocean might act as a sink for CCl_4.

He estimates this sink to be about half as effective as the stratospheric photolysis sink. This leads to residence times in the 25 to 40 y range.[23]

A sink of possible interest for CCl_3F and CCl_2F_2 which has been suggested recently[31] is photodecomposition on the surface of sand particles.[38] Irradiation of CCl_4, CCl_3F, and CCl_2F_2 in the presence of C_2H_6 in vessels containing silica sand results in the formation of chlorine containing products.[38] The formation of these compounds occurs at wavelengths extending to 400 nm; i.e., at wavelengths where the solar radiation penetrates to the earth's surface. It is of interest that, in experiments with CCl_3F, $CHCl_2F$ was detected as a decomposition product.[38]

Because of the detection method used, the experiments described above could not be conducted under an oxygen atmosphere. More recent developments have allowed experiments to be conducted in the presence of oxygen. It was speculated that oxygen might poison the sand surface and therefore reduce the decomposition efficiency. Results of the new experiments[45] in fact show that the apparent efficiency is increased by about a factor of 40. No data are available to indicate if $CHCl_2F$ remains a product in the aerobic experiment.

Subsequently, it has been found that there is a thermal decomposition reaction of halocarbons on the surface of sand.[46] The reaction, which does *not* require radiation, is sensitive to moisture, the rate decreasing rapidly as the moisture is absorbed by the sand. The decomposition efficiencies go in the order $CH_3CCl_3 > CCl_4 > CH_3Cl > CCl_3F > CCl_2F_2$.

Earlier work by Lovelock has suggested that CCl_4 concentrations in the atmosphere are diminished in the vicinity of large bodies of windblown sand.[47] The question of photodecomposition of CCl_3F and CCl_2F_2 on airborne and on surface sand (and other surfaces) clearly merits further study. However, to date, no reliable quantitative estimate of the possible size of the effect is available.

IV. UPWARD MIXING AND PHOTOLYSIS

After release, halocarbons "diffuse" upwards in the atmosphere and a significant fraction reaches the stratosphere.

The concentrations of a variety of chlorocarbons have been measured from below the tropopause to the mid-stratosphere. Figure 5 shows calculated profiles for the chlorocarbons currently included in 1-D models of the atmosphere.[48] The stratospheric chlorocarbon profiles are largely indications of diffusion and photolysis. Since these steps in the model have not been seriously questioned, the measurements, although useful, do not contribute greatly to the overall validation of model calculations.

At sufficiently high altitudes in the stratosphere, chlorofluorocarbon molecules are broken down by sunlight to release chlorine atoms. There is little controversy about this process. The shapes of the concentration vs. altitude plots are consistent with photolysis of CCl_3F and CCl_2F_2 at the expected altitudes in the stratosphere. They are inconsistent with the assumption of no photolysis.

There is one additional question concerning the photolysis processes. It is possible that some of the intermediates in the photolytic decomposition, e.g., CClFO and CF_2O are sufficiently stable that they can build up to significant concentrations in the stratosphere:

$$CCl_3F \xrightarrow{h\nu} Cl + \dot{C}Cl_2F \xrightarrow{O_2} CClFO \quad (1)$$

$$CCl_2F_2 \xrightarrow{h\nu} Cl + \dot{C}ClF_2 \xrightarrow{O_2} CF_2O \quad (2)$$

This question can be investigated by calculating theoretically and measuring experi-

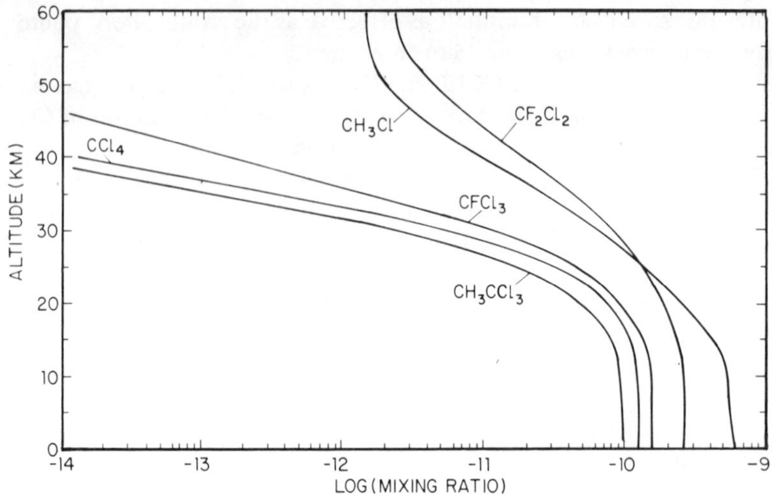

FIGURE 5. Vertical concentration profiles for the chlorocarbons currently included in 1-D models of the atmosphere.[48]

mentally the balance between the starting compounds (CCl_3F, CCl_2F_2), the intermediates ($CClFO$, CF_2O), and the stratospheric end products (HCl, HF).

It is found[49] that HF should be the most abundant flourine containing photolysis product throughout the stratosphere with an essentially constant mixing ratio of 4×10^{-10} above 35 km. The calculated HF mixing ratio is quite insensitive to the choice of eddy diffusion profile.[49] Measured HF values[50-53] range from 1 to 4×10^{-10} with the majority of the measurements at the low end of the range. Concentrations of CF_2O and $CClFO$ are calculated to be an order of magnitude or more lower.

The calculated HF:HCl ratio is characterized by a sharp increase between 18 and 28 km reflecting the susceptibility of HCl to attack by OH. The calculated peak ratio is between 2 and 4 times the observed values.[49,51,52] Measurement and theory can be brought closer together by assuming a relatively dry stratosphere (H_2O mixing ratio 1 ppm) which reduces the OH level.[49]

There are a number of problems in assessing the HF data and HF:HCl ratios which should be noted:

1. The amount of HF coming from $CClF_3$ and CCl_3F is easy to calculate, since it is the dominant stratospheric fluorine species (in the present chemical scheme) and is controlled basically by CFM photolysis and by rainout of the HF. If, in fact, measurement is a factor of 4 lower than theory, there is no way this can be reconciled with current models and new chemistry would probably be required.
2. The measured HF:HCl ratio is lower than theory by a factor of 2 to 4. In the present chemical scheme, the ratio is controlled by the OH concentration. Sze[49] has considered the possibility that the low ratio arises because the actual OH concentration is lower than that calculated in current models. Alternatively, if in fact, there is far more chlorine in the stratosphere than currently assumed, the HF:HCl ratio would be automatically lower than is currently calculated (a doubling of stratospheric chlorine [2 ppb → 4 ppb] would halve the expected HF:HCl ratio).
3. CF_4 has been identified in the atmosphere[54] and roughly estimated to be in the concentration range 0.1 to 1 ppb. This should make some contribution to the stratospheric HF level (although it could be small because of the small absorption

cross section of CF_4). Contributions of HF from sources such as CF_4 would throw measurement and calculations still further out of line.

V. THE TOTAL STRATOSPHERIC CHLORINE BURDEN

The principal identified sources of stratospheric chlorine are CH_3Cl, CCl_4, CCl_3F, CCl_2F_2, and CH_3CCl_3. The contribution of these sources to stratospheric ClX* for mid 1979 estimated using a 1-D model (Figure 5), is as follows:[48]

Source	ClX mixing ratio (ppb)
CH_3Cl	0.56
CCl_4	0.50
CH_3CCl_3	0.29
HCl	0.16
CCl_3F } CCl_2F_2 }	0.62
	2.13 ppb

The lower boundary conditions for these calculations are: known release of CCl_3F and CCl_2F_2 from the CMA survey; ground level mixing ratios of 125 ppt for CCl_4, 100 ppt for CH_3CCl_3, 625 ppt for CH_3Cl, and 1000 ppt for HCl.

For comparison, total experimental stratospheric odd chlorine can be obtained by summing the contributions of the known members of the odd chlorine family, Cl, ClO, HCl, and $ClONO_2$.

HCl measurements by a variety of techniques give mixing ratio values of 0.9 to 3.0 ppb at 35 km. ClO values vary in the range 0.2 to 8 ppb at the same altitude. The contribution of Cl atoms is negligible. An upper limit of 0.8 ppb has been suggested for $ClONO_2$ based on atmospheric infrared spectra.

The maximum experimental odd chlorine mixing ratio in the 30 to 40 km region is thus ∿12 ppb and the minimum 1.1 ppb. While the lower end of the range is near the calculated value (2 ppb), some of the individual ClO measurements exceed the calculated total by as much as a factor of 4. It is clear, therefore, that measurements of total stratospheric chlorine are critical to a testing of the models and also that there may be significant additional sources of stratospheric chlorine yet to be identified. Possible sources would be volcanic action, missing chlorocarbons, or sea salt injected by major storms.

VI. REACTIONS OF CHLORINE IN THE STRATOSPHERE

The fate of the liberated chlorine atom in the stratosphere is by far the biggest gap in our knowledge. This concerns both the quantitative aspects of the known chemistry and, more importantly, the possibility of still unknown gas phase chemistry.

The quantitative aspects of the known chemistry can be refined by laboratory measurements — primarily by measurement of reaction rate constants and photodissociation cross sections. This information can then be used as input to computer models which attempt to simulate the atmosphere. The models predict concentrations of trace species (HCl, ClO, Cl, NO_2, NO, O_3, etc.) in the atmosphere as a function of altitude and these predictions are in turn tested by direct measurement, usually using balloons. Discrepancies are then assessed in terms of possible missing chemistry or inadequacies in the models or the available experimental data.

* ClX: all forms of stratospheric chlorine other than chlorocarbons and CFC's.

Current computer models allow for the effects of:

- Vertical transport
- Diurnal variation of the solar radiation
- Multiple (Rayleigh) scattering of the solar radiation
- Updating of photolysis rates to allow for increased penetration of the solar radiation as the ozone level changes (self-healing effect)
- >100 chemical and photochemical reactions
- >30 chemical species (atoms, molecules, radicals)

Major factors limiting the utility of the models as predictive tools are

- No latitudinal transport
- Inaccuracies in experimental input data (Rate constants, absorption cross sections)
- The possibility of missing chemistry (species and reactions not currently included)
- The possibility of tropospheric sinks
- Possible missing sources of stratospheric chlorine

The basic reactions for ozone depletion by chlorine are $Cl + O_3 \rightarrow ClO + O_2$ and $ClO + O \rightarrow Cl + O_2$ which constitute a catalytic cycle. (Adding the two reactions gives $O_3 + O \rightarrow 2O_2$).

In terms of model validation, the molecules of greatest interest in the halocarbon problem are those containing chlorine. The calculated present day concentrations of the various chlorine species in the stratosphere arising from the known amounts of CH_3Cl, CCl_4, CH_3CCl_3, CCl_3F, and CCl_2F_2 in the troposphere are shown in Figure 6. The profiles are from a diurnal model and are for the afternoon.[48]

We now proceed to a comparison of experiment and model calculation for chlorine containing molecules which have been detected in the stratosphere or have been postulated as possible stratospheric intermediates.

A. HCl

One experimental feature which seems relatively well-established is the range of HCl mixing ratios in the altitude region 15 to 35 km. Measurements have been made by several groups of workers[55-61] and by two independent techniques (infrared spectroscopic and base coated filter methods). For all of these sets of measurements the maximum mixing ratio for HCl in the 15 to 35 km region lies between 0.9 to 3.0 ppb.

A selection of the HCl data is plotted in Figure 7 together with the model calculation curve from Figure 6.

A majority of the measurement sets show maxima in the HCl mixing ratios in the mid-stratosphere. This is to be contrasted with the calculated HCl profile which shows a monotonic increase in HCl concentration with altitude.

The preliminary experimental conclusion is that HCl mixing ratios go through a maximum in the mid to lower stratosphere ($\sim$30 km) and that the value of the maximum is between 1 and 3 ppb, (HCl may be quite variable). It is not unlikely that HCl mixing ratios will increase again above 40 km and mixing ratios much higher than 3 ppb are possible.

The following explanations for the discrepancy between observed and calculated HCl profiles seem possible:

1. The fall-off in HCl mixing ratio with altitude is an experimental artifact or is due to stratospheric variability.

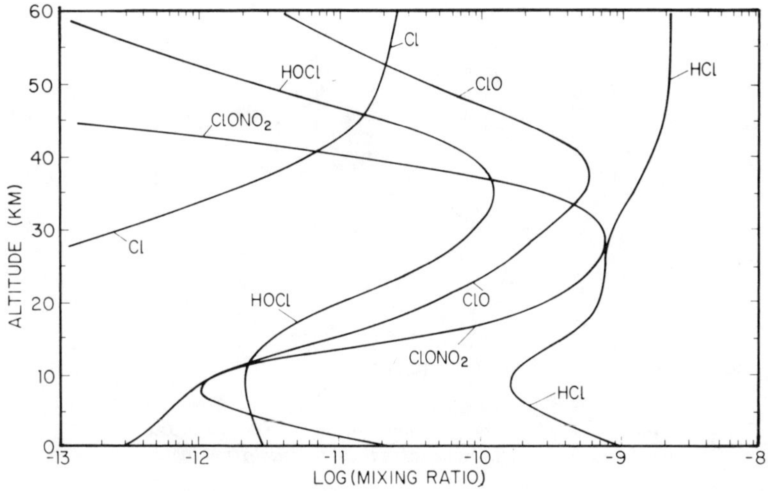

FIGURE 6. Mixing ratio profiles for ClX species from a one dimensional diurnal calculation.[48] The profiles are for 3 p.m.

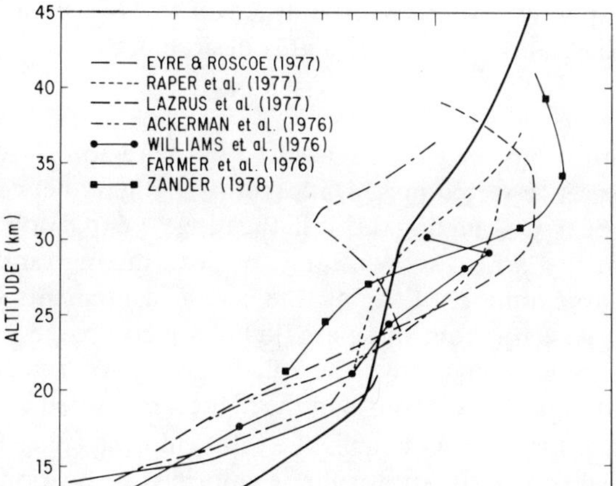

FIGURE 7. Experimental mixing ratio profiles for HCl as a function of altitude together with the theoretical profile[48] from Figure 6.

2. The fall-off in HCl mixing ratio is not reproduced by the calculations because of errors in the rate constants used for the simulations. It is possible, by adjusting rate constants within their error limits, to increase the ClO/HCl ratio in the calculations to the point where the calculations also show a fall-off.
3. The fall-off in HCl mixing ratio arises because important chemistry has been left out of the models.
4. The fall-off in HCl mixing ratio arises because of latitudinal transport effects not included in the one dimensional model.

Preliminary results with a 2-D model suggest that (4) will not provide a viable explanation.[62] Current calculations include HOCl as a stratospheric intermediate; assuming

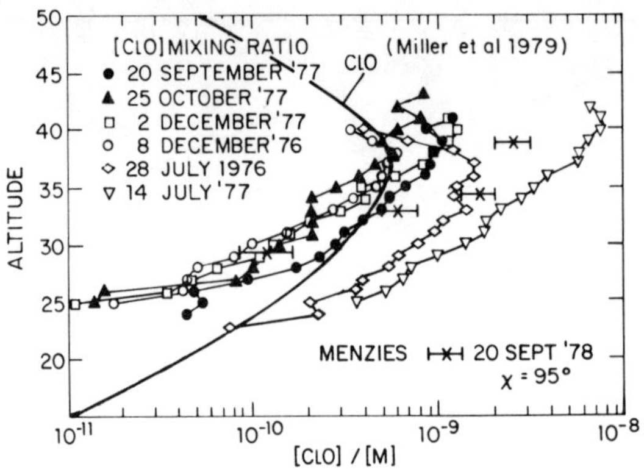

FIGURE 8. ClO Data obtained by *in situ* resonance fluorescence together with the theoretical profile[62] from Figure 6.

that laboratory data on HOCl are adequate for modeling calculations (this may not be the case) the most likely molecule which could contribute to explanation (3) is $HOClO_3$. $HOClO_3$ is photostable but to date no significant kinetic routes to its formation in the stratosphere have been established. The HCl situation needs further study. Experimental data above 40 km are highly desirable.

B. Cl, ClO

Cl and ClO concentrations have been measured in the region 20 to 40 km[63] using *in situ* resonance fluorescence techniques. Data from six flights between July 1976 and December 1977 gave peak ClO mixing ratios in the range 0.6 to 8 ppb. Thus, in contrast to the HCl measurements, which show a range of peak mixing ratios covering only a factor of ∿3 over a large number of flights, the ClO measurements vary by more than a factor of 10. The ClO data from these six flights are compared with the calculated ClO curve from Figure 6 in Figure 8. The calculation gives a peak mixing ratio of ∿0.6 ppb, an order of magnitude lower than the highest measured value.

The highest ClO measurement is 4 times the total chlorine concentration calculated for currently identified sources of stratospheric chlorine (see Section V). There appear to be three possible explanations for this fact:

1. There is a large unidentified temporary holding tank* (analogous to HCl) for chlorine in the stratosphere from which ClO can be released, possibly on a seasonal basis. Molecules such as HOCl and $HOClO_3$, would be possibilities. Explanation (1) would have to be coupled with some mechanism for getting additional chlorine to the stratosphere such as a missing chlorocarbon or a combination with explanation (2).
2. There are large, sporadic injections of chlorine into the stratosphere (volcanoes, major storms, etc.).
3. The ClO measurements are wrong.

A much more interesting feature of the ClO data is that one set of simultaneous measurements of ClO, O_3, and OH reveal normal O_3 and OH concentrations along with a high ClO value. The experimental data of Anderson[63] are shown in Figure 9 together with the calculated profiles of Miller et al.[62] for 2.13 ppb stratospheric ClX.

* A term describing a form of chlorine which is inactive for catalytic destruction of ozone, but which can be reconverted to an active form, for instance, photochemically.

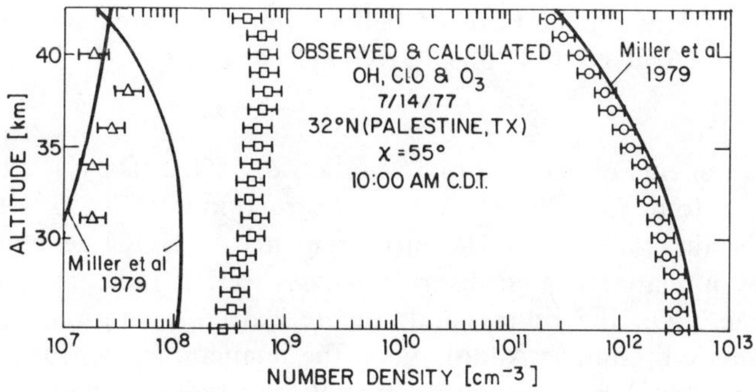

FIGURE 9. Experimental data from simultaneous measurements of ClO, O_3, and OH from Anderson[63] compared with the model calculations of Miller et al.[62]

The high ClO would be expected to deplete the ozone. Possible explanations for this key discrepancy are

1. There is an ozone generating cycle based on chlorine. This is a concept originally formulated by Prasad[64] in terms of the reactions $ClO + O_2(^1\Delta) \rightarrow ClO^* + O_2$; $ClO^* + O_2 + M \rightarrow ClO_3 + M$; $ClO_3 \xrightarrow{h\nu} OClO + O$; and $OClO \xrightarrow{h\nu} ClO + O$ where ClO^* is vibrationally excited ClO.

   Glasgow[65] has suggested the following possible reaction sequence as routes to odd oxygen formation:
   $Cl + O_2 + M \rightarrow ClOO + M$; $ClOO \xrightarrow{h\nu} ClO + O$; $ClO + O_2 + M \rightarrow OOClO + M$; $OOClO \xrightarrow{h\nu} OClO + O$; $ClO + O_2('\Delta) + M \rightarrow ClO_3 + M$; and $ClO + O_3^* \rightarrow OClO + O_2$.

   Sze and McElroy[66] have suggested the reaction $ClOO + O_2(^1\Delta) \rightarrow OClO + O_2$.

   The sequences are admittedly very speculative but would explain high ozone concentrations in the presence of high ClO concentrations.
2. The ozone had not come to steady state following an abrupt change in ClO concentration. It takes less than a day to restore the steady state following a perturbation. Explanations such as chlorine injection from a rocket exhaust or from ablation of a meteor shower have been considered, but seem unlikely.[63]
3. The ClO measurements are wrong.

The observation of normal ozone in the presence of high ClO is clearly the most important current discrepancy between observation and the predictions. Since the data arise from a single set of measurements attempts to verify them are an urgent priority in the experimental program.

Recent measurements of Menzies[67] give partial confirmation of high ClO concentrations; the peak values, measured at sunset, and theoretically extrapolated to noon to allow for diurnal effects, correspond to >3 ppb ClO.

An additional, and more firmly established, feature of the ClO data is that the experimental profiles decrease much more rapidly at altitudes below the peak than do the calculated profiles. Since the amount of ClO expected in this region is controlled by OH in the current chemical scheme, the sharp drop off may be indicative of much lower OH concentrations than the models predict.[68] This in turn could lead to substantial reductions in ozone depletion estimates.

C. $ClONO_2$

Murcray et al.[69] estimated an upper limit for $ClONO_2$ of 2 ppb at 30 km, decreasing to 4 × 10−¹¹ at 15 km, on the basis of balloon-borne infrared solar spectra. Model calculations (Figure 6) give ∿0.8 ppb $ClONO_2$ at ∿30 km.

D. HOCl

Recent measurements of the rate of formation of HOCl (HO_2 + ClO → HOCl + O_2) have shown it to be fast.[70-72] The rate of destruction (HOCl $\xrightarrow{h\nu}$ HO + Cl) is still somewhat uncertain. Laboratory UV measurements on HOCl generated from Cl_2O and H_2O suggest a relatively large absorption cross section and consequent short stratospheric lifetime.[73] On the other hand, theoretical considerations suggest a much longer stratospheric lifetime.[74] Additionally, the temperature dependence has not yet been established. More laboratory work on HOCl is required to determine how important it is in the stratosphere. At the same time, efforts are underway to detect HOCl in the stratosphere using infrared methods.

VII. REMOVAL OF CHLORINE FROM THE STRATOSPHERE

This is the final step in the chain bringing the chlorine back to the surface of the earth. At present this is assumed to occur by downward diffusion of HCl followed by rainout in the troposphere. If there are other stable, water soluble chlorine compounds generated in the stratosphere, they would also contribute to this process.

VIII. OZONE DEPLETION ESTIMATES

Using a 1-D model of atmospheric chemistry and transport which includes both diurnal effects and multiple scattering of the solar radiation, a hypothetical ozone depletion effect of 18.3% is calculated[48] for continued release of CCl_3F and CCl_2F_2 at 750,000 metric tons/year into the indefinite future.

The 18.3% effect is for steady state conditions which would take ∿300 years to achieve. It corresponds to a present day depletion effect of 1.8% and an incremental depletion effect of 0.15%/year. The uncertainties in these numbers are considered in the next section.

IX. UNCERTAINTIES IN ESTIMATES OF OZONE DEPLETION DUE TO CCl_3F AND CCl_2F_2

In this section a number or range of numbers is assigned to each effect in the model. The numbers are multiplicative factors which when multiplied by the current 18.3% ozone depletion figure indicate the range of possible depletion figures due to uncertainties in the particular effect. This is *not* an analytical error analysis (many of the effects are not amenable to an analytical approach), but does give a useful feel for the overall uncertainty picture.

An uncertainty number of 1 is an indication that the effect has been quantified quite precisely (as in the case of production and release). When the numbers are such that $1 > x > 0$ (as in the case of tropospheric lifetimes) it is an indication that the effect cannot increase ozone depletion estimates and also cannot reduce them to zero. When negative values are possible (as in the case of missing stratospheric chemistry) it is an indication that the effect could lead to ozone increases.

The effects are not necessarily all independent, but in most cases, to a good approximation, they are multiplicative.

A. Production and Release

As indicated earlier in the chapter, production and release of CCl_2F_2 and CCl_3F is probably known to <±5%. There is, therefore, a negligible uncertainty introduced into ozone depletion estimates from this effect in the model and we assign it an uncertainty factor of 1.

B. Tropospheric Sinks

The uncertainty in tropospheric lifetime spans at least the range $\infty - 10$ years.[26] Model calculations show that a 10-year tropospheric lifetime would reduce ozone depletion estimates by a factor of ∿10 and we assign an uncertainty factor range of 1 to 0.1 to this effect in the model.

C. Transport

The National Academy of Sciences (NAS) concludes[31] that the uncertainties in ozone depletion estimates due to the one dimensional transport approximation are a factor of 1.7 in either direction, i.e., an uncertainty factor range of 1.7 to 0.6. More recently, the National Aeronautics and Space Administration (NASA) has suggested that the range is larger than this.[75]

D. Experimental Data Used in Model Calculations

The data used in model calculations (reaction rates, photolysis rates, solar flux, temperature distribution in the atmosphere, sources and sink strengths of trace species, etc.) are experimental quantities and subject to experimental uncertainties. The NAS identified[31] seven key reactions as constituting the largest uncertainty in the data base and estimated that they introduced an uncertainty range of 2.5 to 0.4 into the calculations. NASA in a more recent study[75] gives the range as 1.8 to 0.3. It should be noted that the analysis includes thermal reaction rates only. Uncertainties due to photolysis rates or any of the other factors listed above will extend the ranges.

E. The Chemical Scheme in the Stratosphere

The question of completeness of the stratospheric chemical scheme is the area of greatest uncertainty in the model. By its nature it is difficult to quantify since we are asking, "Have we left any important chemistry out of the scheme?" In principal, as discussed in Section VI B, unknown chemistry could result in ozone generation due to atmospheric chlorine (just as current known chemistry allows oxides of nitrogen to generate ozone in the atmosphere below 30 km). Examples of new chemistry are the introduction of $ClONO_2$ into the chemical scheme. The NAS found[31] that this reduced ozone depletion estimates by a factor of 1.85. More recently HOCl has been suggested as a possibly important stratospheric intermediate.

In summary, it is clear that, at one extreme, the effect of CCl_3F and CCl_2F_2 on the ozone layer could be negligible. It should be noted that, at the other extreme, ozone trend analysis can put an *upper limit* on the size of the effect. The current best statistical estimate for the level at which an ozone depletion effect would already have been detected (95%±) is less than half the value calculated by the chemical models for the present day atmosphere. The conclusion from this study is that either the current chemical models are wrong (i.e., the actual steady state ozone depletion will be <9%) or there is an offsetting trend (such as increasing CO_2) which is partially cancelling the CFM effect.

X. APPENDIX

Since writing this chapter there have been many developments in the science of halocarbon behavior in the atmosphere. It would be impossible to cover these developments adequately in a short appendix. However, it is worth listing some of the major advances.

2-D chemical models of the atmosphere are now available which contain chemistry as complete as that in advanced one dimensional models.[77] This has made, and will continue to make, a large impact on our ability to validate the current picture of atmospheric chemistry. A major reason is that theory and experiment can now be compared as a function of latitude and effects such as the Noxon cliff[78] (sharp drop off in column NO_2 at high latitudes in the winter) investigated theoretically for the first time.

The quantitative aspects of the known gas phase stratospheric chemistry have continued to be refined. As suggested earlier in the chapter, the agreement between measurement and theory for ClO below 30 km has improved due to revisions in rate constants[79] which reduce the calculated OH concentration in this region. This leads to a concomittant substantial reduction in the calculated steady state ozone depletion. A significant contribution to this overall change arises from remeasurement of the absorption cross section for HO_2NO_2.[80] Slower photolysis of HO_2NO_2 leads to larger calculated stratospheric concentrations for this molecule and makes it one of the important "new" species in the chemical scheme along with $ClONO_2$.

As noted above, the rate constant revisions have greatly decreased calculated ozone depletion from fluorocarbons at steady state. The value of 18.3% given in the chapter becomes 5 to 7% in the more recent calculations.[81,82]

Additionally, more attention is now being paid to the calculation of multiple perturbations on stratospheric ozone.[83] Two important effects, in addition to the CFM effect, arise from increasing CO_2 in the atmosphere and from the effect of exhaust gases from *conventional* jet aircraft. When all these effects are lumped together, the present day calculated depletion of ozone drops from the 1.8% calculated in 1979[48] to essentially zero.

Ozone trend analysis is providing an interesting complementary approach to the resolution of the question of stratospheric ozone perturbations. Statistical analysis of the ozone data record from a large number of Dobson ozone stations, by three independent groups,[84-86] has shown that there was no abnormal trend in ozone detectable during the 1970s relative to historical behavior (the trends found in the respective studies were $1.7 \pm 2.0\%$, $0.8 \pm 1.3\%$, and $1.0 \pm 1.2\%$). Within the stated uncertainty limits, the results are consistent with the depletion calculations described in the previous paragaraph and the approach offers the possibility of an "early warning system" to alert us to potential threats to stratospheric ozone in the future.

REFERENCES

1. A. D. Little, Inc., Preliminary Economic Impact Assessment of Possible Regulatory Action to Control Atmospheric Emissions of Selected Halocarbons, Publ. No. EPA-450/3-75-073, Environmental Protection Agency, Washington, D.C., 1975.
2. Manufacturing Chemists Association Technical Panel: World Production and Release of Chlorofluorocarbons 11 and 12 through 1979, May 23, 1980.
3. Manufacturing Chemists Association Technical Panel: 1975 World Production and Sales of Fluorocarbon 22, 1976.
4. McCarthy, R. L., Bower, F. A., and Jesson, J. P., The fluorocarbon-ozone theory. I. Production and release of CCl_3F and CCl_2F_2 (Fluorocarbons 11 and 12) through 1975, *Atmos. Environ.*, 11, 491, 1977.
5. Bauer, E., A Catalog of Perturbing Influences on Stratospheric Ozone, Institute of Defense Analysis Paper P1340, Washington, D.C., September 1978.
6. Gabally, I. E., Man-made carbon tetrachloride in the atmosphere, *Science*, 193, 573, 1976.
7. McConnell, J. C. and Schiff, H. I., Methyl chloroform: impact on stratospheric ozone, *Science*, 199, 174, 1978.

8. **Singh, H. B.**, private communication, 1979.
9. **Howard, C. J. and Evenson, K. M.**, Rate constants for the reactions of OH with ethane and halogen substituted ethanes at 296°K, *J. Chem. Phys.*, 64, 4303, 1976.
10. **Singh, H. B., Fowler, D. P., and Peyton, T. O.**, Atmospheric carbon tetrachloride: another man-made pollutant, *Science*, 192, 1231, 1976.
11. **Parry, H. D.**, Ozone depletion by chlorofluoromethanes? Yet another look, *J. Appl. Meteor.*, 16, 1137, 1977.
12. U.S. International Trade Commission Reports.
13. **Neely, W. B. and Plonka, J. H.**, Estimation of time averaged hydroxyl radical concentrations in the troposphere, *Environ. Sci. Tech.*, 12, 317, 1978.
14. **Neely, W. B.**, private communication, 1979.
15. **Zafiriou, O. C.**, Reaction of methyl halides with seawater and marine aerosols, *J. Marine Res.*, 33, 73, 1975.
16. **Lovelock, J. E.**, Natural halocarbons in the air and in the sea, *Nature (London)*, 256, 193, 1975.
17. **Palmer, T. Y.**, Combustion sources of atmospheric chlorine, *Nature (London)*, 263, 44, 1976.
18. **Countryman, C. M.**, PSW-19, PSW Station USDA Forest Service, Berkeley, California, 1964.
19. **Cadle, R. D., Fernald, F. G., and Frush, C. L.**, Combined uses of lidar and numerical diffusion models to estimate the quantity and dispersion of volcanic eruption clouds in the stratosphere, *J. Geophys. Res.*, 82, 1783, 1977.
20. **Self, S.**, private communication to Bauer, E., quoted in A Catalog of Perturbing Influences on Stratospheric Ozone, Institute of Defense Analysis Paper P1340, September 1978.
21. **Stolarski, R. S. and Cicerone, R. J.**, Stratospheric chlorine: a possible sink for ozone, *Can. J. Chem.*, 52, 1510, 1974.
22. **Naughton, J. J., Lewis, V., Thomes, D., and Finlayson, J. B.**, Fume compositions found at various stages of activity at Kilauea Volcano, Hawaii, *J. Geophys. Res.*, 80, 2963, 1975.
23. **Singh, H. B., Salas, L. J., Shigeishi, H., and Scribner, E.**, Atmospheric halocarbons, hydrocarbons, and sulfur hexafluoride: global distributions, sources and sinks, *Science*, 203, 899, 1979.
24. **Singh, H. B., Salas, L. J., Shigeishi, H., and Scribner, E.**, Global Distribution of Selected Halocarbons, Hydrocarbons, SF_6, and N_2O, Publ. No. EPA-600/3-78-100, Environmental Protection Agency, Research Triangle Park, NC, 1978.
25. **Rasmussen, R. A.**, private communication, 1979.
26. **Meakin, P., Gumerman, P. S., Glasgow, L. C., and Jesson, J. P.**, The fluorocarbon-ozone theory. III. Fluorocarbon mixing and photolysis. The effects of eddy diffusion and tropospheric lifetime on CCl_3F and CCl_2F_2 tropospheric mixing ratios, *Atmos. Environ.*, 12, 1271, 1978.
27. **Pack, D. H., Lovelock, J. E., Cotton, G., and Curthoys, C.**, Halocarbon behavior from a long time series, *Atmos. Environ.*, 11, 329, 1977.
28. **Rasmussen, R. A.**, Interlaboratory comparison of fluorocarbon measurements, *Atmos. Environ.*, 12, 2505, 1978.
29. **Jesson, J. P., Meakin, P., and Glasgow, L. C.**, The Fluorocarbon-ozone theory. II. Tropospheric lifetimes — an estimate of the tropospheric lifetime of CCl_3F, *Atmos. Environ.*, 11, 499, 1977.
30. **Lovelock, J. E., Maggs, R. J., and Wade, R. J.**, Halogenated hydrocarbons in and over the Atlantic, *Nature (London)*, 241, 194, 1973.
31. National Academy of Sciences, *Halocarbons: Effect on Stratospheric Ozone*, National Academy of Sciences, Washington, D.C., 1976.
32. **Cunnold, D. M., Alyea, F. N., and Prinn, R. G.**, A. Methodology for Determining the Atmospheric Lifetime of Fluorocarbons (preprint 1978).
33. **Glasgow, L. C., Jesson, J. P., and Ward, R. B.**, Urban non-urban relationships of halocarbons, *Atmos. Environ.*, 12, 962, 1978.
34. **Singh, H. B.**, American Chemical Society Fall Meeting, Miami Beach: 1978.
35. **Cronn, D. R.**, American Chemical Society Fall Meeting, Miami Beach 1978.
36. **Rasmussen, R. A.**, American Chemical Society Fall Meeting, Miami Beach 1978.
37. **Bruner, F.**, private communication, 1979.
38. **Ausloos, P., Rebbert, R. E., and Glasgow, L. C.**, Photodecomposition of chloromethanes adsorbed on silica surfaces, *J. Res. Natl. Bur. Stand.*, 82, 1, 1977.
39. **Singh, H. B., Salas, L., Shigeishi, H., and Smith, A. H.**, Atmospheric Fates of Halogenated Compounds: Second Year (Phase I) Summary Report, EPA Grant R-80380202, SRI Project 4487, Stanford Research Institute, Menlo Park, California, 1977.
40. **Singh, H. B.**, Atmospheric halocarbons: evidence in favor of reduced average hydroxyl radical concentration in the troposphere, *Geophys. Res. Lett.*, 4, 101, 1977.
41. **Singh, H. B.**, Preliminary estimation of average tropospheric HO concentrations in the Northern and Southern Hemispheres, *Geophys. Res. Lett.*, 4, 453, 1977.

42. **Lovelock, J. E.**, Methyl chloroform in the troposphere as an indicator of OH radical abundance, *Nature (London)*, 267, 32, 1977.
43. **Chang, J. S. and Penner, J. E.**, Analysis of global budgets of halocarbons, *Atmos. Environ.*, 12, 1867, 1978.
44. **Crutzen, P. and Fishman, J.**, Average concentrations of OH in the troposphere, and the budgets of CH_4, CO, H_2, and CH_3CCl_3, *Geophys. Res. Lett.*, 4, 321, 1977.
45. **Rebbert, R. E.**, WMO Symposium Toronto, Canada, June 1978.
46. **Ausloos, P. and Rebbert, R. E.**, private communication, 1979.
47. **Lovelock, J. E.**, Presented at the Third European Geophysical Society Meeting, Amsterdam, September 1976.
48. **Miller, C., Filkin, D. L., and Jesson, J. P.**, NASA Atmospheric Modelling Comparison, Harpers Ferry, June 3-8, 1979.
49. **Sze, N. D.**, Stratospheric fluorine: a comparison between theory and measurement, *Geophys. Res. Lett.*, 5, 781, 1978.
50. **Mroz, E. J., Lazrus, A. L., and Bonelli, J. E.**, Direct measurement of stratospheric fluorine, *Geophys. Res. Lett.*, 4, 149, 1977.
51. **Farmer, C. B. and Raper, O. F.**, The HF:HCl ratio in the 14-38 km region of the stratosphere, *Geophys. Res. Lett.*, 4, 527, 1977.
52. **Zander, R. G., Roland, G., and Delbouille, L.**, Confirming the presence of hydrofluoric acid in the upper stratosphere, *Geophys. Res. Lett.*, 4, 117, 1977.
53. **Buijs, H. J., Vail, G., and Tremblay, G.**, Simultaneous Measurement of the Volume Mixing Ratios for HCl and HF in the Stratosphere, Report to CMA, Project 75-90, Washington, D.C., 1977.
54. **Gassmann, M.**, Freon-14 in the "ultra-clean" krypton and in the atmosphere, *Nat. Sci. J.*, 61, 127, 1974.
55. **Lazrus, A. L., Gandrud, B. W., Woodward, R. N., and Sedlacek, W. A.**, Direct measurements of stratospheric chlorine and bromine, *J. Geophys. Res.*, 81, 1067, 1976.
56. **Lazrus, A. L., Gandrud, B. W., Greenberg, J., Bonelli, J., Mroz, E., and Sedlacek, W. A.**, Midlatitude seasonal measurements of stratospheric acidic chlorine vapor, *Geophys. Res. Lett.*, 4, 587, 1977.
57. **Ackerman, M., Drimount, D., Girard, A., Gottignies, M., and Muller, C.**, Stratospheric HCl from infrared spectra, *Geophys. Res. Lett.*, 3, 81, 1976.
58. **Farmer, C. B., Raper, O. F., and Norton, R. H.**, Spectroscopic detection and vertical distribution of HCl in the troposphere and stratosphere, *Geophys. Res. Lett.*, 3, 13, 1976.
59. **Eyre, J. R. and Roscoe, H. K.**, Radiometric determination of stratospheric HCl, *Nature (London)*, 266, 243, 1977.
60. **Raper, O. F., Farmer, C. B., Troth, R. A., and Robbins, B. D.**, The vertical distribution of HCl in the 20 to 40 km region of the stratosphere, *Geophys. Res. Lett.*, 4, 531, 1977.
61. **Williams, W. J., Kosters, J. J., Goldman, A., and Murcray, D. G.**, Measurement of the stratospheric mixing ratio of HCl using infrared absorption technique, *Geophys. Res. Lett.*, 3, 383, 1976.
62. **Miller, C., Filkin, D. L., and Jesson, J. P.**, unpublished results, 1979.
63. **Anderson, J. G., Grassl, H. J., Shetter, R. E., Margitan, J. J.**, *Stratospheric Free Chlorine Measured by Balloon Borne in situ Resonance Fluorescence*, preprint, 1979.
64. **Prasad, S. S.**, Some aspects of the stratospheric Cl-ClO-Cl cycle: possible roles of ClO*, $ClNO_3$, and HOCl, *Planet Spac. Sci.*, 24, 1187, 1976.
65. **Glasgow, L. C.**, private communication, 1978.
66. **Sze, N. D. and McElroy, M. B.**, Report to CMA Fluorocarbon Technical Panel, Washington, D.C., 1979.
67. **Menzies, R. T.**, Remote measurement of ClO in the stratosphere, *Geophys. Res. Lett.*, 1979.
68. **Sze, N. D.**, private communication, 1978.
69. **Murcray, D. G., Goldman, A., Williams, W. J., Murcray, F. H., Bonomo, F. S., Bradford, C. M., Cook, G. R., Hanst, P. L., and Molina, M. J.**, Upper limits for stratospheric $ClONO_2$ from balloon borne infrared measurements, *Geophys. Res. Lett.*, 4, 227, 1977.
70. **Birks, J., Leck, T. J., and Shoemaker, B.**, Formation of HOCl in the Stratosphere, 4th Rocky Mountain Regional Am. Chem. Soc. Meet., Boulder, CO, June 5-7, 1978.
71. **Stimpfle, R., Perry, R. A., and Howard, C. J.**, Laser Magnetic Resonance Spectrum of ClO and the Kinetics of the Reactions of ClO with NO_2 and HO_2, 4th Rocky Mountain Regional Am. Chem. Soc. Meet., Boulder, CO, June 5-7, 1979.
72. **Reimann, B. and Kaufman, F.**, *J. Chem. Phys.*, 1978.
73. **Molina, M. J.**, Ultraviolet absorption Spectrum of HOCl, Paper presented at World Meteorological Meeting, Toronto, June 26-30, 1978.
74. **Jaffe, R. L. and Langhoff, S. R.**, Theoretical study of the photodissociation of HOCl, *J. Chem. Phys.*, 68, 1638, 1978.

75. Chlorofluoromethanes and the Stratosphere, NASA Ref. Publ. 1010, Washington, D.C., 1977.
76. **St. John, D. S.**, private communication, 1979.
77. **Miller, C., Filkin, D. L., Owens, A. J., Steed, J. M., and Jesson, J. P.**, A two-dimensional model of stratospheric chemistry and transport, *J. Geophys. Res.*, 1981.
78. **Noxon, J. F.**, Stratospheric NO_2, 2, global behavior, *J. Geophys. Res.*, 84, 5067, 1979.
79. NASA/CODATA Recommendation for Stratospheric Chemical Reaction Rates, 1981.
80. **Molina, L. T. and Molina, M. J.**, UV absorption cross sections of HO_2NO_2 vapor, *J. Photochem.*, 14, 97, 1981.
81. **Steed, J. M., Owens, A. J., Miller, C., Filkin, D. L., and Jesson, J. P.**, Two-Dimensional Model Calculations of Potential Ozone Perturbation by Chlorofluorocarbons, in preparation, 1981.
82. **Wuebbles, D. J.**, private communication, 1981.
83. **Penner, J. E.**, The Effect of Increased CO_2 on Atmospheric O_3, American Geophysical Union Fall Meeting, San Francisco, CA., December 8-12, 1980.
84. **Bloomfield, P., Thompson, M. L., Watson, G. S., and Zeger, S.**, Frequency Domain Estimation of Trends in Stratospheric Ozone, Technical Report No. 182, Series 2, Dept. of Statistics, Princeton University, January 1981.
85. **Reinsel, G., Tiao, G. C., Wang, M. N., Lewis, R., and Nyehka, D.**, *Atmos. Environ.*, 1981, in press.
86. **St. John, D. S., Bailey, S. P., Fellner, W. H., Minor, J. M., and Snee, R. D.**, *J. Geophys. Res.*, 1981, in press.

Chapter 3

NITROGEN FERTILIZERS

C. C. Delwiche

TABLE OF CONTENTS

I. GENERAL

A. Nitrous Oxide and Denitrification

Atmospheric nitrous oxide (N_2O) can contribute indirectly to the formation and destruction of ozone in the stratospheric ozone envelope,[1-5] although its contribution is highly dependent upon other simultaneous perturbations.[6] Nitric oxide (NO) resulting from the photochemical destruction of N_2O is one of the agents. NO can also contribute to the photochemical formation of ozone at lower altitudes and so a net increase in atmospheric N_2O can result in both an effect on stratospheric ozone and a lowering of the mean altitude of the ozone column.

The present low concentration of N_2O in the atmosphere (a mixing ratio of approximately 3.5×10^{-7} on a volume basis) presumably is a dynamic expression of processes of its formation and destruction. Any process or reaction which contributes to an increased N_2O concentration will influence the concentration of stratospheric ozone.

The biological reaction most commonly cited for N_2O production is denitrification in which nitrate ions (NO_3^-) or other nitrogenous compounds serve as electron acceptor for the biological oxidation of organic compounds under circumstances of low oxygen supply. Any action of management which results in increased nitrate ion in the soil may result in increased denitrification and liberation of N_2O to the atmosphere.

The production of N_2O, along with N_2, in the denitrification reaction, has long been recognized,[7] but the quantitative relationship between the two is still uncertain. There are other biological processes which can produce or utilize N_2O, and any manipulation of plant, animal, and microbial environments can influence N_2O production and utilization. Since such manipulation is an expected result of human activity, it is appropriate to establish these effects quantitatively if possible.

B. Past, Present, and Future Patterns of Nitrogen Fertilizer Use

Nitrogen deficiency is by far the most commonly encountered deficiency (other than water) in agricultural crops, and nitrogen fertilization of some type has been used historically wherever there has been agriculture involving tillage (as contrasted with grazing or simple food gathering economies). Until the development of industrial fertilizers this was accomplished with varying degrees of success by a combination of management practices including the use of "green manure" crops,* the return of organic residues to the soil, the employment of a rotation involving a fallow period, and to a very limited degree, the occasional use of naturally occurring inorganic nitrogenous deposits. Only in the period following World War I, when effective industrial methods were developed for the large-scale fixation of nitrogen, have industrial sources of nitrogenous fertilizers been employed. Since that time there has been a dramatic increase in their use. In 1950, approximately 3.5×10^6 tons fertilizer nitrogen were used, and by 1974 this had increased to 4×10^7 tons per year. It is estimated that by the year 2000 approximately 1.7×10^8 tons of fertilizer nitrogen will be required unless some alternative technology is devised.[8] The reasons for this great expansion in the use of nitrogenous fertilizers and the likelihood that the expansion will continue at least for the next few decades are complex and closely related to a number of society's problems. World population growth and the need for more food to support an expanded population is only part of the reason because fertilizer use has expanded much more rapidly than has population. The growth of industrial activity in virtually every society and a parallel demand for labor have resulted in a general shift of population from rural to urban areas almost worldwide. This population shift and industrial development have stimulated the increased mechanization of agriculture. The improved

* A crop which is plowed back into the soil as a means of fertilization.

economic status of the average population, particularly in industrialized countries, has raised costs of labor, further stimulating mechanization and intensification of agricultural production. Aside from the demand for nitrogenous fertilizers which this intensification implies, the greater efficiency (in logistics and labor) resulting from the use of high analysis fertilizers, particularly anhydrous ammonia, has encouraged the change to industrial sources of fixed nitrogen. At the same time the shift of population to urban centers has caused a general ''opening'' of the production and decay cycle. The products of agriculture are exported to urban centers and food and fiber wastes reappear remote from their site of production where they now become pollutants and are generally lost from the production cycle, adding to the requirement for nitrogen and other fertilizers. Even in many of the less industrialized countries, this exodus to the city has occurred, and is due in part to the intensification of production from the use of industrially fixed nitrogen. To some extent, the use of industrially fixed nitrogen further opens the production-utilization chain, demanding yet additional fertilizer.

The suggestion that the process of urbanization be reversed, so as to bring the waste products closer to the site of production or that the waste products of the urban area be retransported back to the soil, is perhaps a simplistic solution for several reasons. Cost in labor and energy of returning and distributing these wastes probably exceeds their value as fertilizer, particularly when only the nitrogen content is considered. In addtion, the energy content of most organic residues is sufficiently high that if technology were available for the use of this material as fuel (pyrolysis or direct combustion) the nitrogen which could be fixed by the energy thus produced would exceed that present in the waste material. Organic residues also potentially carry with them both biological and chemical contaminants which would be undesirable if returned to the soil.

Thus, a new technology which salvages both the energy and the mineral constituents of these wastes, although it might appear to be further compounding technology, probably would be a wiser solution. There has been some success in the application of sewage plant effluent water and sludge to forests, golf courses, and agricultural land. This avoids the problem of eutrification, but concentration of heavy metals and refractory organic materials, as well as the possible survival of pathogens, limit the long-term promise of such disposal. Additionally, the quantity of these materials and their concentration in urban areas make such disposal impractical (and energy consuming) in most cases. The evolution of waste disposal systems has been based upon dilution and dispersion but present limitations of water (for dilution) and space (for dispersion) make these concepts obsolete — if ever they were valid. Regardless of what solutions are finally reached for managing wastes, increased demand for industrially fixed nitrogen are almost certain. Some of the demand will be met by the use of legumes and other nitrogen-fixing crops, and probably more energy-efficient means will be devised for industrial nitrogen fixation. Present commercial processes involving the steam reforming of natural gas, or naptha, and the combination of hydrogen so obtained with atmospheric nitrogen, are energy expensive, requiring approximately 4.5×10^7 Joules of energy per kg of nitrogen fixed.[9] Other industrial processes are equivalently energy expensive.

II. FLUX PATTERNS

A. Flow of Nitrogen Through the Soil-Plant System

To help estimate the extent to which applied nitrogen fertilizer might eventually be lost to the atmosphere as nitrous oxide, it would be useful to devise a complete and quantitative flow chart accounting for all processes transferring nitrogen. Although we have some reasonably good models for some native ecosystems, it is particularly

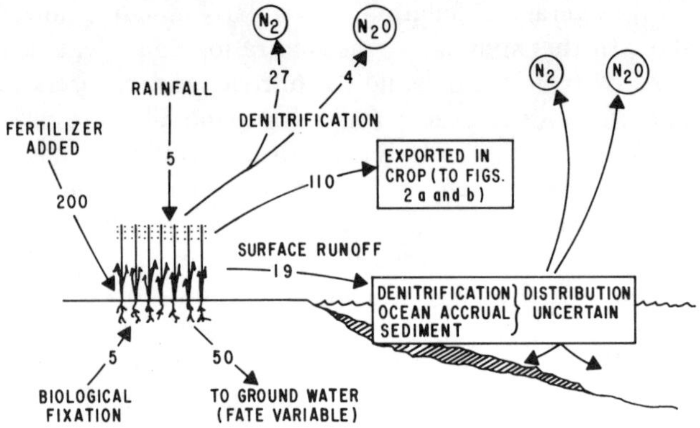

FIGURE 1. Estimated flow pattern of nitrogen in a hypothetical agricultural system. Transfer rates are in Kg N per ha per year. Assumptions: Crop is maize with a yield of 300 Kg grain per year exported, stover remaining on field. Rainfall and irrigation 1200 mm annually. Erosional losses from the soil, a fine sandy loam, are assumed to be 800 Kg/year, constituting the main mechanism for transferring nitrogen from soil to ocean.

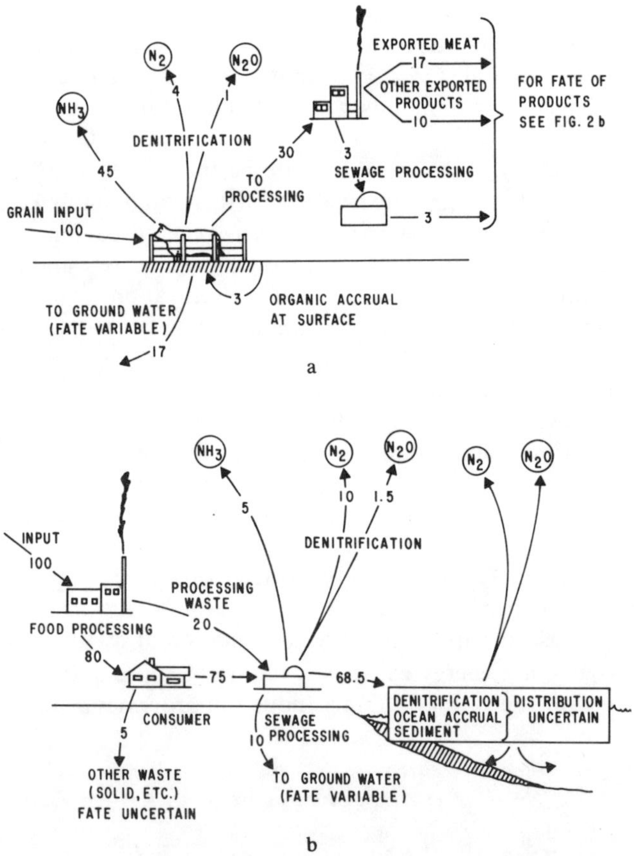

FIGURE 2. Estimated flow pattern of nitrogen, as percent of primary input, imported by an urban center comparing (a) maize going through a feed lot and (b) maize consumed as processed products. In either case the loss from the field is complete and the nitrogen is delivered to drainage waters, lost to the atmosphere by denitrification, or volatilized and returned to the soil by rainfall.

difficult to devise one for agricultural systems. There is no "typical" agricultural unit and different crops and different management regimes will behave very differently in a quantitative sense. Figure 1 outlines the various pools and transfer mechanisms which are important and gives some estimates for an intensive crop, in this case, maize.

Estimates of the fraction of applied nitrogen fertilizer which is recovered in the crop vary over a wide range from 10% for a heavily fertilized citrus crop to 70% or more with less heavily fertilized cereal crops or maize.[10] In lysimeter studies, nitrogen applications of 100 kg N per hectare per year to an annual grass and clover forage gave recoveries of approximately 60% in the harvested crop. Of the remainder, about 3% was lost by leaching and 14% unaccounted for, presumably having been lost in denitrification.[11] The flow chart in Figure 1, however, represents an example of the heavier fertilization rates typical of those used on many intensive crops.

The form in which nitrogen is lost also varies. Under some conditions N_2 can be the sole product of denitrification, whereas under others, N_2O is produced almost exclusively. Nitric Oxide (NO) is also a possible intermediate in the denitrification reaction, but because of its high reactivity, it is unlikely that large quantities of it escape from the soil. The transfer rates given in Figure 1 assume that 15% of the nitrogen lost in denitrification is released to the atmosphere as N_2O. This quantity is about 5 to 6 kg N per hectare per year in this model, and is therefore a contribution to the atmospheric N_2O reservoir.

B. The Flow of Nitrogen Through the Food Chain

Because of the shift of population to urban centers and the consequent shift in the flow of the waste products of agricultural production, the procedures for disposal of waste products have become more significant in determining the fate of applied nitrogenous fertilizers. Figure 2 outlines the pattern of this nitrogen flow and more particularly N_2O production as an additional factor influencing atmospheric ozone concentrations. As with the production system there is no "typical" route for the flow of food or feed through the utilization chain. Figure 2 shows one of the more common disposal systems in which a brief anaerobic digestion is followed by aerobic processes designed primarily to lower the biological oxygen demand (BOD) to permit the discharge of waste waters to streams or other waters. In such a system, possibly 80% of the nitrogen is eventually discharged as NO_3^- to a river or estuary.[12] Since populations are concentrated on riverways or estuaries, a model of this sort typifies a large fraction of the total sewage disposal system. NO_3^-, since it is commonly accompanied by phosphate and other essentials to plant and microbial growth may favor eutrophication of the waters involved. Eventually, the nitrogen finds its way to sediments in streams and estuarial (and coastal) deposits. In these sediments, conditions are frequently anoxic and the nitrogen may be bound in organic sediments of varying stability. In these sediments or near their surface, additional microcycles operate resulting in some oxidation of organic materials, the concomitant oxidation of ammonium ion (NH_4^+) to nitrate ion, and the possibility of denitrification in anoxic zones of the sediment. This is an additional potential source of nitrous oxide, but the sediments also are a potential sink for nitrous oxide. Many sediments are undersaturated with nitrous oxide relative to the atmosphere and some profiles of estuarial waters demonstrate a net uptake of N_2O.

Because of the related problem of release of carbon dioxide to the atmosphere by fossil fuel combustion, it is possible that the additional eutrophication of streams and estuaries and the deposition of organic material in sediments is a positive thing tending to counteract the excessive release of carbon dioxide to the atmosphere. Where detrital residues of this integrated process are deposited and how long they persist therefore become unanswered questions of considerable import. The shift in the pattern of nitrogen fertilizer utilization and increased fossil fuel utilization are interrelated and data are not available for the interpretation of the significance of this interrelationship. Probably the net result is an increased deposition of organic materials in estuaries and coastal waters. If the deposits of organic material are long-lived (in a geological sense) then to some extent they tend to ameliorate two problems. The model presented pessimistically assumes that the resultant deposits are short-lived and that therefore there is a net yield of N_2O to the atmosphere. The significance of this yield is subject to interpretation later.

III. FIELD EXPERIENCE

A. Influence of Management Practices and Fertilizer Placement

It is desirable to maximize the effectiveness of applied fertilizers in meeting the objective of increased productivity. The unpredictability of rainfall intensity and timing, as well as the characteristics of ion uptake by plants, imply that the timing of fertilizer usage and the form and location in which it is applied to the soil, affect the extent to which applied fertilizers actually reach the plant and result in increased plant growth and productivity.

Three factors have been exploited particularly to maximize advantages of nitrogen fertilizer use: (1) the location in the soil (or on foliage) where the nitrogen fertilizer is applied, (2) the time at which it is applied, and (3) the manner in which it is made available to the plants. "Slow release" fertilizers designed to limit the rates at which nutrients are solublized and made available to the plants, have received much attention. A typical method is to coat granules of urea with sulfur, limiting the rate at which they can be dissolved and released to the soil and therefore limiting the rate at which they can be lost by transmission through the active root layer by water from rainfall or irrigation.

Nitrification inhibitors which slow the oxidation of ammonium ion to nitrate ion by nitrification have also been useful.[13] Ammonium ion, a cation, is more tenaciously held on the exchange complex of the soil than is nitrate ion, an anion. This limits the extent to which nitrogen can be moved through the soil by irrigation or rainwater, increasing the possibility of retention and uptake by plants.

B. Ratio of N_2O and N_2 Production in Denitrification

If the excessive use of nitrogenous fertilizers resulted only in an increased denitrification with the loss to the atmosphere of nitrogen gas, the only significance of the total process would be the energy expenditure in nitrogen fixation. This, although considerable, is small compared with the total energy budget, much of which is subject to other controls and less urgent than food production. Because N_2O is one of the products of denitrification the question is more complex. The ratio of N_2O to N_2 produced in denitrification, as has been indicated above, is uncertain. N_2O is in some respects an illogical product of denitrification. In an anoxic environment where organic substrate is limiting, the production of nitrogen gas in denitrification thermodynamically yields more energy than does the production of N_2O. If organic substrate is abundant, as it frequently is in anoxic situation, the production of nitrogen gas still would be more advantageous from a thermodynamic standpoint than would be the production of N_2O. We know, however, that N_2O is produced, and the explanation must be kinetic rather than thermodynamic. Because denitrification results in the utilization of an acidic anion (NO_3^-) and the production of nitrogen gas which is neutral or nitrous oxide which is the (theoretically at least) anhydride of a weak acid, there is a shift of pH of the medium in which the reaction is occurring from the acidic to one more alkaline. The reaction should be enhanced in an acidic environment and more limited

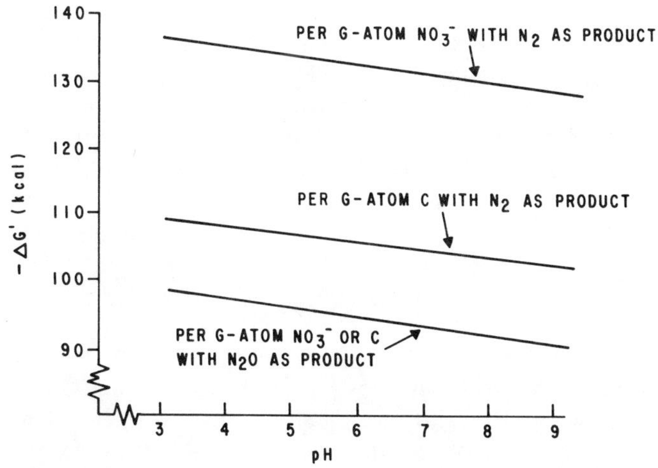

FIGURE 3. Free energy yield in denitrification as a function of pH calculated from standard tables using a carbohydrate substrate and with N_2O or N_2 as gaseous product.

in one with alkaline reaction. A plot of the free energy yield in denitrification with the production of N_2 or N_2O is shown in Figure 3. N_2O production should be more likely in soils of higher pH than those of lower pH. Experience bears this out and N_2O is a larger fraction of the total under alkaline conditions. Nevertheless, field studies have indicated that N_2O is produced in both alkaline and acidic soils.

Studies to determine the ratio of N_2 to N_2O production in soils are complex and laborious, and although it is easy to get some indication of the ratio of the two gases in laboratory studies, it is difficult to verify these under field conditions. Best indications are that N_2O production usually is between 5 and 20% of the total nitrogenous gas release in denitrification.[14,15] The value used in compiling the model of Figure 1 was 15%. Any large deviation from this value, one way or the other, would influence estimates of the atmospheric significance of the use of nitrogenous fertilizers.

IV. N_2O PRODUCTION IN NITRIFICATION

Nitrification, the biological oxidation of ammonium ion to nitrite ion and then to nitrate ion has the overall stoichiometry shown by the reactions below:

$$NH_4^+ + 1\tfrac{1}{2}\,O_2 \rightarrow NO_2^- + H_2O + 2H^+$$

$$\Delta G'_{298} \text{ (at pH 7)} = -70.8^* \qquad (1)$$

$$NO_2^- + \tfrac{1}{2}\,O_2 \rightarrow NO_3^-$$

$$\Delta G'_{298} = -20.1 \qquad (2)$$

In the overall sequence from ammonium ion to nitrate ion, nitrogen undergoes a net valence change of 8 electrons from a valence of −3 in ammonium ion to +5 in nitrate ion. The reaction is shown here in two steps, the first from NH_4^+ to NO_2^- and the second from NO_2^- to NO_3^-, with different groups of organisms being involved.

* H^+ is produced in this reaction and thus the reaction is favored in an alkaline environment. This effect may be expressed as a value for $-\Delta G'$ which varies with pH.

Classically, it is viewed as an autotrophic process with *Nitrosomonas* carrying on the first step and *Nitrobacter* the second. There are other organisms capable of nitrification, some of them heterotrophic.[16,17] As indicated by the values given for ΔG, both reactions are energy-yielding, at least for the autotrophic organisms carrying them out. Whether biologically useful energy is realized in heterotrophic nitrification is not certain but most probably it is. The overall sequence of theoretical intermediates in the process is given below:

$$\underset{\text{Ammonium}}{NH_4^+} \rightarrow \underset{\text{Hydroxylamine}}{HONH_3^+} \rightarrow \underset{\text{Nitroxyl}}{[HNO]} \rightarrow \underset{\text{Nitrite}}{NO_2^-} \rightarrow \underset{\text{Nitrate}}{NO_3^-} \quad (3)$$

The reaction is depicted as going in two-electron steps, although intermediate compounds involving a one-electron shift are possible. Nitroxyl is shown here in brackets because free nitroxyl radical probably does not exist but some compound at this valence level undoubtedly is involved. Hyponitrous acid (HONNOH), the dimer of nitroxyl is one of several possible compounds at this valence level. It can be decomposed spontaneously to yield N_2O, which has been found in nitrification processes under some circumstances. Although in laboratory studies a quantitative recovery of nitrite ion can be made, this is not always the case and presumably some nitrogen has been lost in the process, possibly as N_2O.

Because hydrogen ion is a product in the oxidation of ammonium ion to nitrate ion, the energy yield in the process is a function of pH. This relationship is shown graphically in Figure 4. The next step, the oxidation of nitrite ion to nitrate ion, is also slightly pH-dependent because of the somewhat greater dissociation constant of HNO_3 compared with that of HNO_2. Expressing it otherwise, HNO_3 is a "stronger" acid than is HNO_2.

These pH effects are not great compared with the overall energy yield in the oxidation of NH_4^+ to NO_2^- or NO_2^- to NO_3^-, but they are expressed in the optimal pH for the growth of these organisms, in the neighborhood of pH 8. Under some conditions the termination of the reaction at the level of hyponitrous acid and the ejection of N_2O to the atmosphere could therefore be advantageous for *Nitrosomonas* provided the energy of the reaction is recovered, although it would be disadvantageous to *Nitrobacter*.

With resting cell cultures of *Nitrosomonas europaea* as much as 25% of the total NH_4^+ oxidized has been recovered as N_2O. Under field conditions the production of N_2O appears to be much less, usually less than 1% of the applied NH_4^+.[18]

V. N_2O PRODUCTION AND UTILIZATION IN MARINE ENVIRONMENTS

Several characteristics of the ocean and of estuaries and some fresh water bodies[19] suggest that processes of the nitrogen cycle, particularly where the production of N_2O is concerned, should have quite different characteristics from those observed on land. These include the high solubility of N_2O in water compared with N_2, the frequent presence of anoxic sediments, water volumes which are low in dissolved oxygen, and (at least in oceans and estuaries) pH values which are higher than those commonly encountered in soils.

In both the Atlantic and Pacific oceans, supersaturation values of N_2O have been observed suggesting that these may be sources of N_2O.[20-23] Supersaturation with N_2O has also been observed in estuaries but under-saturation and sometimes the complete absence of N_2O have also been observed in estuarial waters and in sediments.[24] Deep

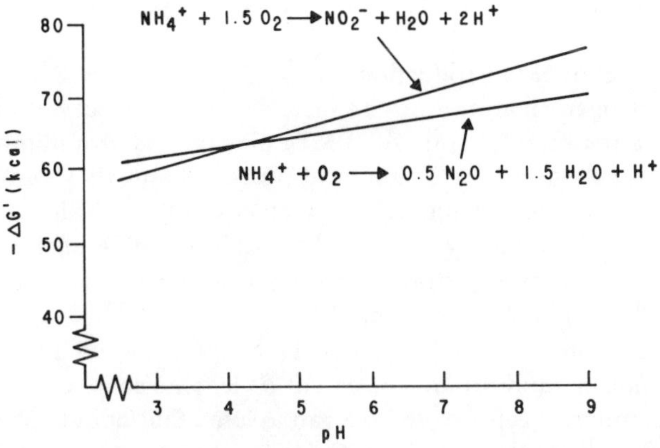

FIGURE 4. Free energy yield in nitrification as a function of pH with N_2O or NO_2^- as product.

ocean sediments are accessible only with difficulty for N_2O determination and little information is available on these. The conclusion is unavoidable, however, that N_2O is being formed in some zones and utilized in others and the most likely consequence is that the oceans as a whole are a net source of N_2O, possibly a large one. Some of this could originate in a micro cycle operating at the interface between sediments and water with nitrification taking place in aerobic waters and denitrification with consequent N_2O production in the anoxic sediments immediately below.

A more likely source in the deep ocean could be nitrification below the hypolymnion where photosynthesis is no longer taking place, detritus is being oxidized and oxygen concentration is limited. The energy yield of nitrification with the production of N_2O is compared to the energy yield of nitrification with production of nitrite ion:

$$HN_4^+ + O_2 \rightarrow ½\ N_2O + 1½\ H_2O + H^+$$

$$\Delta G'_{298}\ \text{(at pH 7)} = -54.62^* \qquad (4)$$

$$NH_4^+ + 1½\ O_2 \rightarrow NO_2^- + H_2O + 2H^+$$

$$\Delta G'_{298}\ \text{(at pH 7)} = -65.07^* \qquad (5)$$

The energy yield per mole NH_4^+ oxidized is greatest when NO_2^- is the product. Comparing the energy yield per mole oxygen consumed, however, the energy yield is greatest with N_2O as the final nitrogenous product, although it is not known what fraction of this energy is actually utilizable by the organism. Thus, under conditions of low oxygen supply N_2O production would seem to be favored. Moreover, there is the added advantage that the production of NO_2^- makes available an energy substrate for yet another autotrophic reaction in its oxidation to NO_3^-. This would place still greater demand on a limited oxygen supply, and so the liberation of NO_2^- to the water by *Nitrosomonas* would appear to be an unwarranted generosity. Comparative observations of O_2 and N_2O in waters overlying an anoxic volume tend to be in agreement with such a possibility, however they could be explained by the simultaneous operation of nitrification and denitrification reactions with N_2O as the final product.

* H^+ is produced in this reaction and thus the reaction is favored in an alkaline environment. This effect may be expressed as a value for $-\Delta G'$ which varies with pH.

VI. ESTIMATES OF RATES AND QUANTITIES

A. The Balance Sheet of N_2O Production

Estimates of nitrogen fixation rates by historic natural processes are uncertain,[25,26] but starting with a reasonable figure of 105 Tg (Teragrams or millions of metric tons) nitrogen fixed annually by historic biological processes and 10 Tg annually by atmospheric reactions and further assuming that a steady state prevails, an equivalent 115 Tg nitrogen from these sources is denitrified annually. If 10% of this appears as N_2O, the annual yield of N_2O from historic processes is about 18 Tg equivalent to 11.5 Tg N. This is shown diagramatically in Figure 5.

Legume crops contribute an additional 80 Tg N annually and although this component probably is not in equilibrium, the extent of legume crop usage has not changed greatly in the past quarter century and we can assume that about 75% of this component is denitrified with a 10% yield of N_2O for 9.5 Tg* N_2O equivalent to 6 Tg nitrogen.

Nitrogen fixed annually by industrial processes and combustion (40 Tg and 18 Tg, respectively) is of recent origin and probably not in secular equilibrium but we can assume that 1/3 of it is denitrified, the remainder appearing as NO_3^- in ground and surface waters or buried in sediments as organic N. The denitrified portion would yield 3 Tg N_2O equivalent to 2 Tg N. The total input of N_2O from these three denitrification sources is 30.5 Tg equivalent to 19.5 Tg N.

Present estimates of the residence time of N_2O in the atmosphere range from 8 to 70 years.[1,3,27] (Resident time is pool size divided by the rate of formation or removal assuming a steady state exists.) Thus the estimates which we make here of the annual rate of N_2O production from denitrification prior to human intervention is about 0.64% of the atmospheric pool giving a residence time of over 150 years (0.0064^{-1}) calculated from the present pool size. Clearly it would be valuable to find a reliable sample of preindustrial air.

The human influence has been comparatively recent so conditions are not in a steady state. If the system *were* so tightly coupled that present atmospheric N_2O reflected directly N_2O production from all of these sources, the calculated residence time would be about 90 years. Thus the large atmospheric N_2O pool suggests a *long* residence time. To the extent that atmospheric N_2O concentrations vary, a *short* residence time is suggested.

Key uncertainties remain which could explain such a discrepancy, and each of which would effectively increase atmospheric N_2O emissions:

1. The estimate of total nitrogen fixation by biological processes may be low.
2. The fraction of nitrogen appearing as N_2O in the denitrification reaction may be underestimated.
3. There may be a short circuit in the nitrogen cycle with some of the N_2O being reduced to NH_4^+ and reappearing as N_2O either as a result of nitrification or denitrification.
4. The contribution of N_2O by nitrification may be greater than has been assumed (although this would also require additional nitrogen fixation).

Of these possibilities, perhaps all are real. Past estimates of nitrogen fixation in the ocean have been small because it has been assumed that the ocean is phosphorus-limited and that therefore no large fixation of nitrogen would be necessary to support oceanic life processes. But the comparatively high values of supersaturation of ocean

* Teragrams, millions of metric tons and megatons are equal.

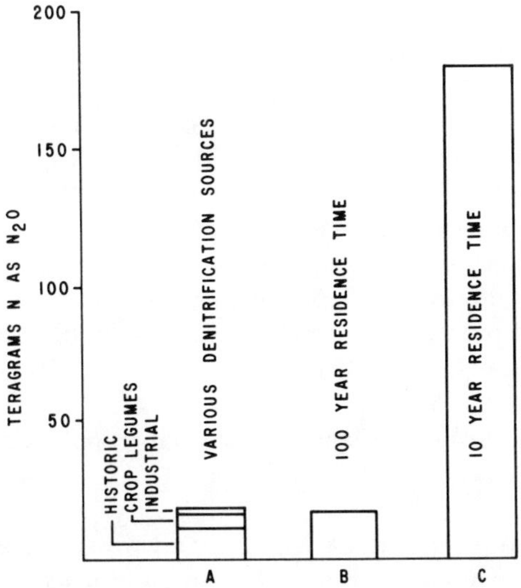

FIGURE 5. Comparison of annual N_2O production (expressed as Tg N) as estimated here (A) with that which would be required to match observed atmospheric concentrations with assumed atmospheric residence times of 100 years (B) and 10 years (C).

waters with N_2O suggest that there may be a loss of nitrogen from the ocean as N_2O either from nitrification or denitrification reactions, and that additional nitrogen has to be fixed in order to accommodate this loss.

B. Natural Vs. Human Sources

The annual supply of industrially-fixed nitrogen, the use of legume crops, and other processes resulting in "new" fixed nitrogen as estimated here is approximately equal to that which we have assumed to have been fixed by "natural" processes. As a first approximation it can be assumed that the annual input of nitrous oxide to the atmosphere has also been doubled by these procedures. Actually because the production of N_2O is favored by high concentrations of ammonium ion it is possible that the human influence is even greater, although other processes as for example the burial of nitrogen in sediments, tend to have the opposite effect. There is now an annual input of new fixed nitrogen from human activities of approximately 140 Tg, greater than that from historic processes. The real significance of this change hinges on a number of uncertainties. These include our uncertainty as to the sensitivity of the nitrogen cycle to atmospheric N_2O concentration, the question as to how much of the incremental nitrogen from human sources is denitrified and what fraction remains in new organic matter or buried sediments, the uncertainty as to the fraction of denitrified nitrogen which appears as N_2O, the true residence time of N_2O in the atmosphere, and the significance of the ocean as compared with terrestrial activities in the total nitrogen cycle and in N_2O production and utilization. All of these are unknowns at present and we are left with the uneasy intuition that human influence on the nitrogen cycle has been large and that the consequences as they apply to the N_2O cycle and stratospheric ozone may be significant.

VII. RESEARCH NEEDS

Evaluation of the relationship of fertilizer use to ozone concentration (and location) in the stratosphere is dependent upon several variables. Uncertainty regarding these variables and comparative significance of each dictate the most logical directions which future research should take. The questions include:

1. What effect will a given increment in atmospheric N_2O have on ozone concentration and the altitude of the ozone "layer" in the stratosphere?
2. What fraction of the N_2O in the atmosphere comes from the soil and what fraction from the sea or other waters?
3. What fraction of fertilizer nitrogen is eventually converted to N_2O and where?
4. Assuming that additional N_2O production from human sources is undesirable (not proved) what are the best methods available to limit this source, both in the field and in sewage disposal systems?

As a consequence of this analysis added research emphasis in several areas is indicated. These include:

1. Detailed examination of both agricultural and waste management procedures to determine the fate of "new" fixed nitrogen and the extent to which it results in a yield of N_2O to the atmosphere.
2. An examination of processes involving nitrogen in the ocean, other water bodies, and in sediments.
3. A more quantitative study of processes involving N_2O in the atmosphere and a determination of the extent to which these influence the concentration of the stratospheric ozone.
4. The search for methods for closing the present chain of food production and waste management in order to keep fixed nitrogen tightly in the system, thus minimizing the need for newly fixed nitrogen and minimizing the loss from the system of N_2O and minimizing energy requirements for nitrogen fixation.
5. The continuation and expansion of the search for methods to "target" applied nitrogen to the crop, possibly by modifying the fertilizer, its means and timing of application, or modifying the crop itself, thereby increasing the efficiency with which the fertilizer is used.
6. An examination of conditions which tend to influence the ratio of N_2O to N_2 in denitrification assuming that when N_2 is produced in the process it is relatively harmless whereas the production of N_2O may have more serious consequences.

REFERENCES

1. **Hahn, J. and Junge, C.**, Atmospheric nitrous oxide: a critical review, *Z. Naturforsch. Teil A*, 32, 190, 1977.
2. **Duewer, W. H., Wuebbles, D. J., Ellsaesser, H. W., and Chang, J. S.**, NO_x catalytic ozone destruction: sensitivity to rate coefficients, *J. Geophys. Res.*, 82, 935, 1977.
3. **Junge, C.**, The cycle of atmospheric gases—natural and man made, *Q. J. R. Meteorol. Soc.*, 98, 711, 1972.
4. **Crutzen, P. J. and Ehhalt, D.**, Effects of nitrogen fertilizers and combustion on the stratospheric ozone layer, *Ambio*, 6, 112, 1977.
5. **Johnston, H. S.**, Analysis of the independent variables in the perturbation of stratospheric ozone by nitrogen fertilizers, *J. Geophys. Res.*, 82, 1767, 1977.
6. **McElroy, M. B., Wofsy, S. C., and Yung, Y. L.**, The nitrogen cycle: perturbations due to man and their impact on atmospheric N_2O and O_3, *Philos. Trans. R. Soc. London, Ser. B*, 277, 159, 1977; Logan, J. A., Prather, M. J., Wolfsy, S. C., and McElroy, M. B., *Philos. Trans. R. Soc. London, Ser. A*, 290, 187, 1978.
7. **Delwiche, C. C. and Bryan, B. A.**, Denitrification, *Ann. Rev. Microbiol.*, 30, 241, 1976.

8. **Cooke, G. W.**, Fertilisers in 2000 A. D., *Phosphorus Agric.*, 53, 1, 1969.
9. **Blouin, G. M. and Davis, C. H.**, Energy requirements for the production and distribution of chemical fertilizers in the United States, Conference Workshop of the Southern Regional Education Board, Energy in Agriculture, Atlanta, October 1-3, 1975, 51.
10. **Broadbent, F. E. and Carlton, A. B.**, Field trials with isotopically labeled nitrogen fertilizer, in *Nitrogen in the Environment,* Nielsen, D. R. and MacDonald, J. G., Eds., Academic Press, New York, 1978, 1.
11. **Jones, M. B., Delwiche, C. C., and Williams, W. A.**, Uptake and losses of ^{15}N applied to annual grass and clover in lysimeters, *Agron. J.*, 69, 1019, 1977.
12. **Schroeder, E. D.**, *Water and Wastewater Treatment,* McGraw Hill, New York, 1977.
13. **Huber, D. M., Warren, H. L., Nelson, D. W., and Tsai, C. Y.**, Nitrification inhibitors — new tools for food production, *Bioscience,* 27, 523, 1977.
14. **Nelson, D. W. and Bremner**, Gaseous products of nitrite decomposition in soils, *Soil Biol. Biochem.*, 2, 203, 1970.
15. **Rolston, D. E. and Marino, M. A.**, Simultaneous transport of nitrate and gaseous denitrification products in soil, *Soil Sci. Soc. Am. J.*, 40, 860, 1976.
16. **Obaton, M., Amarger, N., and Alexander, M.**, Heterotrophic nitrification by *Pseudomonas aeruginosa, Arch. Mikrobiol.*, 63, 122, 1968.
17. **Verstraete, W. and Alexander, M.**, Heterotrophic nitrification in samples of natural ecosystems, *Environ. Sci. Tech.*, 7, 39, 1973.
18. **Yoshida, T. and Alexander, M.**, Nitrous oxide formation by *Nitrosomonas europaea* and heterotrophic microorganisms, *Soil Sci. Soc. Am. Proc.*, 34, 880, 1970.
19. **Kaplan, W. A., Elkins, J. W., Kolb, C. E., McElroy, M. B., Wofsy, S. C., and Duran, A. P.**, Nitrous oxide in fresh water systems: an estimate for the yield of atmospheric N_2O associated with disposal of human waste, *Pure Appl. Geophys.*, 116, 423, 1978.
20. **Hahn, J.**, The North Atlantic Ocean as a source of atmospheric N_2O, *Tellus,* 26, 160, 1974.
21. **Seiler, W. and Schmidt, U.**, Dissolved nonconservative gases in seawater, in *The Sea,* Goldberg, E., Ed., Wiley-Interscience, New York, 1974, 219.
22. **Yoshinari, T.**, Nitrous oxide in the sea, *Marine Chem.*, 4, 189, 1976.
23. **Cohen, Y. and Gordon, L. I.**, Nitrous oxide in the oxygen minimum of the eastern tropical North Pacific: evidence for its consumption during denitrification and possible mechanisms for its production, *Deep-Sea Res.*, 25, 509, 1978.
24. **Cohen, Y.**, Consumption of dissolved nitrous oxide in an anoxic basin, Saanich Inlet, British Columbia, *Nature (London),* 272, 235, 1978.
25. **Delwiche, C. C. and Likens, G. E.**, Biological response to fossil fuel combustion products, in *Global Chemical Cycles and Their Alteration by Man,* Stumm, W., Ed., Dahlem Konferenzen, Berlin, 1977, 73.
26. **Hardy, R. W. F. and Havelka, U. D.**, Nitrogen fixation research: a key to world food? *Science,* 188, 633, 1975.
27. **Schütz, K., Junge, C., Beck, R. and Albrecht, B.**, Studies of atmospheric N_2O, *J. Geophys. Res.*, 75, 2230, 1970.

Section 2: Effects and Research

Chapter 4

OZONE CHANGE: BIOLOGICAL EFFECTS*

D. S. Nachtwey and R. D. Rundel

TABLE OF CONTENTS

* This paper reflects the views of the authors and in no way is to be construed to reflect the official or unofficial view, policy, or plans of the National Aeronautics and Space Administration.

I. INTRODUCTION

The studies presented in other chapters in this book argue that several human activities may reduce the average thickness of the stratospheric ozone layer. It is clear that any ozone layer reduction will lead to some increase in solar ultraviolet (UV) radiation reaching the Earth's surface if all other factors (clouds, haze, aerosols) remain the same. The type of UV radiation that will increase is the so-called UV-B* radiation (290 to 320 nm). This radiation is known to be biologically damaging at sufficiently high doses, but it is apparently fairly well-tolerated at current solar irradiance levels. The major question involved in assessing the potential biological impacts of ozone layer reduction is, will the increased UV-B exceed the tolerance limits for organisms?

It must be admitted at the outset that current knowledge does not allow an unequivocal answer to the major question even if we knew for certain how much the ozone layer will be reduced under any set of circumstances. Nonetheless, this chapter will present some of the available information and will explore in some detail the uncertainties involved in assessing the potential biological impacts of ozone layer reduction. Although the chapter has a rather negative tone, it is hoped that exploring some of the problems associated with obtaining an answer to the major question will serve as a guide to pitfalls for researchers in the field and will point out to regulators just how uncertain is our information about the effects of UV-B on the biosphere.

Before proceeding, it is first necessary to discuss some of the basic physical aspects of increased solar UV irradiance and the concept of biologically effective dose. Some of these aspects have been previously published in various isolated papers and may be well-known to many readers, but they are repeated below in simplified form for those readers who may be unfamiliar with the fundamental processes involved in ozone layer reduction/increased biologically effective UV.

II. FUNDAMENTAL PHYSICAL ASPECTS OF SOLAR ULTRAVIOLET IRRADIANCE

The direct flux of solar UV radiation reaching the Earth's surface on a cloudless day depends on the air mass (consisting of air, variable amounts of trace gases such as ozone and aerosol particulates) traversed, which in turn varies with the solar zenith angle (the angle of the sun relative to the vertical), which varies with time of day, season, and latitude (Figure 1). The mechanism by which the ozone layer partially shields the Earth is illustrated schematically in Figure 2. Figure 2 shows the solar UV irradiance above the atmosphere, I_0. The atmospheric ozone layer absorbs some of this incoming UV in an exponential manner: $I = I_0e^{-kx}$, where k is the absorption coefficient for ozone and x represents the pathlength. The amount of absorption depends on the wavelength (λ) of UV radiation, and the concentration of ozone along the pathlength of the radiation through the atmosphere. As shown in Figure 1a, the actual pathlength AP is related to the vertical column length PB by the secant of the solar zenith angle θ. For simplification of calculation, the amount of ozone in a vertical column from the ground to the top of the atmosphere is expressed as the thickness that the entire ozone column above the Earth would occupy if it were compressed at standard temperature and pressure. This thickness, symbolized w, is expressed in units

* The term UV-B was originally designated by Coblentz and Stair (cited in Meyer and Seitz)[1] as that portion of the UV spectrum from 280 to 315 nm; current usage has pushed the lower limit to 290 nm and the upper limit to 320 nm to make it more consistent with the damaging UV radiation around the absorption edge of ozone. UV-B is distinguished from UV-A (320 to 400 nm) which is generally benign and UV-C (<290 nm) which is very biologically damaging, but absent in sea level radiation, because of absorption in the upper atmosphere.

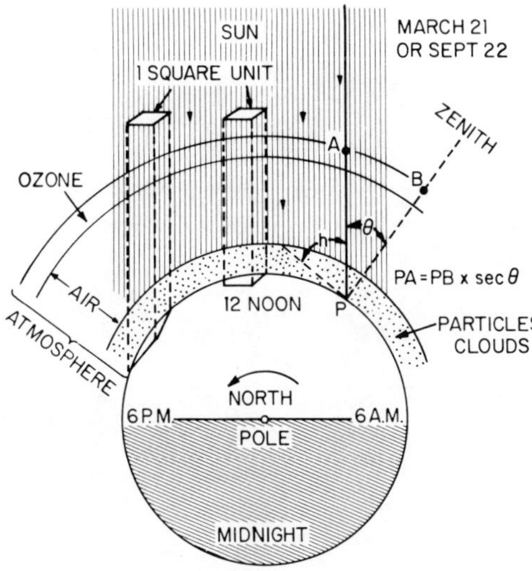

(a)

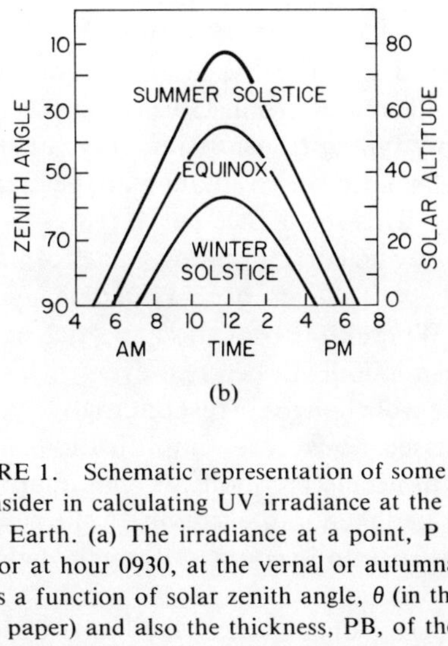

(b)

FIGURE 1. Schematic representation of some factors to consider in calculating UV irradiance at the surface of the Earth. (a) The irradiance at a point, P (on the Equator at hour 0930, at the vernal or autumnal equinox) is a function of solar zenith angle, θ (in the plane of the paper) and also the thickness, PB, of the attenuating atmosphere. (For diagramatic purposes, the ozone layer, air molecule layer, and particulate, or aerosol, layers are separated.) The radiation must go through a pathlength or air mass, AP, to reach P; AP = PB × sec θ. Also illustrated is the fact that the same amount of radiation passing perpendicularly through one square unit will be spread out over a greater area when the solar zenith angle is high (solar elevation, h, is low) than when the solar zenith angle is low, e.g., at 12 noon. (b) Schematic of the change of solar zenith angle with time of day and season at 34° N. latitude. (c) and also (d) illustrate the change of solar zenith angle with season of the year. Note in (c) that the day length (length of arc above horizon) changes with the seasons.[51]

of atmosphere centimeters (atm.cm). At mid-latitudes, the thickness averages about 0.32 atm.cm.

The combination of these concepts leads to an equation that allows the calculation

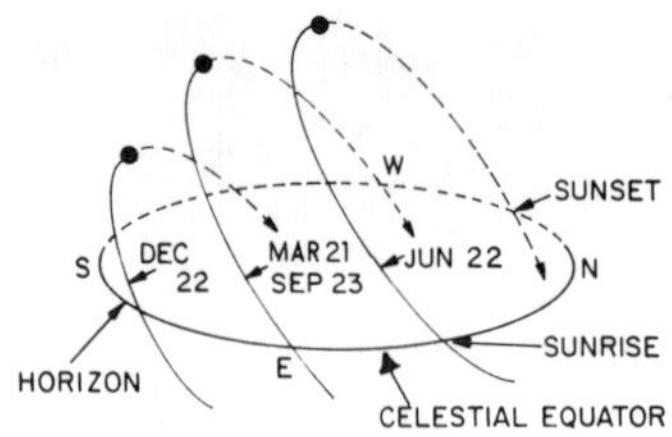

(c)

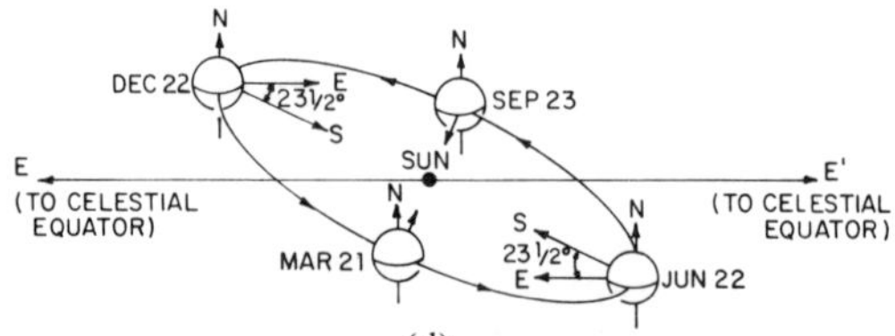

(d)

FIGURE 1

of the transmission T = I/I_0 for the direct flux at any wavelength through the ozone layer for any solar zenith angle: $T_\lambda = e^{-k\lambda w \sec\theta}$.

It may be seen in Figure 2 that the ozone absorption coefficient, k (λ), increases steeply as the wavelength decreases from 350 nm down to 280 nm. This increasing absorption leads to a steeply falling transmission curve at shorter wavelengths (Figure 2). Figure 2 shows the solar spectral irradiance at the Earth's surface with the sun directly overhead ($\theta = 0°$). This irradiance curve represents the product of values of the extraterrestrial irradiance and corresponding values of the transmission curve.

A reduction in the thickness of the atmospheric ozone layer (w) will lead to a change in the transmission curve. Where the absorption coefficient is large, as at short wavelengths, the change in transmission will be great even with small changes in w and the solar spectral UV irradiance will change correspondingly.

The fundamentals discussed above are simplifications. In addition to the ozone layer, the aerosol and air molecule components of the atmosphere and the multiple scattering of photons must be taken into account. In the figures and tables that follow these aspects have been taken into account, although their discussion is deferred to Section IX.

Figure 3 illustrates the effect on the clear-sky, solar spectral UV irradiance at the ground produced by various percentage reductions in the ozone layer thickness. It should be emphasized that this illustration is but one example from an infinite number of possible examples; the curves would be different for different initial ozone layer thicknesses, different solar zenith angles, and clouds and aerosols.

The present day natural thickness of the ozone layer varies substantially. Figure 4 illustrates the daily variations at Tallahassee, Florida, as an example.[2] Actually, even on a given day, the ozone layer thickness can vary by 10% with the passage of a meteorological front. Because of the ozone variation, even at a fixed solar zenith angle and a given latitude, the UV irradiances vary correspondingly. The average thickness at one location varies seasonally (25% between spring and fall), and yearly (5% in the course of a decade). Moreover, the annual average ozone layer thickness also varies geographically (both latitudinally and longitudinally); for example, between Minnesota and Texas, it can vary as much as 30%. Thus the current ozone layer thickness as a function of time at any latitude and longitude can be characterized as a wavy curve of peaks and valleys. If all else remains equal, a reduction in the average ozone layer thickness will lead to a lowering of the entire curve. The accompanying peaks of UV irradiance as well as average UV irradiances, will then be correspondingly higher.

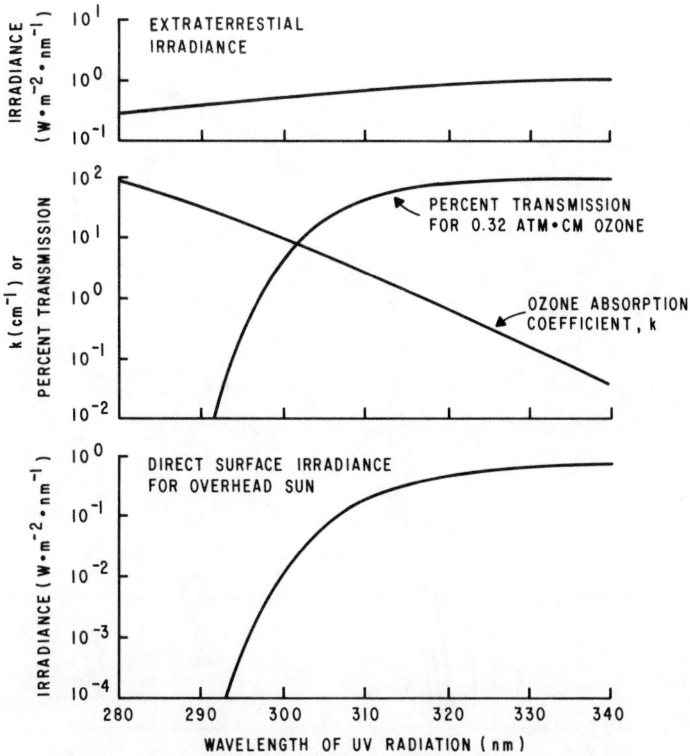

FIGURE 2. Schematic representation of the influence of ozone on the direct UV radiation at the Earth's surface. See text.

III. THE CONCEPT OF BIOLOGICALLY EFFECTIVE UV-B RADIATION

Not all wavelengths of UV radiation are equally effective in producing a photobiological effect. In order to produce an effect, the particular photon must be absorbed by some critical target molecule(s) within the organism. The probability that a photon of a particular wavelength is absorbed depends upon the molecule's electronic configuration, which determines its absorption spectrum. The relative effectiveness of photons of different wavelengths in producing a given biological response (action spectrum) reflects the absorption spectrum of the target molecule and the probability that the absorption leads to a biological effect. Although action spectrum studies indicate that a variety of molecules may be involved in the biological effects of UV radiation, almost all such action spectra have in common an increasing effectiveness with decreasing wavelength.

Figure 5a schematically illustrates a generalized action spectrum of the type mentioned above and illustrates its use in quantifying biologically effective solar UV-B radiation and the consequences of ozone layer reduction. In Figure 5a the schematic action spectrum is superimposed on the calculated spectral irradiances for the sun at 30° zenith angle with standard ozone (0.32 atm.cm; 0% reduction) and with a 50% reduction from this base level. The product of a point (at any given wavelength) on the action spectrum curve E (λ) and a comparable point on a solar spectral irradiance curve I (λ) will give a value which reflects both the relative effectiveness of the photons at that wavelength and the number of photons at that wavelength reaching the earth. These product curves for the biologically effective UV-B irradiances with standard

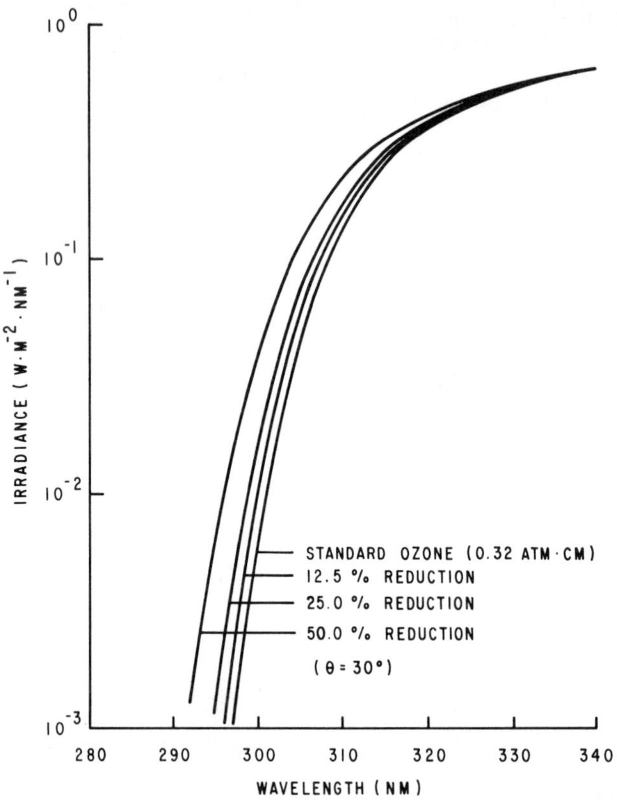

FIGURE 3. Spectral irradiance from the Sun at $\theta = 30°$ with various reductions in the ozone layer thickness from the standard thickness.

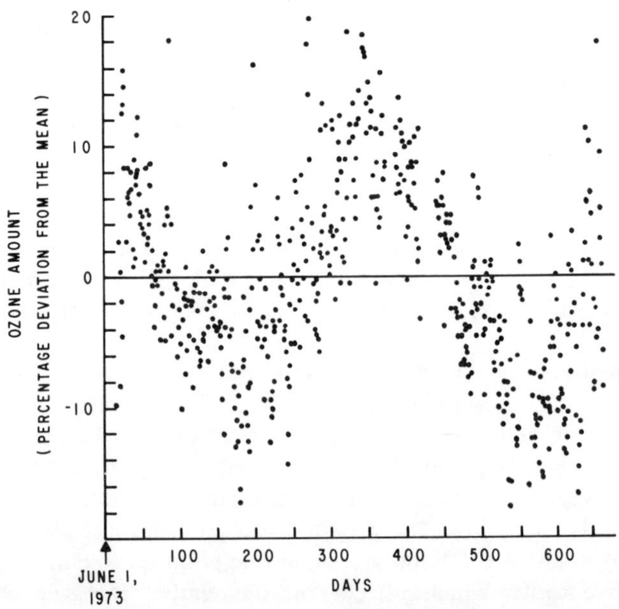

FIGURE 4. Daily total ozone measurements (Dobson Meter) at Tallahassee, Florida, expressed as percentage deviation from the mean value of 0.305 atm.cm.

ozone and with 50% reduced ozone are also shown. Figure 5b shows the product curves on linear coordinates. The area under each product curve represents the total biologically effective power from the sun under the two conditions: effective UV-B = $\int E(\lambda)\, I(\lambda)\, d\lambda$.

Although any given ozone layer decrease will increase UV-B radiation, such increases will not be equal throughout the UV-B spectral interval: at ~290 nm the increase may be an order of magnitude; at 320 nm the increase may be very small (Figure 5a). For purposes of assessing the impact of ozone layer reduction, the important factor is not the increase in UV-B radiation *per se*, but the increase in biologically effective UV-B radiation, i.e., the irradiance weighted by some biological effectiveness function.*

For assessment purposes, the value of interest is not the biologically effective irradiance, but the biologically effective dose (irradiance × exposure time). In general, this UV-B(BE) dose is calculated by integrating the biologically effective irradiance as a function of time of day, while specifying the ozone layer thickness, and specifying the latitude and season, which affect maximum solar zenith angle and day length. (Refer to Section IX for further details.)

IV. MOLECULAR BIOLOGY OF ULTRAVIOLET RADIATION EFFECTS

A large extant literature documents the fact that UV radiation is capable of affecting a wide variety of organisms. Most of the information, however, has been obtained with UV-C radiation (less than 290 nm, but primarily 254 nm) under laboratory conditions. Analyses have shown that these data obtained with UV-C cannot be readily extrapolated to a *quantitative* assessment of the biospheric impact of increased solar UV-B radiation in nature.[3] Similar studies of the effects of UV-B radiation *per se* were, until quite recently, relatively rare. Nonetheless, *qualitatively,* most of the effects demonstrated with UV-C radiation are found to occur with UV-B radiation. For example, the absorption of a UV-B or UV-C photon can excite molecules to an extent where they can undergo a chemical change. A chemical change in highly specific biological macromolecules constitutes a damaging lesion that can interfere with the function of the molecule. The macromolecules of greatest importance are deoxyribose nucleic acid (DNA), ribose nucleic acid (RNA), and proteins. Probably the most biologically important target molecule for UV radiation effects is DNA. Sets of DNA molecules in each cell carry genetic information that specifies the structure of RNA and through RNA the structure of proteins. The sets of DNA molecules are usually present in only one or two copies, whereas the RNA and protein molecules are present in multiple copies. Damage to the latter two types of macromolecules can thus be overcome by replacement with new copies if the DNA is intact, but damage to the DNA molecules if unrepaired can inactivate the cell.

The DNA molecule consists of two antiparallel helical strands, each of which is made up of a chain of links of a phosphoric acid attached to a five-carbon sugar (deoxyribose). Attached to each sugar molecule and extending into the center of the helix is one of four types of bases: two purines — adenine and guanine; and two

* Different authors use different designations for biologically effective UV radiation: DUV for Damaging UV; BHUV for Biologically Harmful UV; UV-B_{ee} for erythemally effective UV; UV-B_e or UV-B(BE) for Biologically Effective UV. The use of UV-B with some modifier indicates integrated spectral irradiances weighted with some effectiveness function. In this Chapter, UV-B(BE) is used as a general term when discussing the *quantitative* estimation of the increased biologically effective dose-rate and dose resulting from ozone layer reduction. UV-B will still be used when discussing qualitative aspects of ozone reduction and where accuracy doesn't require the use of the more specific term.

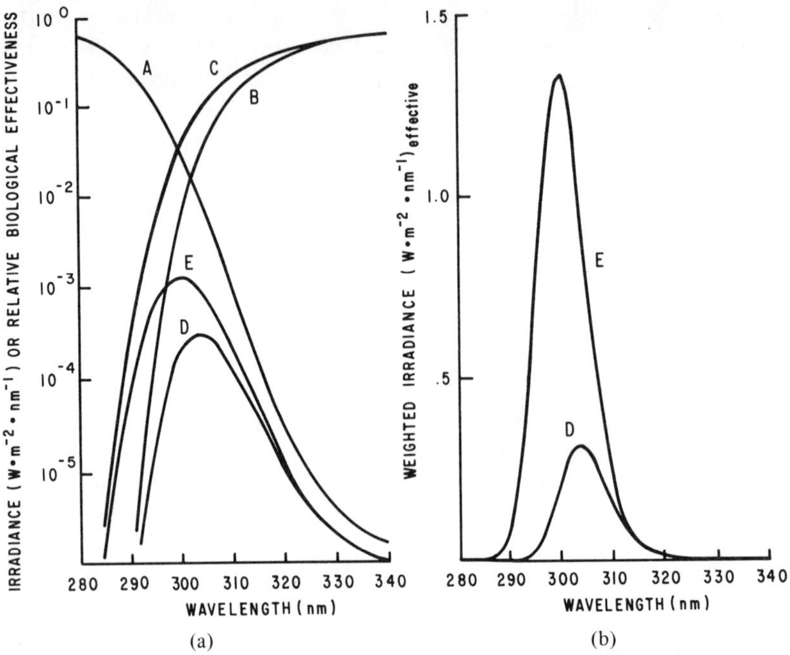

FIGURE 5. Schematic illustration of the concept of biologically effective radiation. Figure 5a A. Relative biological effectiveness of different wavelengths (weighting function). B. Solar irradiance at θ = 30° with the standard ozone layer thickness of 0.32 atm.cm. C. Solar irradiance at θ = 30° with a 50% reduction from the standard. D. The product of values on Curves A and B, which yields the biologically effective irradiance for standard ozone, i.e. $W \cdot m^{-2} \cdot nm^{-1}$ weighted by the biological effectiveness weighting function, $(W \cdot m^{-2} \cdot nm^{-1})_{eff}$. E. The product of values on Curves A and C, the biologically effective irradiance for 50% reduced ozone.
Figure 5b. Curves D and E from Figure 5a plotted on linear coordinates.

pyrimidines — cytosine and thymine. The hereditary information in DNA is contained in the sequence of these bases on one of the strands. The other strand contains complementary information because adenine (A) is always apposed to thymine (T) and guanine (G) is always apposed to cytosine (C). Duplication of the DNA molecule preparatory to cell division involves the splitting of the double stranded helix into two parent strands with new sugar-phosphate-bases (nucleotides) being laid down in appropriate apposition to the parent strand and then being joined together by an enzyme, DNA polymerase, to form a new chain.

Some of the types of damage that UV-C and UV-B can induce in DNA are illustrated in Figure 6.

1. Water, split into H and OH, can add to the broken double bonds of a pyrimidine (hydration).
2. Adjacent pyrimidines on a strand can link together to form a cyclobutyl pyrimidine dimer: T with T, T with C, and C with C. Thymine-thymine dimers (TT) are the most common type produced by 254 nm UV radiation but as the wavelength of UV increases, the relative proportions of TT, TC, and CC dimers change.
3. These pyrimidines can also link across complementary DNA strands (not shown).
4. Single strand breaks in the sugar-phosphate backbone can occur.
5. Alteration of nucleotide bases by (1), (2), or (3) leads to breakage of hydrogen bonds and localized denaturation, which causes structural distortion of the DNA molecule.[4]

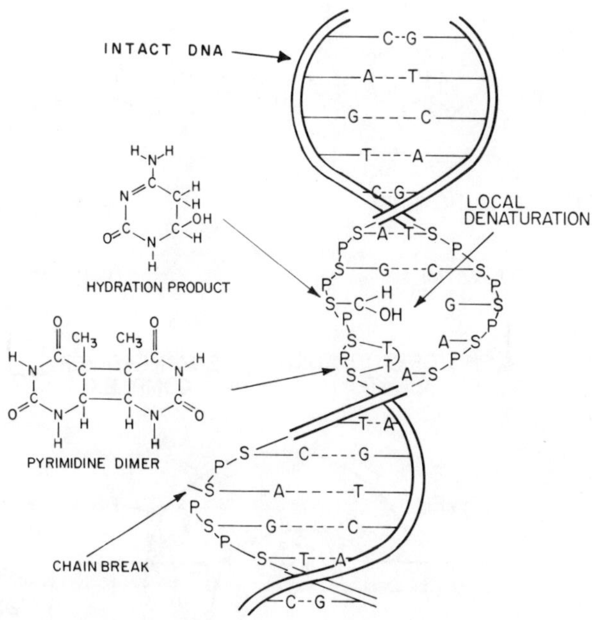

FIGURE 6. Various alterations in DNA produced by UV-B or UV-C radiation. From Giese, A. C., Hanawalt, P. C., and Smith, K. C., *Molecular Photobiology—Inactivation and Recovery,* Academic Press, New York., 1969.

Lesions of the types shown if unrepaired or misrepaired can interfere with DNA replication, can delay cell division and thus growth, can lead to changes in the sequence of nucleotide bases thereby causing a mutation, and can kill cells. Furthermore, the doses of solar UV radiation required to induce these lesions are not large. As will be discussed in more detail in the next section, the noon hour solar UV-B dose at midlatitudes is capable of producing large numbers of these lesions in any appropriately exposed cell.

V. MECHANISMS FOR COPING WITH UV RADIATION

Accepting for the time being that current levels of solar UV-B radiation can produce the types of lesions outlined above, we now consider various mechanisms by which organisms can cope with such damage.

A. Molecular Repair Systems

All organisms have evolved several enzymatic mechanisms for repairing lesions of the types shown in Figure 6, whether they have been induced by UV-C or UV-B. At least three repair mechanisms are recognized: photoreactivation (Figure 7A), excision-resynthesis-repair (Figure 7B), and postreplication repair (Figure 7C).[4] The first requires an enzyme plus visible light (violet to green) or near UV radiation (~340 to 400 nm) to supply the energy to break apart *in situ* a pyrimidine dimer. This very efficient enzyme system is very widely distributed in organisms. However, it appears to play only a minor, if any, role in repair of UV damage in cells from placental mammals.[5,6] The second repair system (excision-resynthesis-repair) consists of a cutting away of the damaged portion of one strand of the DNA double helix and replacing it with new building blocks (nucleotides), using the remaining strand as a template. The third enzyme system (postreplication repair) involves the replication around the lesions in DNA

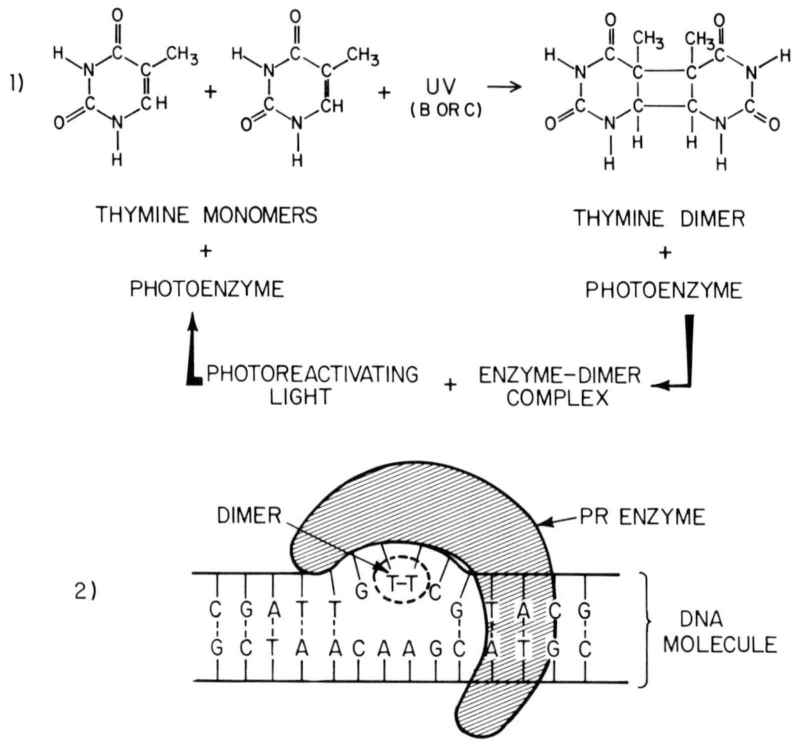

FIGURE 7a. Photoreactivation. (1) Thymine dimer formation between neighboring thymine residues in DNA after exposure to UV-B or UV-C radiation; similar dimers are formed in smaller numbers between other neighboring pyrimidine residues in DNA. (2) Diagrammatic representation of possible structural distortion of DNA as a result of dimer formation. The distortion specifically attracts the photoreactivating enzyme (shown as a shaded comma-like structure), which on absorption of light splits the thymine dimer (monomerization), restoring the DNA to its original state. (From Giese, A. C., *Living With Our Sun's Ultraviolet Rays*, Plenum Press, New York, 1976. With permission.)

with subsequent repair using undamaged strands of the DNA duplex to provide the information for putting in the correct building blocks. The second and third types of repair systems are collectively called dark-repair systems. One of the enzymes of the dark-repair systems is also involved in repair of UV-induced single-strand breaks in DNA.

The importance of these repair systems is apparent when one considers that mutant bacteria lacking some of them are killed by about a minute of exposure to sunlight[7] while wild type bacteria would require about 500 times the equivalent UV dose.

Although these repair mechanisms are very effective after low doses of UV radiation, they are not 100% effective, otherwise biological effects involving damaged DNA (e.g., mutation, cell death) would not be observed. For example, 5 J/m^2 of 254 nm UV will produce about 500,000 thymine-containing ($\hat{T}T$ or $T\hat{C}$) dimers per mammalian cell. After about 24 hr, more than half of these will have been repaired by the excision repair system; by 72 hr the dimer damage is essentially completely repaired.[6] Even so, some cells die. After 10 $J \cdot m^{-2}$ of 254 nm UV, 10^6 dimers are induced/mammalian cell. About 80% of the dimers in a population of cells are excised but 90% of the cells are killed. Higher doses lead to progressively less repair and more cell killing. Similar studies with comparable effective doses of UV-B radiation indicate a similar relationship: if a million dimers/cell are induced, only about 80% will be repaired; any dose yielding more than about a million dimers/cell will be followed by less and less repair.[8]

Doses of solar UV-B radiation that can produce more than a million dimers/cell

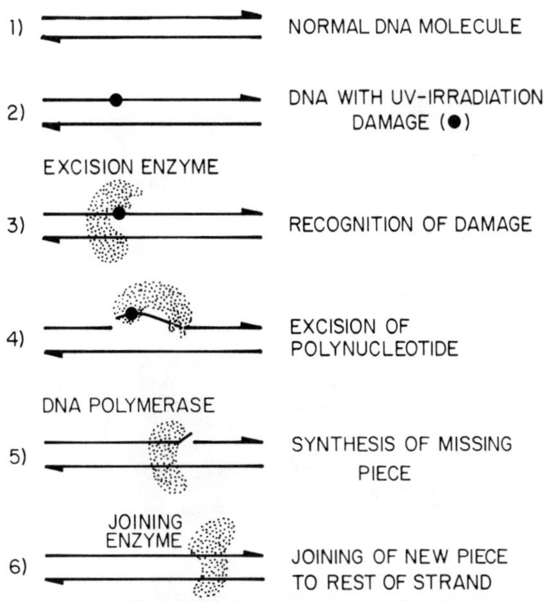

FIGURE 7b. Excision resynthesis repair. Possible mechanism of excision repair by "cut and patch" of UV-radiation-induced damage in a DNA molecule. DNA, shown in a simplified model with anti-parallel strands, (1) is injured by UV-B or UV-C radiation as shown by the black dots (2). When the injury is recognized (3), the excision enzyme snips it out as a polynucleotide (4). It is then replaced by a piece of DNA synthesized under action of the DNA-polymerase using the information on the complementary strand(5). The new piece is then attached(6) by the joining enzyme (ligase). The differing shapes of the enzymes are for diagrammatic purposes only; their real shapes are not known. (From Giese, A. C., *Living With Our Sun's Ultraviolet Rays,* Plenum Press, New York, 1976. With permission.)

currently reach the Earth's surface. Therefore, additional mechanisms of coping must be involved for life as we know it to exist. In the following paragraphs, some of these mechanisms are discussed.

B. Avoidance

Many organisms live almost permanently under ground, rocks, leaves, or in litter, or inside fruits and vegetation. Others live in the dark depths of water where essentially no UV-B penetrates. Some terrestrial organisms have nocturnal habits and avoid the sun in the daytime. Obviously, UV radiation effects on organisms that totally avoid sunlight are not a source of concern. However, many organisms only partially avoid UV-B radiation, being exposed to various levels of solar radiation at least at some time in their life cycles: aquatic organisms live in a variable solar UV-B radiation field depending upon their depth in the water. Many plants live only in the shade of other plants. Many animals are exposed to direct solar UV-B radiation during much of the day but seek shade during midday. In this regard though, it should be emphasized that shade-dwelling animals and plants are still exposed to some UV-B radiation, that which filters through gaps in the leaf canopy or is reflected and re-reflected from leaves or particles in the atmosphere. Ozone layer reduction will increase the UV-B(BE) to which these partially exposed organisms are subjected if all other factors remain the same. Thus avoidance alone, if it is cued by other factors (UV-A radiation, visible light, heat, moisture, etc.), may not suffice to protect organisms from UV-B increase.

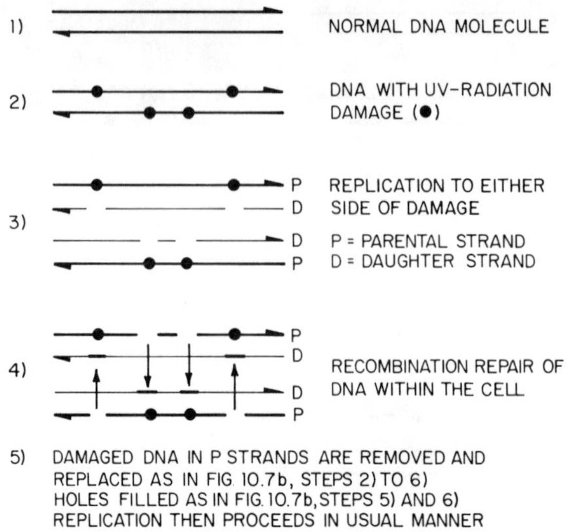

FIGURE 7c. Post-replication repair. A simplified model for post-replication repair of UV-damaged DNA molecules (shown as antiparallel lines with arrows): (1) normal DNA molecule; (2) dots indicate damage in DNA; (3) DNA synthesis (replication) under direction of DNA polymerase (not shown) proceeds on either side of the damage in the parental strands, leaving gaps in the daughter strands. After replication (3) the missing pieces are added to the incomplete DNA daughter strands. (D) by recombination with corresponding uninjured pieces of the parental DNA strands (P) within the cell. The remaining pieces of the parental DNA strands (4) may be repaired by enzymatic excision (by the DNA polymerase) of damaged regions in the DNA and by insertion of polynucleotides in these gaps and in those resulting from recombination repairs, in the same manner as in Figure 7b. It is possible that in some cases the gaps are repaired without recombination. (From Giese, A. C., *Living With Our Sun's Ultraviolet Rays,* Plenum Press, New York, 1976. With permission.)

C. Protective Screening

As protection against the solar UV-B radiation to which exposed organisms are subjected, there are a variety of organismal screening agents: (1) nonliving materials like fur, or hair, feathers, horny layers of epidermis in vertebrates; chitinous exoskeletons in arthropods; calcareous, or siliceous shells in other invertebrates; bark, waxy cuticles, and cellulose walls in plants; etc., (2) Pigments within living cells, e.g., melanin, xanthophylls, anthocyanins, etc., and (3) Noncritical substances within living cells such as chlorophyll, proteins, and RNA, which, although they have very important functions within cells, are present in large numbers and are usually replaceable if they become damaged.

The above screening agents can absorb UV-B photons and thus shield critical cellular targets such as DNA. The energy of the photon absorbed by these screens is usually degraded to harmless molecular motion (heat), but some chemical damage can occur. However, most damaged molecules can be broken down and recycled, or the damaged structure, e.g., horny epidermis, can be sloughed and replaced with new horny epidermis. Such consequences have a cost in energy but they protect against permanent damage to the organism.

D. Redundancy of Hereditary Information

Despite the above protective measures, some solar UV-B radiation may penetrate through to critical target molecules such as DNA. DNA contains the genetic information for making all of the constituents of cells and combining them into ordered cellular structures, organs, and organisms. Since DNA is so important to living organisms, there is frequently a redundancy of genetic information within the cells. Thus, even unrepaired damage to DNA is not necessarily detrimental if a redundant section of undamaged DNA can take over.

E. Cell Division Delay, Possibly for Repair Time

A prominent feature in the response of almost all cells to UV radiation is a delay of cell division. Certain evidence[9,10] indicates that cell division delay may represent an evolved mechanism by which organisms are allowed time for repair of UV-induced damage before it is "fixed" by cell division to become lethal.

F. Miscellaneous Mechanisms

In addition to the above general mechanisms, some organisms have evolved other ways to cope with solar UV-B damage:

1. Many cell types are more sensitive to the lethal effects of UV radiation during cell division than at any other time in their cell cycle.[11] Such extra sensitivity may result from the damage being "fixed" in some way before it can be repaired. Some cells in some organisms show cyclic daily patterns with cell division at night. Many unicellular algae have evolved a complicated reproductive pattern in which they grow quite large during the day (by photosynthesis) and then proceed through a series of rapid cell divisions at night to yield 4, 8, 16 and sometimes 32 cells. This complex pattern of reproduction may have evolved as a mechanism for coping with the increased sensitivity to UV radiation prior to and during cell division.
2. Calkins[12] has pointed out that reproduction by surviving unicellular organisms could replace those killed by UV-B radiation. As long as the dose of UV-B radiation does not exceed a certain "replacement limiting dose" (RLD) for each species of organism, then, according to his model, populations could survive. (The computation of the RLD accounts for cell division delay as well as reproductive death.)
3. Some plants with UV-B-radiation-damaged leaves may lose the leaf through the orderly process of senescence and abscission and replace it with a new one. This has an energy cost, but the organism survives.
4. Some organisms under unfavorable conditions form UV-radiation resistant spores and germinate at a later season when conditions are more favorable.[13]

VI. FUNDAMENTAL QUESTIONS FOR ASSESSING THE CONSEQUENCES OF INCREASED UV-B(BE)

With the knowledge that organisms in general have a variety of mechanisms for coping with current levels of solar UV-B(BE) radiation, we now consider the situation of an increase in UV-B(BE) radiation from ozone layer reduction. Figure 8 illustrates several National Academy of Sciences/National Research Council (NAS/NRC) projections of the time course of average ozone layer reduction (left hand ordinate) resulting from various scenarios of release of chlorofluorocarbons.[14] The right hand ordinate shows the percentage increases in annual average DNA-damage weighted UV-B(BE) radiation dose (at 40° N latitude) corresponding to the percentage ozone layer reductions. (Note that the right hand ordinate scale is nonlinear.) For all of the scenarios shown, the average ozone layer thickness continues to be reduced until well into the 22nd century at which time a constant steady state level of reduction is reached (see

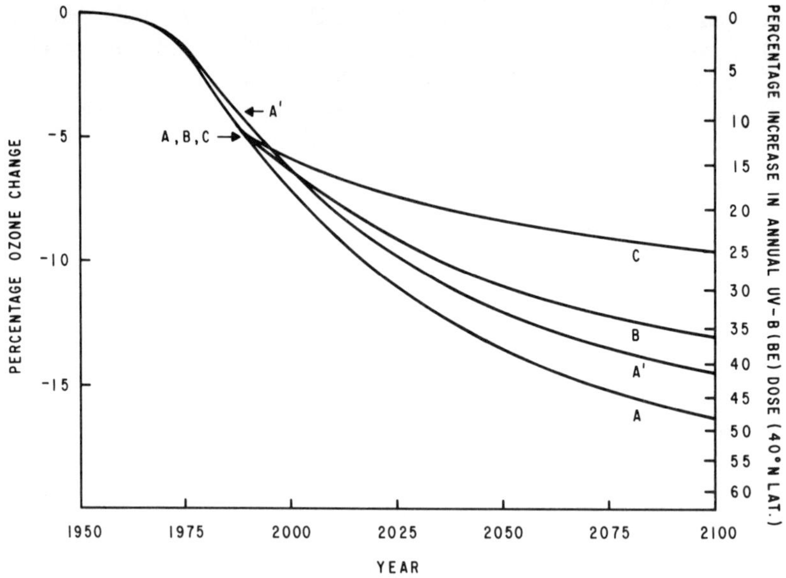

FIGURE 8. Projections of time-dependent decreases in the stratospheric ozone layer and increases in annual UV-B(BE) dose resulting from various scenarios for the release of chlorofluoromethanes (CFMs). Percentage ozone layer reduction is indicated on the left-hand ordinate; increased annual UV-B(BE) dose (DNA-damage weighted) at 40° N. latitude is indicated on the right-hand ordinate (note nonlinear scale).

		Steady state values	
Scenario	CFM release rate	% O_3 Reduction	Annual % UV-B(BE) increase
A	Constant at 1977 value	−18.6%	58.8%
A′	Same as A but multiplied by 0.89 to correct for feedback effects and possible tropospheric sinks	−16.5	50.1
B	Constant at 1977 value until 1983, when reduced by 25%; constant thereafter at reduced value.	−14.9	43.9
C	Constant at 1977 value until 1983, reduced then by 25% until 1988, when reduced an additional 25%; constant thereafter at the doubly reduced rate	−10.6	28.9

Redrawn from NAS Report[14] with numerical values for percentage ozone layer reductions from J. Chang (Personal communication).

legend for Figure 8). Reaching a steady state level depends upon a constant CFM release rate. If CFMs were to be released at an increasing rate, the ozone layer thickness would continue to fall.

For illustrative purposes we will concentrate on Scenario B, which yields about a 15% ozone layer reduction at steady state. Half of this reduction is expected to occur by about the year 2010. The increased UV-B(BE) corresponding to these reductions are 44% and 19%, respectively.

To assess the consequences of the increased UV-B(BE) radiation, the following fundamental questions must be answered:

1. Is all UV-B(BE) radiation detrimental to some extent or is there some limited tolerance of UV-B(BE) radiation without significant damage?
2. If there is a tolerable level of UV-B(BE) radiation, are organisms in nature currently existing at the limits of their tolerance?
3. If so, what will be the quantitative effect of exceeding that tolerance level?
4. If not, how much more UV-B(BE) can they tolerate?

VII. LACK OF AVAILABLE INFORMATION FOR ANSWERING FUNDAMENTAL QUESTIONS

We cannot unequivocally answer the fundamental questions on the basis of available information. This ignorance should not be surprising given that almost all of our information on UV-B radiation effects has been obtained in the last six or so years. It has been obtained from the short-term (18-month) biological studies portion of the Climatic Impact Assessment Program (CIAP) of the U.S. Department of Transportation plus a few subsequent studies from a National Aeronautics and Space Administration (NASA) program and the Environmental Protection Agency (EPA) Biological and Climatic Effects Research Program (BACER) now called the Stratospheric Impact Research and Assessment (SIRA) Program.

The difficulty in obtaining experimental answers to the fundamental questions lies in logistics, technology, experimental design, and the variability of biological systems. The direct way to obtain empirical answers is to expose organisms to expected increases in UV-B radiation in a reasonably natural manner. This is easier said than done. There are literally millions of species of organisms that are exposed to solar UV-B radiation. These organisms exist in an overwhelming variety of environmental conditions. Responses to UV radiation have been shown to differ with different conditions of visible light, temperature, moisture, nutrients, pH, etc.

Given this problem in assessment, the logical approach is to expose representative organisms, selected on the basis of economic or ecological value, to simulated increases in UV-B radiation under different environmental conditions that approach, as much as possible, the conditions under which these species normally grow.

However, the second problem of technology arises. The problems in simulation of enhanced UV-B irradiation are discussed in detail in Section X. In brief, with the best simulators presently available, there are unnatural characteristics. Most investigators employ fluorescent sunlamps filtered with plastic sheets to approximate the solar UV-B spectral irradiance. The match is not good and this has led to many problems (see Section X). Moreover, in laboratory settings, the amount of visible light and UV-A radiation available for photoreactivation repair is usually more than an order of magnitude less than that from the sun.

Even in experiments performed in the field with supplemental UV-B from filtered fluorescent sunlamps, technical problems arise. For example, the fluorescent fixtures themselves shade the plants from photosynthetically active visible radiation and thus may contribute to the reduction of biomass accumulation. Infestations of insects attracted by UV-A can compromise experimental findings.

Because of the technological problems mentioned above, many if not most of the studies performed to date are, unfortunately, only marginally applicable to an assessment of the ozone layer reduction problem. In fairness to the investigators, it should be emphasized that environmental photobiology is in an early evolutionary state. All of the environmental parameters that interact with an organism's response have not always been recognized. Failure to recognize some factors vitiates the applicability of the results to an assessment. Moreover, it is difficult for mechanistically oriented biologists to recognize that demonstrating a lack of effect of an agent is as important as demonstrating an effect — the establishment of a threshold or tolerance limit being a lack of an effect.

A review of all of the environmental photobiology studies with a discussion of their weaknesses would be tediously long and of limited value for quantitatively assessing the consequences of a 15% ozone layer reduction. The following brief summaries of effects, based on the NAS/NRC findings[15] may suffice. Certain of the studies mentioned will be discussed further in a more general context in Section VIII.

A. Agricultural Plants

Tests of more than 100 species or varieties of species in controlled environment growth chambers indicate that productivity (biomass increase) in approximately 20% of the plant types is quite sensitive to daily UV-B doses of the order of those delivered by Florida sunshine at present ozone levels. Twenty percent of the plant types were resistant to doses four times greater than this. The remaining 60% showed some intermediate sensitivity. These tests would suggest that a significant fraction of the present agricultural varieties are presently under UV radiation stress and would suffer decreased production with a 15% ozone layer reduction.

However, tests of 15 species and varieties, irradiated in the field under conditions of high visible light and UV-A-radiation, appeared to be more UV resistant than plants grown in the laboratory environmental growth chambers. Where it is possible to make dosage-response comparisons of the same type plants, the difference in resistance is on the order of fourfold. If this result turns out to be general, the higher sensitivity indicated in the growth chambers does not represent what would happen in the field and the expected consequences of ozone depletion become considerably reduced. Nevertheless, some species (sugar beets, tomatoes, mustard, corn) still appear to be affected, although the field experiments were necessarily less well-controlled and the results less clear-cut than were the growth chamber studies.

Resistant varieties of some of the species have been identified. This offers the possibility of selecting plants better adapted to UV-B stress. Efforts could be made to combine the UV resistance trait with the other characteristics desired in these plants (high crop yields and resistance to certain diseases, etc.) through systematic breeding. However, in much of the agriculture of undeveloped countries, introduction of new varieties would be difficult, and any emergence of more UV-resistant varieties might have to depend on the slower process of natural selection after the ozone decrease had occurred.

B. Nonagricultural Land Ecosystems

Perturbations of nonagricultural ecosystems could ultimately be as important as more direct effects on human health or human food supplies, but there is almost no information with which to assess the consequences of increased solar UV-B radiation on such ecosystems. Because the few nonagricultural plants that have been tested show about the same range of sensitivity as crop plants, reduced productivity of some forest and grazing land plants might occur. However, change in the species composition of these systems seems a more likely response: enhanced solar UV-B has been shown to alter competitive balance between species and lead to a change in community structure. Such a change would alter the character of the vegetation, and might qualitatively change the products provided. However, because UV sensitivities and their range of variation in wild species are known in so few cases, and the theoretical modeling of ecosystems is presently so primitive, such changes cannot be predicted confidently.

C. Domestic and Wild Animals

The very incomplete information suggests that the impact of ozone depletion on domestic and wild animals would not be large. An increase in cancer eye affecting some cattle breeds could be expected, but the fraction of cattle affected would be small (< 1%). Most of the increase would occur in older breeding cattle.

D. Aquatic Ecosystems

Experimental studies on over 60 aquatic microorganisms, protozoa, algae, and small invertebrates indicate that most of them are sensitive to current water-surface levels of UV-B radiation. Because they currently thrive in nature, attenuation of UV-B by natural waters, especially in productive areas, may play an important role in their survival. However, the degree of penetration of UV-B into such waters is not well-characterized.

UV-B irradiation of young anchovies by means of filtered fluorescent sun lamps indicates that in surface waters this commercially important species lives normally near its UV tolerance limit. UV-B(BE) increases similar to those expected with a 15% ozone layer reduction would, as a "worst case", kill over 50% of the anchovies in the top 10 m of the clearest ocean water, or else would require them to substantially readjust their usual water depth to diminish UV exposure. Such avoidance, if it were to occur, would result in a substantial shift in ecological relationships.

Similar studies show that mackerel are more UV resistant than anchovies. The "worst case", corresponding to organisms exposed near the water surface, indicates essentially no losses.

Analogous studies of crab and shrimp larvae indicate that they too are near their UV-B(BE) tolerance limit, considering their location in the water column (near the surface) and their reproductive/development season (late winter — early spring). A 15% ozone layer reduction might substantially shorten their reproductive/development season with unknown, but probably detrimental, consequences to productivity.

This very limited sampling gives little perspective on the UV sensitivity range of commercially important aquatic organisms. If we presume that such a small sample is roughly representative, an appreciable fraction of fishery species young could be vulnerable to the UV-B(BE) levels expected with a 15% ozone reduction. However, until more is known about the penetration of UV-B(BE) into the waters where they live, one can only speculate about potential vulnerability. It is likely that the most productive waters and seasons for anchovies are characterized also by high contents of chlorophyll that attenuate UV-B(BE).

VIII. POTENTIAL EFFECTS ON THE BIOSPHERE IN GENERAL OF OZONE LAYER REDUCTION/INCREASED UV-B(BE)

Although we cannot answer the fundamental questions on the basis of available information, a general discussion of the questions and potential answers can perhaps provide a useful perspective on the problem.

The first fundamental question, viz., is all UV-B radiation detrimental to some extent ...?, is perhaps the one fraught with the greatest import for assessment. If all UV-B radiation is currently producing damage in exposed parts of organisms and the organisms are coping with this damage by a constant expenditure of energy for repair or replacement of damaged molecules, cells, or nonliving structures, then any increase in UV-B(BE) radiation will likely increase this expenditure of energy. A diversion of input energy to such coping activity and away from productive activity may be considered detrimental. For example, decreased energy for productive activity could decrease the total biomass of an ecosystem, all else being equal. This view suggests that any increase in average UV-B(BE) radiation will have some effect on exposed parts of organisms wherever they may be — in direct sun, in shade, or under water — all else being equal. Moreover, from photochemical laws, we can expect that, at low damage levels, the additional number of damaged molecules will be directly proportional to increased UV-B(BE) flux. If the amount of energy expended for coping activity is large compared to that for productive activity, then the effect of increased UV-B(BE) radiation could be large; conversely, if the amount is small, the effect could be small. At present we have no information for any organism on the fraction of its energy flow used in coping with UV-B(BE) radiation.

The weight of evidence[16] so far suggests that some UV-B(BE) radiation is tolerable by at least some organisms. Some plants and insects can withstand daily irradiation with very high doses of simulated solar UV-B(BE) (corresponding to a 40 to 50% reduction in ozone) without apparent effect on growth in biomass or rate of development. On the other hand, some plants are drastically affected by such high doses. If a substantial fraction of the organisms of the world show some effects at all levels of UV-B(BE) radiation, including current levels, then an increased average UV-B(BE) radiation can be expected to have additional effects on them and the ecological system of which they are a part. The only way to assess such additional effects quantitatively is to obtain dosage-response curves for each different organism to see how great the response may be for a given increased average UV-B(BE) radiation dose rate. If, on the other hand, a large fraction of the organisms of the world can tolerate current or greater levels of UV-B(BE) without detriment, then the problem can be examined from a different viewpoint.

It is not unreasonable to expect some level of tolerance in many, if not most, organisms or populations of organisms. The energy flow into repair of damage to, say, cow's eyes and unpigmented skin is doubtlessly a very small fraction of the total energy flow, so a moderate increase in UV-B(BE) radiation would not likely decrease production. Similarly, this concept likely holds for other partially protected vertebrates and invertebrates. In the case of plants, damage to the photosynthetic apparatus in *individual* leaves leading to a reduction in leaf size and/or plant size could be compensated in a *population* of plants by decreased competition between the smaller plants for soil nutrients and water, thereby allowing more plants to grow per unit area. Thus, other factors in nature besides UV-B(BE) radiation may limit the amount of total biomass production.

Let us assume then that for most populations of organisms there is some tolerance of UV-B(BE), and address the other three fundamental questions listed above. If we assume that the current UV-B(BE) irradiances are being tolerated in nature, it may be instructive to examine just what is is that is being tolerated. As discussed previously, the natural ozone levels fluctuate, the solar zenith angle varies with latitude and throughout the day and year, and clouds come and go. All of these factors lead to widely varying UV-B(BE) irradiances to which organisms are exposed. As mentioned earlier, the UV-B(BE) fluctuations can be represented by a curve with peaks and valleys. If all else remains equal, an average ozone reduction would lead to a raising of the entire UV-B(BE) irradiance curve, i.e., it will lead to an increased average UV-B(BE) flux for any particular time of year or location and will lead to increased peak fluxes.

In experiments in which many different kinds of organisms have been UV-irradiated (both UV-B and UV-C), one frequently finds a sigmoid dose-response curve. The response starts out increasing slowly with increasing dose and then rises rapidly. Extrap-

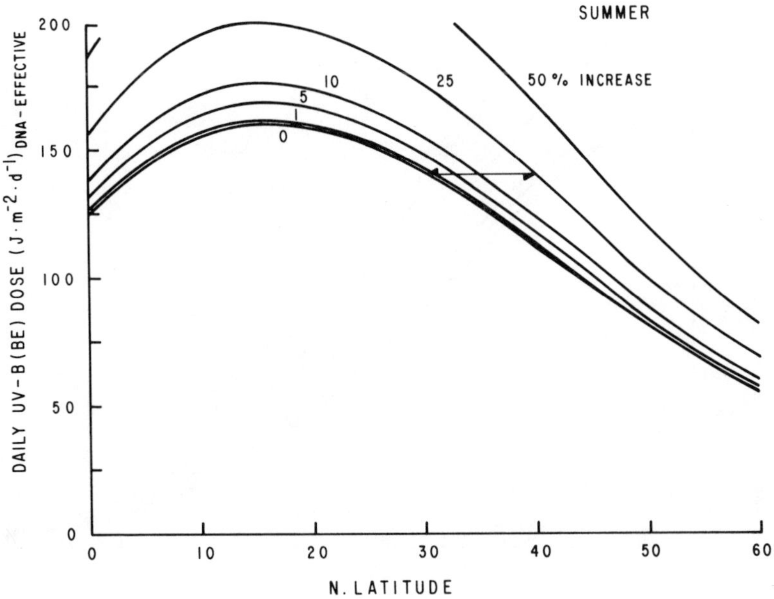

FIGURE 9. Daily clear-sky UV-B(BE) dose (DNA-damage weighted) for the summer solstice as a function of N. latitude for 0, 1, 5, 10, 25, and 50% increases in UV-B(BE) radiation. (The average ozone layer thicknesses for the season and latitude have been taken into account.) The arrow between 30° and about 40° represents the "zone of potential impact" resulting from a 25% increase in UV-B(BE).

olation of the steeply rising portion of the dose-response curve ot the abscissa (dose) yields what has been termed a quasithreshold dose, i.e., a dose below which little or essentially no effect (killing, inhibition of growth, etc.) is obtained. A logical explanation for such a curve is that the defenses and repair systems of the organisms can keep pace with damage up to a point but beyond this point, the defenses are overwhelmed. Buttressing this concept are experimental studies showing that mammalian cells[8] and plant cells[17] given low doses of UV-C radiation (254 nm) can repair essentially all of a particular type of damage (pyrimidine dimer lesions in DNA) but those given higher doses cannot and therefore subsequently die. It is reasonable to expect that in nature cells and the organisms composed of cells also have some quasithreshold dose below which little or no effect is seen.

The fundamental questions 2, 3, and 4 concern where this quasithreshold dose lies relative to current solar UV-B(BE) doses. These questions can be approached theoretically: Conrad[18] has discussed biological adaptability to naturally varying, potentially detrimental factors in general. He presents the idea that adaptability can be measured by calculating the maximum "uncertainty of the environment." He goes on to predict on theoretical grounds involving costs of adaptability that "adaptability will tend, in the course of evolution, to assume its lowest possible value. In other words, it should tend to decrease until it equals the uncertainty of the environment." Employing Conrad's reasoning, the natural extreme UV-B(BE) radiation dose would be the crucial factor for predicting the limit of tolerance or quasithreshold dose. The extreme UV-B(BE) radiation dose will occur under current conditions when the ozone level has fluctuated to its minimum and when the sky is clear and the solar elevation is at its maximum, which depends upon latitude. If the limit of UV-B(BE) radiation tolerance for a species is constant for all members of the species in nature, then a species that can just tolerate the current extreme UV-B(BE) radiation dose at say 30° latitude should be able to tolerate the extreme dose at all higher latitudes, all else being equal.

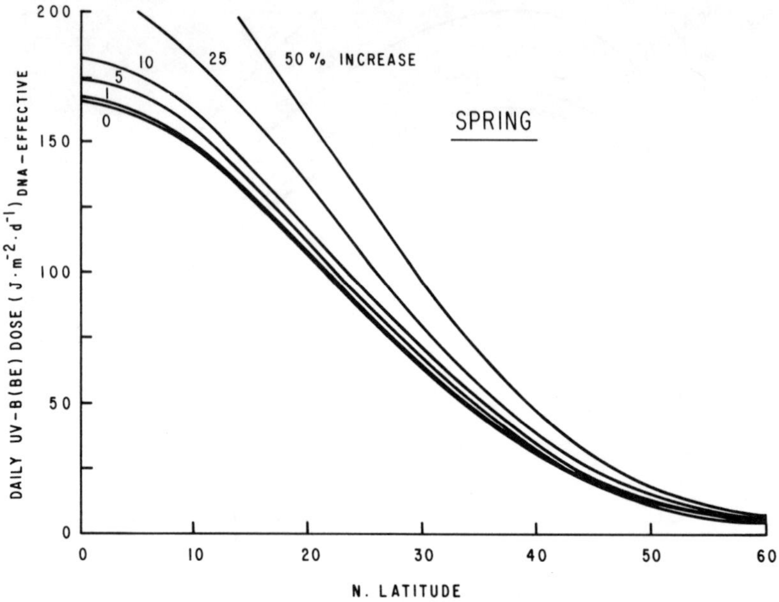

FIGURE 10. Same as Figure 9 but for the Spring Equinox.

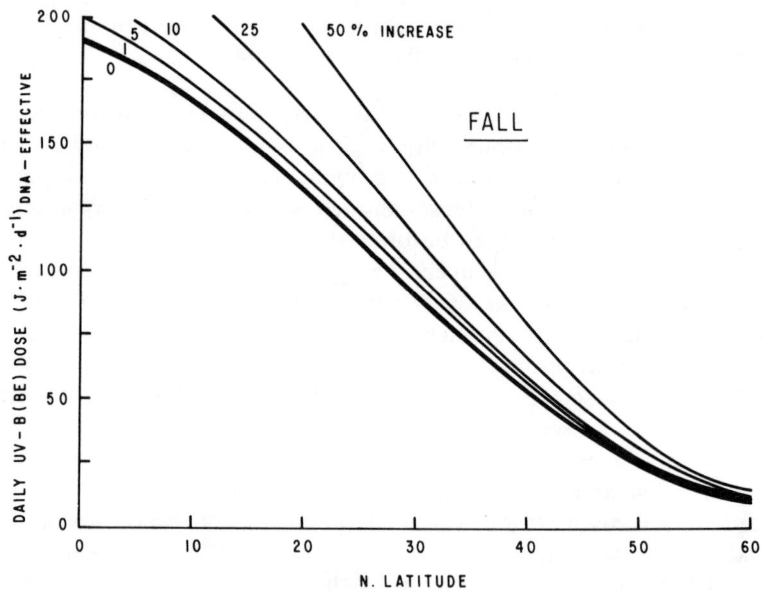

FIGURE 11. Same as Figure 9 but for the Fall Equinox.

With ozone layer reduction, the normal extreme UV-B(BE) radiation dose (tolerance dose) will occur at a higher latitude than normal. Below this latitude it will be exceeded. Thus increased UV-B(BE) radiation can potentially impact organisms in the zone from the latitude where the extreme dose is currently received up to the latitude where it will be received with the ozone layer reduction — the "zone of potential impact."

Figure 9 shows summer solstice, clear sky, daily UV-B(BE) doses in $J/m^2/day$ weighted by the DNA-damage weighting function (see Section IX) as a function of latitude. The curve labeled 0% represents the current dose and the other curves represent various percentage increases in daily UV-B(BE) doses. With the assumption that the extreme UV-B(BE) doses are directly proportional to these clear sky doses (calculated with the average ozone layer thickness for the season and latitude), the zone of potential impact can be determined for organisms whose lower latitude limit is known. For example, the horizontal arrow extending from 30° N latitude on the 0% curve to almost 40° N latitude on the 25% increase curve represents the zone of potential impact from a 25% increase in UV-B(BE) dose for organisms whose lower latitude limit is 30° N latitude. This 10° zone represents 1111 km. Similar families of curves for the spring and fall equinoxes are shown in Figures 10 and 11. Note that the highest daily dose in the Northern Hemisphere occurs at the equator on the fall equinox when the ozone layer thickness is lowest and the sun at noon is directly overhead. A 25% increase in UV-B(BE) dose would potentially impact organisms that can just tolerate the extreme daily dose at the equator; almost 12° on both sides of the equator — a total zone of about 2700 km — would be involved.

It should be emphasized that the foregoing analysis yields zones of *potential* impact; it should not be construed as implying that all sensitive organisms will be eliminated from the zone. Actually it is not presently known whether any single factor, be it UV-B(BE) radiation or any other environmental factor, limits the geographical range of any particular organism. More than likely, competitive interaction of organisms, responding to all of the many environmental factors, leads to geographic limitation.[19] Nonetheless, UV-B(BE) radiation at high intensity is a stressful agent. An increase may influence the competitive balance in at least part of the zone of potential impact and thus lead to a change in the species composition of communities there. Fox and Caldwell[20] have demonstrated that enhanced UV-B(BE) radiation can indeed influence competition between agricultural plants and their associated weeds.

The foregoing dealt with increases in the extreme UV-B(BE) doses. In addition to extreme doses and average doses, the frequency of days with particular doses may also be important. Organisms existing near their limit of tolerance may do so by tolerating the majority of the daily doses to which they are exposed in a particular ecological niche but may suffer some sublethal detriment at the highest doses with recovery occurring when the daily dose fluctuates back to a lower level.

With variations in ozone layer thickness, clouds, smog, and haze, the daily UV-B(BE) doses at a fixed latitude also vary widely. Figure 12 shows cumulative frequency distributions of the daily UV-B(BE) doses measured by a Robertson-Berger Meter (R-B Meter) in arbitrary units (see Section IX) during the summers of 1974 to 1977 at various representative locations.[21] Because the R-B Meter has a UV response spectrum that is somewhat insensitive to changes in the ozone layer thickness (see Section IX), the results presented are not as representative of DNA damaging UV-B(BE) radiation changes as would be desirable. Nevertheless, the results illustrate the wide variability in daily UV-B(BE) radiation doses to which organisms in different regions are exposed.

Curves such as those in Figure 12 can be used to assess the consequences of ozone layer reduction/increased UV-B(BE) radiation. Figure 13 curve A shows the cumulative frequency distribution of daily R-B Meter counts during summer in Oakland, CA, as an example. A ~5% decrease in ozone, with all else being equal, will lead to about a 10% shift of the entire curve to higher counts (Figure 13 Curve B). If organisms can tolerate radiation doses only up to the dose they receive, say 80% of the time under unperturbed circumstances, (i.e., 8.5×10^3 counts), then 20% of the time they may be detrimentally affected. With a 10% increase in UV-B(BE) radiation, this threshold dose or greater will occur 42% of the time instead of 20% of the time (see arrow from 20% point). If they can tolerate 90% of the doses without setback, then with a 10% increase in UV-B(BE) radiation the threshold dose (8.75×10^3 counts) will be exceeded 36% of the time instead of 10% of the time — a more than tripling of the number of

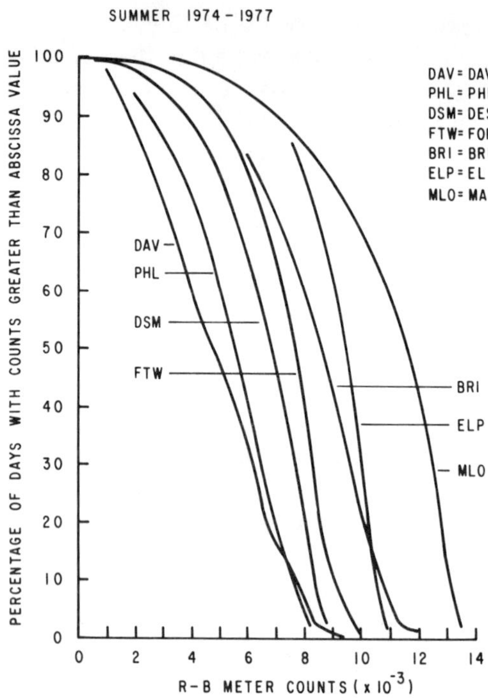

FIGURE 12. Frequency of summer days during 1974 to 1977 with Robertson-Berger meter counts greater than the value shown on the abscissa for various representative localities.[21]

detrimental days. If only 1% of days are currently nontolerable, these will increase 27-fold. Thus, with a 10% increase in UV-B(BE), the number of days during which the dose exceeds the tolerability threshold will increase nonlinearly by considerably more than 10%.

Germane to this point is the fact that, in some cases, plant growth,[22] and shellfish and finfish reproduction[23,24] are restricted to certain times of the year, usually spring. The environmental factors influencing this period of development involve temperature, moisture, day length, nutrients, and possibly other factors. The amount of UV-B(BE) radiation early in the growing season, when young UV-B-radiation-sensitive plants are developing, may not exceed the threshold under unperturbed conditions. However, given a shift in the distribution curve, the UV-B(BE) irradiance may now exceed threshold for some days, thus shortening the reproduction/development period.

The above considerations indicate that a 44% increase in UV-B(BE) has a strong potential for adversely altering the normal situation even though the sensitivities and limits of tolerance of most organisms are not known.

IX. UNCERTAINTIES ASSOCIATED WITH ASSESSMENT OF INCREASED UV-B(BE)

Up to this point we have discussed the potential effects of UV-B(BE) on a qualitative to semiquantitative basis. In order to consider the consequences of ozone layer reduction on a quantitative basis — the only basis on which an accurate assessment can be approached — when the future studies provide more adequate information, it is necessary to analyze quantitatively the uncertainties associated with the calculation of UV-B(BE) doses.

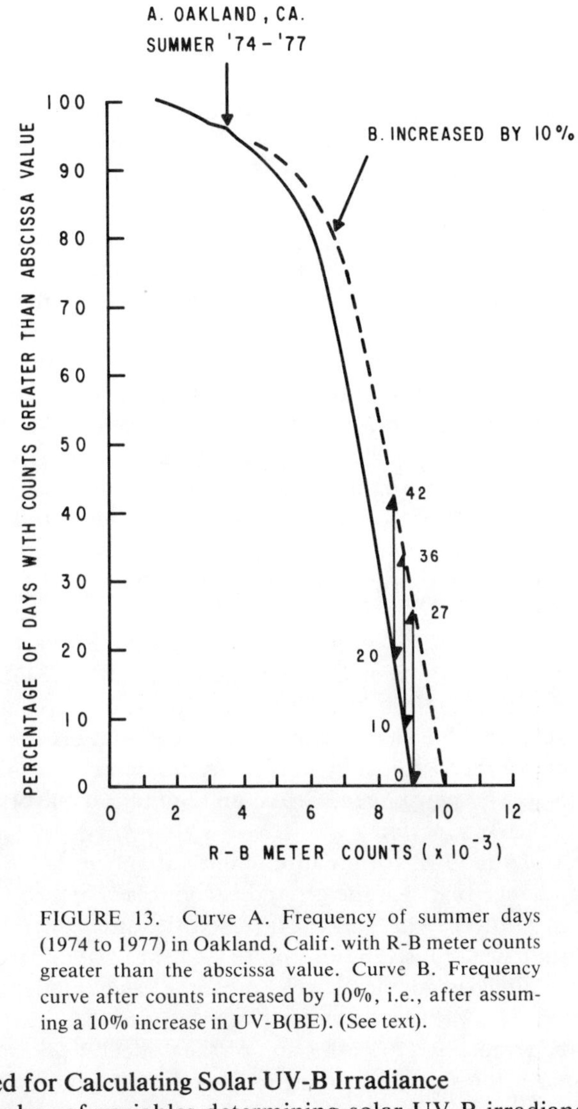

FIGURE 13. Curve A. Frequency of summer days (1974 to 1977) in Oakland, Calif. with R-B meter counts greater than the abscissa value. Curve B. Frequency curve after counts increased by 10%, i.e., after assuming a 10% increase in UV-B(BE). (See text).

A. Methods Used for Calculating Solar UV-B Irradiance

The great number of variables determining solar UV-B irradiances (see Section II) makes it imperative that the irradiances be calculated; no set of measurements of the type required for assessment exists. The accurate calculation of the transmission of radiation through the atmosphere requires that account be taken of scattering as well as absorption. This means solving the full set of equations for radiative transfer, an expensive and time-consuming proposition. Furthermore, the large number of variables upon which radiative transfer depends (e.g., wavelength, solar zenith angle, time of year, latitude, ozone and particulate content of the atmosphere, etc.) rules out a full radiative transfer calculation for each possible case.

A. E. S. Green and co-workers have approached this problem by constructing empirical equations with adjustable parameters, which can be fitted to experimental data or to the results of radiative transfer calculations. These empirical equations may then be used to interpolate to any case of interest.

In their first attempt at such a procedure[25] they utilized data gathered over a period of years at Davos, Switzerland, by Bener.[26] Unfortunately, Davos is at latitude 46° 48′N, and thus the minimum solar zenith angle is about 23.4°, and in addition Davos is at 1590 m above sea level. Thus, the empirical equations developed by Green, Sawada, and Shettle fit to Bener's data (termed the GSS-Bener model) were based on extrapolations to sea level and overhead sun, which introduced significant uncertainties.

Green, Sawada, and Shettle tried to improve their formulation by fitting to the rather simplified, radiative transfer calculations of Shettle and Green.[27] This empirical model (termed the GSS-multiscatter model) was an improvement over the previous model, since only interpolation, not extrapolation, was required.

Both the GSS-Bener and GSS-multiscatter models have been utilized, by Green, Mo, and Miller[28] and by other workers, to calculate UV-B irradiance for a number of cases. Unfortunately many of the reports have not made clear which parameters were employed.

Recently, Green, Cross, and Smith[29] have developed a new empirical model (termed the GCS model) based on the elaborate radiative transfer calculations of Halpern, Dave, and Braslau[30] and Dave and Halpern.[31] In these calculations special care was given to modeling the effects of the atmospheric aerosol content. They actually made calculations for ten different aerosol models, spanning a range of aerosol contents and refractive indices. Their model C-1, which uses a typical mid-latitude aerosol column content (about 2×10^7 particles/cm^2) with a typical scattering refractive index (1.50) and a typical absorption refractive index (0.01), is probably the best "standard" case. Figure 14 shows a comparison of the results of Dave and co-workers for model C-1 (assuming an ozone content of 0.30 atm.cm and a ground albedo of zero), compared with the various parameterizations of Green and co-workers. The GSS-Bener formulation gives irradiances that are too low, especially for near-overhead sun; the mean error over the spread of zenith angles is 30%. The GSS-multiscatter formulation is much better for near overhead sun, but fails badly for large solar zenith angles. Its mean deviation is 28%. The new Green, Cross, and Smith (GCS) formulation, with a mean error of 8%, is clearly superior.

The improvement of the GCS formulation comes about in the following manner. The total global irradiance reaching the ground is comprised of two parts: direct irradiance, consisting of photons which have traversed the atmosphere without interaction, and diffuse irradiance, consisting of photons which have undergone at least one scattering event. The direct irradiance $I_{dir}(\lambda, \theta)$ can be easily modelled with standard Beer's law absorption, $I_{dir}(\lambda, \theta) = I_o(\lambda) \exp(-\tau/\cos\theta)$ where $I_o(\lambda)$ is the solar irradiance at the top of the atmosphere, τ is the sum of the optical depths, for ozone, particulates, and air and θ is the solar zenith angle. This formulation for the direct irradiance (including a correction for the fact that the Earth's atmosphere is spherical rather than planar) is used in all three Green models.

The diffuse component is much more difficult to model. Its behavior is definitely not described well by Beer's law. Because the diffuse component is from about 40% (for overhead sun) to nearly 100% (at sunrise or sunset) of the total irradiance, it must be accurately modeled to yield correct total irradiances. Green, Sawada, and Shettle[25] tried to force diffuse irradiance to fit an equation of the same general functional form as the previous equation. Green, Cross, and Smith,[29] however, adopt a quite different technique. They have constructed empirical equations for the two ratios:

$$M(\lambda) = I_{diff}(\lambda, 0)/I_{dir}(\lambda, 0) \text{ and } S(\lambda) = I_{diff}(\lambda, \theta)/I_{diff}(\lambda, 0)$$

where M represents the ratio of diffuse to direct irradiance for overhead sun and S represents the ratio of diffuse irradiance at same solar zenith angle θ to that for overhead sun ($\theta = 0°$). The diffuse irradiance can then be calculated from the $I_{diff}(\lambda, \theta) = S(\lambda, \theta) M(\lambda) I_{dir}(\lambda, 0)$. Thus the GCS formulation is more flexible and yields more accurate irradiances for all wavelengths and solar zenith angles. We recommend use

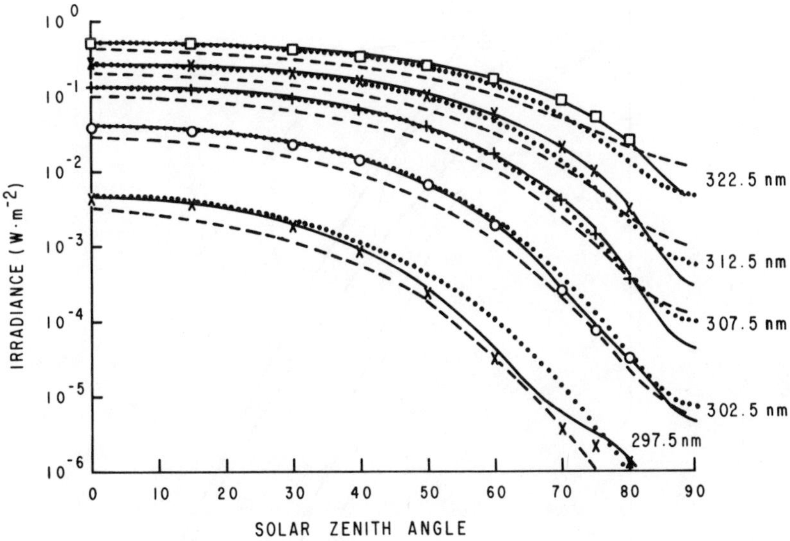

FIGURE 14. Comparison of irradiances calculated by three analytic models of Green and coworkers with the data points resulting from the radiative transfer calculations of Dave and co-workers. (See text.) ---- GSS-Bener; •••• GSS-multiscatter; —— GCS.

of the GCS formulation for future assessments of the biospheric impact of ozone layer reduction. Through its use most of the uncertainty that is attributable to solar irradiance calculations becomes so small relative to the other sources of uncertainty that it becomes essentially negligible. All values in tables and figures in this chapter are based on the GCS formulation. We also recommend that authors specifically indicate which of the several Green models they choose to utilize.

B. Uncertainties Associated with Calculation of Biologically Effective UV-B Radiation

Given a particular solar irradiance model and a particular weighting function, the calculation of the biologically effective dose impinging on a horizontal plane during a cloudless day is fairly straightforward, albeit mathematically cumbersome. As discussed in Section IX A, the uncertainty associated with the choice of solar irradiance model may be on the order of 30%. The uncertainty introduced by ignoring clouds and haze has not been adequately characterized, but one may assume that, if current patterns of cloudiness remain the same after some ozone layer reduction, the increased UV-B(BE) dose with clouds will be proportional to the increased UV-B(BE) dose calculated without clouds. The influence of inclinations to the sun and sky other than that of a horizontal plane has been examined by Burt and Luther.[32] They found that the latitudinal gradient of erythemal doses is reduced for vertically oriented receivers but that the percentage increase in UV-B(BE) dose resulting from a particular percentage reduction in the ozone layer is only weakly dependent on receiver orientation.

The major uncertainty associated with the calculation of UV-B(BE) dose is introduced by the choice of weighting function. The remainder of this section explores this uncertainty in some detail.

Different investigators use different weighting functions. Figure 15 shows plots of several of the analytical equations used as weighting functions by various investigators in assessing the biological impact of ozone layer reduction. Some of these are based upon experimentally determined action spectra, others are contrived "average action spectra." The plethora of weighting functions used by different investigators results

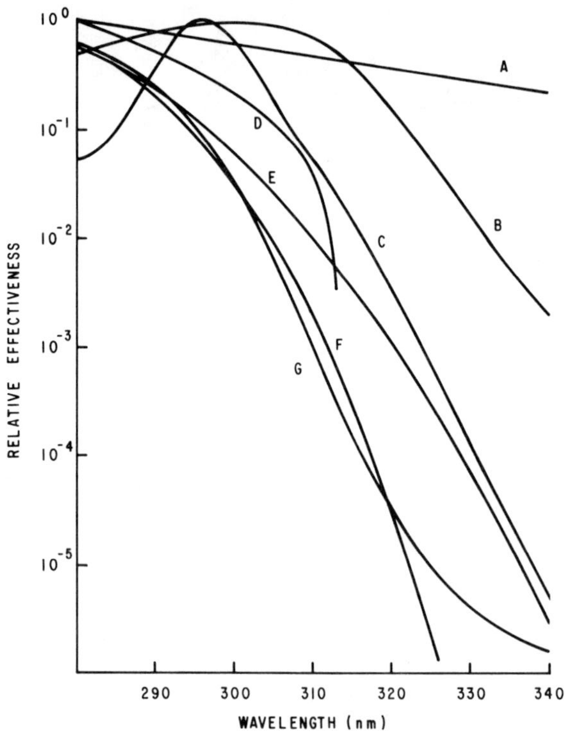

FIGURE 15. Weighting functions (see text). A. Photosynthesis inhibition. B. Robertson-Berger Meter. C. Erythema. D. Caldwell generalized plant action spectrum. E. AΣ 21. F. AΣ 9. G. DNA-damage.

partly from historical precedents and evolution toward more biologically reasonable choices, but mainly from our ignorance of just which is the appropriate choice.

The research on UV-B biological effects supported by the CIAP effort of the U.S. Department of Transportation was influenced by the emphasis on human skin cancer, which justified the use of a human erythemal (sunburn) action spectrum for the weighting function (Figure 15, Curve C). The program report[16] includes extensive tables of erythemal doses as a function of latitude, season, etc. These tables became the starting point for various investigations into biological effects other than human skin response. It was believed that, although other organisms may not sunburn, use of an erythemal weighting function yielded a reasonable, relative UV-B(BE) radiation measurement. This use was reinforced by the availability of the fairly inexpensive Robertson-Berger Sunburning Ultraviolet Meter or R-B Meter*, which was reputed at that time to have a response spectrum (Figure 15, Curve B) sufficiently similar to the erythemal action spectrum to allow practical dosimetry for biological studies, thus obviating the need for expensive spectroradiometers. As environmental photobiology has evolved since the CIAP effort, it has become increasingly recognized that the choice of the weighting

* The spectral response of the R-B Meter roughly approximates the relative human erythemal response spectrum and the signal output in counts results from the repetitive charging of a capacitor to a certain point at which it is discharged. All extant meters have been "standardized" by electronic adjustment to yield the same number of counts/day as a standard meter exposed under the same solar irradiance conditions at Philadelphia, Pa. "Calibration" against untanned caucasian skin indicated that the amount of solar radiation producing a minimal erythema produced 400 counts on the R-B meter. Thus 400 counts, or one sunburn unit (SU), was considered equivalent to a minimal erythemal dose in light-skinned humans.

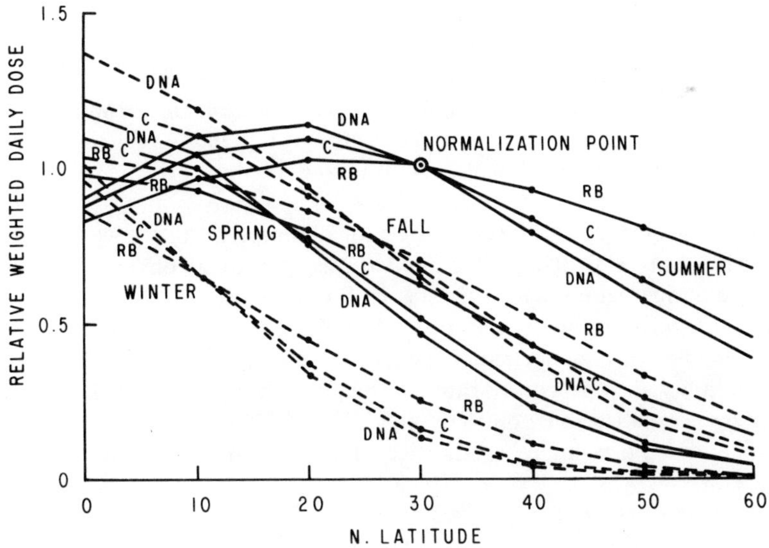

FIGURE 16. Influence of weighting function on calculated relative current daily UV-B(BE) doses as functions of N. latitude. All weighted doses have been normalized to that at 30° N. latitude for the summer solstice.

functions used in determining biologically effective doses profoundly affects the assessment of the consequences of ozone layer reductions and that the erythemal action spectrum and R-B meter response spectrum may not be the most appropriate bases for the weighting function. As will be detailed below, the question of which is the most appropriate cannot be readily answered at this time. For most biological effects studied, the action spectrum is not known.

In lieu of a known action spectrum for a particular biological effect, various investigators use generalized weighting functions based on a variety of measured action spectra using comparable biological material. For example, Caldwell[33] examined a number of action spectrum studies with plant material and developed a generalized plant action spectrum (Figure 15, Curve D). Setlow[34] compiled the results of a number of action spectrum studies with bacteria and viruses and generated a generalized DNA-damage action spectrum (Figure 15, Curve G). In unpublished studies conducted for the EPA-SIRA Program (formerly the EPA-BACER Program) investigators at the U.S. Department of Agriculture (USDA) Agricultural Research Center, Beltsville, Maryland, employed a simplified weighting function, termed AΣ9,[35] (Figure 15, Curve F) which approximates the Setlow DNA-damage action spectrum. Investigators at the University of Florida[36] employed an average weighting function, termed AΣ21 (Figure 15, Curve E) that falls between the DNA-damage action spectrum and the Caldwell action spectrum in the short wavelength portion of the UV-B region.

Figure 15 also shows a weighting function (Curve A) based on an action spectrum study of photosynthesis inhibition in isolated chloroplasts.[37] This study is the only one known that has critically examined the spectral influence of UV (and visible) radiation on photosynthesis independently from influences on cell growth and metabolism. Recent studies by Smith and Baker[38] have shown that photosynthesis by phytoplankton exposed to solar radiation through various cutoff filters is inhibited in a manner that is consistent with the Jones and Kok[37] action spectrum. Significantly, about 50% of the inhibitory effect results from radiation greater than 390 nm, i.e., in the visible region of the spectrum, and 25% results from the UV-A radiation between 340 to 390 nm. Thus, 75% of the photosynthetic inhibition by solar radiation results from wavelengths that are not expected to change with ozone layer reduction. Only the remaining 25% of the inhibitory effect is subject to increase by such reduction. Moreover, as one may infer from the shallowness of the curve, the increase in photosynthesis inhibition by any given ozone layer reduction will be relatively small. If this situation is applicable to plants in general, then ozone layer reduction is not likely to produce much of a significant additional impact on photosynthesis over that already being experienced under current irradiances.

Figure 16 illustrates the influence of the choice of representative weighting functions on the current daily UV-B(BE) doses at various seasons at different latitudes. (The appropriate average ozone layer thicknesses for the season and latitude, from Table 1, have been included in the calculations.) Although, the absolute weighted UV-B(BE) doses calculated with any particular weighting function vary widely because of the choice of parameters, one can normalize them for comparative purposes. In Figure 16, the single normalization point for all UV-B(BE) dose curves is shown at 30° N latitude in summer, a point representative of the maximum current daily UV-B(BE) dose in the U.S.

It may be noted first that organisms are currently being exposed to a wide range of UV-B(BE) doses at different latitudes and seasons, and that the range is greater for some weighting functions than others. It may also be noted that the normalization holds for only the one point in space and time so biological effects measured with dosimetry expressed in units using one weighting function cannot be easily extrapolated to an assessment of effects for which a different weighting function is used. Moreover, given the wide variability in current levels of UV-B(BE) (using any weighting function) the investigator attempting to obtain useful information for assessing the effects of increased UV-B(BE) is faced with a difficult problem: what reference point should be chosen for assessing some percentage increase in UV-B(BE) resulting from some specific ozone layer reduction?

In general, investigators have chosen a "reference day" appropriate to the biological effect studied (e.g., the reproductive/development season) and the latitude of the region where the experiment is conducted. Some investigators have then expressed the responses to UV-B(BE) doses in terms of the standard. The latitudinal variability of UV-B(BE) dose illustrated in Figure 16 precludes using results expressed in such a manner for assessment of potential effects at other latitudes. A preferable means of expressing results would be in terms of biological response per unit UV-B(BE) dose, for example, percent inhibition per DNA-damage-weighted ($J \cdot m^{-2}$), and then comparing this dosage-response to current and enhanced UV-B(BE) doses in the same units.

The substantial difference between the curves for the R-B meter and for the DNA-effectiveness weighting function are particularly noteworthy. It may be noted that the R-B meter is less sensitive to the latitudinal gradient of UV-B change. This occurs because much of the "dose" sensed by the R-B meter results from longer wavelengths that change less with ozone layer thickness and zenith angle than do the shorter wavelengths. Conversely, the DNA-damage-weighted doses show a much steeper gradient because the weighting emphasis is on the shorter wavelengths, which change greatly. The differences in the gradients have important implications. There is preliminary evidence that the dosage-response for skin-cancer-induction in hairless mice by UV-B radiation is consistent with a DNA-damage weighting function.[39] However, most epidemiological studies have correlated human nonmelanoma skin cancer incidence with UV-B(BE) dose as measured by the R-B meter at different cities on the latitudinal gradient. From such correlations, projections of increased skin cancer for particular ozone layer depletions have been made. If the DNA-damage weighting function is the more appropriate one for skin cancer, it may be seen that the latitudinal gradient will be much steeper, which will introduce an additional uncertainty in the already highly uncertain assessment of the consequence of ozone layer reduction on skin cancer incidence.

Because there have been a large number of studies with microorganisms and cultured cells that implicate DNA as the most sensitive cellular target molecule for UV radiation effects such as killing, mutation, and cell-division delay, one might ask, "Why not just settle on a DNA action spectrum for use as a weighting function and proceed from there?" In fact, the DNA-damage weighting function is currently favored by most investigators. However, there are a number of reasons for questioning whether it is the best or only weighting function to use.

1. Action spectrum studies, historically, have not been conducted to ascertain a weighting function for polychromatic radiation; they have been performed using monochromatic radiation to obtain a general idea of the nature of the molecule involved in some photobiological effect. For this purpose, the reciprocal of the dose of radiation at a particular wavelength required to produce a given biological effect is plotted relative to that at the most effective wavelength, generating a relative effectiveness curve as a function of wavelength. The resulting curve is compared to the absorption spectrum of some molecule that might be expected to be involved in the biological response. Emphasis is generally placed on the shape and position of peaks and high valleys. Rarely, do investigators emphasize the tail that falls off several orders of magnitude toward longer wavelengths — a tail that contributes significantly to dose when the very high solar UV-A irradiances are factored in. Setlow[34] is the most notable exception; he has followed the action on DNA down about six orders of magnitude from the peak.
2. Even with the well-characterized Setlow DNA-damage action spectrum as a basis for a weighting function, one cannot uncritically apply it to mammalian cells or multicellular organisms in which overlying structures, pigments, and expendable molecules may screen the DNA from UV-B photons and thus alter the expected action spectrum, even if DNA is the target molecule. For example, Rothman and Setlow[40] performed careful action spectra studies on the killing of V-79 Chinese hamster cells and the induction of lesions in DNA (thymine dimers). They found that the shape of the action spectra were similar but not identical to that for bacterial and viral inactivation: the mammalian cells were about twice as sensitive to radiation at wavelengths greater than 290 nm relative to the peak sensitivity at 265 nm, than were the bacteria and viruses.
3. Other molecules (unknown) which absorb appreciably more photons at longer wavelengths than do DNA molecules can apparently lead to photooxidation of DNA, mutation and death of bacterial cells.[41] This photodynamic action involving other molecules changes the action spectrum relative to that expected for a pure photochemical effect on DNA. The numbers and types of these unknown photodynamic molecules could easily be different in different cell types yielding unique action spectra for different cell types.
4. Action spectrum studies usually employ monochromatic radiation, frequently an isolated line from a mercury vapor arc source. There is a strong possibility that radiation from a polychromatic source such as the sun or fluorescent sun lamps may lead to synergistic biological effects not detectable with monochromatic radiation. For example, Rothman and Setlow[40] using monochromatic radiation, found an excellent correlation between pyrimidine dimer formation in DNA and cell killing as a function of wavelength. They concluded that dimer formation is the most important type of damage in cell killing over a broad range of UV wavelengths. On the other hand, Elkind et al.[42] in a comparative study of the killing of three mammalian cell types with monochromatic 254 nm UV and polychromatic UV from unfiltered fluorescent sunlamps (FS20) found distinct dif-

ferences in the shapes of the survival curves, indicating at least partially different modes of killing by the two types of radiation. From their analysis they conclude that if the thymine dimer is the principal lethal lesion following 254 nm UV irradiation, it is not the only or perhaps even the principal lesion following sunlight-like radiation. As follow-up, Elkind and Han[43] compared the induction of single-strand DNA breaks (including alkali-labile lesions) in mammalian cells by sunlamp UV-B radiation and 254 nm UV radiation. At comparable dimer-inducing doses, there are more single strand breaks in DNA produced by UV-B than by 254 nm UV radiation. Moreover, they found that following exposure, V-79 cells handle 254 nm UV-induced DNA breaks differently from sunlamp UV-induced breaks: after 254 nm UV irradiation the number of breaks increases for about 20 min and then either decreases (at low doses) remains about the same (at moderate doses) or increases gradually (at high doses); after sunlamp UV, the breakage does not promptly increase and the breaks generally persist. Possibly the sunlamp UV produces additional photochemical damage that interferes with excision repair of nearby thymine dimers. Alternatively, broad band sunlamp UV (and by extension solar UV) might directly inhibit enzymatic repair systems in addition to producing DNA damage. Tyrrell and co-workers have demonstrated that very high, but nonlethal, doses of 365 nm UV radiation can inhibit photoreactivating enzyme[44] and excision-repair[45] of 254 nm UV-induced dimers in DNA of bacteria. The action spectra for these effects have not been determined. Although the effective doses of 365 nm UV are in the megajoule per square meter range, this amount of UV-A radiation is not unreasonably out of line with that found in nature: the daily dose of UV-A radiation (320 to 380 nm) in summer is between 1 and 2 $MJ \cdot m^{-2}\ d^{-1}$ at all latitudes up to 60° (the highest calculated). Moreover, Nachtwey[46] has demonstrated that irradiation of a model unicellular freshwater alga *(Chlamydomonas)* with UV-B + UV-A from fluorescent sunlamps with nonlethal doses, given at levels comparable to that in sunlight can inhibit the photoreactivation process. Again, the action spectrum is not known, but the results indicate that UV-B is more effective than UV-A in producing the effect.

5. In addition to the above mentioned interactions involving DNA damage and repair by different wavebands, one must consider that fundamentally different cellular processes with different action spectra may be involved in a biological response to broadband UV radiation. For example, in the reduction of plant growth, UV radiation both reduces photosynthesis and inhibits normal cell division. Photosynthesis inhibition probably does not involve DNA as the affected chromophore whereas cell division inhibition probably does. Thus, reduction of plant growth likely results from the integrated combination of two relatively independent effects each with a different action spectrum.

The results discussed above suggest that polychromatic radiation can have effects not recognizable with monochromatic radiation and that an action spectrum obtained with monochromatic radiation may not be applicable as a weighting function for polychromatic radiation. Thus, the appropriate weighting function to employ for assessment purposes cannot be specified. A great deal of additional effort in the area of action spectroscopy, including studies of synergisms between wavebands, is required before an accurate assessment can be approached.

So far we have discussed only the influence of the weighting function on current levels of UV-B(BE) and some of the implications. The weighting function also quantitatively influences the increase in UV-B(BE) resulting from a particular percentage ozone layer reduction. As would be expected from the wide variability of the UV-

Table 1
MEAN TOTAL COLUMN CONTENT OF ATMOSPHERIC OZONE FOR EACH LATITUDE IN THE NORTHERN HEMISPHERE

Latitude (N)	Spring (atm. cm)	Summer (atm. cm)	Fall (atm. cm)	Winter (atm. cm)
0°	0.260	0.256	0.244	0.241
10°	0.268	0.261	0.253	0.247
20°	0.287	0.273	0.261	0.260
30°	0.313	0.292	0.270	0.284
40°	0.352	0.314	0.281	0.318
50°	0.395	0.333	0.299	0.357
60°	0.419	0.346	0.308	0.373
70°	0.430	0.349	0.307	0.370
80°	0.435	0.347	0.299	0.364
90°	0.436	0.339	0.290	0.361

Source: Bener[26] after J. London.

B(BE) dose discussed above, the relationship of Δ% daily UV-B(BE) dose/- Δ% ozone layer thickness (the radiation amplification factor, RAF) varies with latitude and season as well as weighting function. Table 2 shows the RAF values for 1% ozone layer reductions as a function of weighting function, season and latitude. Table 3 shows the range as a function of latitude and the percentage spread. It may be noted that the appropriate RAF varies from the value of two by a significant amount even for a given weighting function. The oft-repeated statement that a 1% ozone layer reduction leads to a 2% increase in UV-B(BE) is clearly only a rough rule-of-thumb; it doesn't hold precisely for any weighting function.

In Table 2, it is noteworthy that, with any weighting function, the highest percentage increases in UV-B(BE) dose occur at higher latitudes where there are lower current UV-B(BE) doses and presumably less current and potentially increased biospheric impacts from UV-B(BE) radiation. On the other hand, near the equator, where the greatest current UV-B(BE) irradiation occurs, the percentage increases in UV-B(BE) dose are lower although the increase in absolute UV-B(BE) dose will still be high.

For ozone layer reductions greater than 1%, the RAF for any weighting function increases nonlinearly. The RAF for any percentage ozone layer reduction, RAF (X%), can be roughly scaled up from that for 1% ozone layer reduction [RAF(1%)] by the relationship $\text{RAF}(X\%) = \text{RAF}(1\%)[1/(1-X\%/100)]$. This expression yields values that are within about 10% of the correct RAF (X%) for $X<20\%$ and latitudes <50° N.

Figure 17 shows the relative increases (decreases) in daily UV-B(BE) dose as a function of the ozone layer thickness with different weighting functions. To facilitate comparisons, the curves have been normalized to unity at the current ozone layer thickness (0.292 atm.cm) for 30° N latitude at the summer solstice, the same normalization point employed in Figure 16. Figure 17 reveals that the relative increase in daily UV-B(BE) dose for a given percentage ozone layer reduction is strongly dependent on the weighting function.

A more accurate general empirical relationship for calculating RAF (X%) values for different seasons and latitudes using the DNA-damage weighting function is presented in Table 4. This empirical relationship was derived by a non-linear least squares fitting routine[47,48] to "data" generated by the GCS model for 1, 5, 10, 16, and 25% ozone layer reductions from the average ozone layer thickness for the appropriate latitude band and season (Table 1). Note that for winter, the expression provides values only up to 50° N.

Table 2
RADIATION AMPLIFICATION FACTORS FOR 1% OZONE DEPLETION, RAF (1%), EMPLOYING DIFFERENT WEIGHTING FUNCTIONS

N. latitude/ season		Setlow DNA	Caldwell	R-B meter	Erythema	AΣ21	AΣ9
	Sp	2.21	1.61	0.73	1.61	1.28	1.93
	Su	2.23	1.66	0.75	1.63	1.28	1.95
0°	F	2.18	1.56	0.71	1.58	1.26	1.90
	W	2.21	1.61	0.73	1.60	1.27	1.92
	An	2.21	1.61	0.73	1.60	1.27	1.92
	Sp	2.23	1.65	0.75	1.63	1.29	1.95
	Su	2.23	1.63	0.74	1.62	1.28	1.94
10°	F	2.21	1.60	0.73	1.60	1.27	1.92
	W	2.25	1.69	0.76	1.64	1.28	1.97
	An	2.23	1.63	0.74	1.62	1.28	1.94
	Sp	2.28	1.74	0.78	1.67	1.31	2.00
	Su	2.24	1.65	0.75	1.63	1.29	1.95
20°	F	2.24	1.66	0.75	1.63	1.28	1.95
	W	2.30	1.85	0.80	1.69	1.31	2.03
	An	2.25	1.70	0.76	1.65	1.29	1.97
	Sp	2.33	1.89	0.83	1.72	1.34	2.07
	Su	2.27	1.72	0.77	1.66	1.30	1.99
30°	F	2.27	1.75	0.78	1.68	1.30	2.00
	W	2.33	2.11	0.87	1.72	1.34	2.12
	An	2.29	1.79	0.80	1.68	1.31	2.02
	Sp	2.39	2.14	0.90	1.76	1.38	2.17
	Su	2.31	1.81	0.81	1.70	1.33	2.04
40°	F	2.32	1.88	0.82	1.70	1.32	2.05
	W	2.26	2.50	0.94	1.70	1.36	2.21
	An	2.32	1.90	0.84	1.71	1.34	2.07
	Sp	2.38	2.48	0.97	1.78	1.42	2.28
	Su	2.35	1.93	0.84	1.73	1.35	2.09
50°	F	2.35	2.10	0.87	1.73	1.35	2.13
	W	1.94	2.76	0.98	1.61	1.32	2.17
	An	2.35	2.04	0.88	1.74	1.36	2.12
	Sp	2.28	2.84	1.03	1.76	1.43	2.35
	Su	2.38	2.08	0.88	1.75	1.37	2.15
60°	F	2.30	2.36	0.92	1.71	1.36	2.19
	W	1.44	2.09	0.89	1.37	1.13	1.82
	An	2.36	2.17	0.91	1.75	1.37	2.17

The relationships between RAF (X%) and latitude, season, and weighting function indicate that a considerable degree of uncertainty is involved in the statement, "an X% ozone layer reduction leads to a 2X% increase in UV-B(BE) radiation." Such uncertainty is compounded when one considers the UV-B(BE) dosage/biological response aspects in assessing the biospheric impact of ozone layer reduction. The following example may clarify the nature of the complication: Biggs et al.[36] found that the total yield of advanced mature Walter tomatoes from plants subjected in the field to natural solar radiation plus a gradient of enhanced UV-B radiation from cellulose-acetate-filtered fluorescent sunlamps (see Section X for details) declined in a linear manner with increasing added UV-B irradiance. Their data were recalculated using

Table 3
RANGES AND PERCENTAGE SPREAD OF RAF (1%) VALUES FOR 0 to 60° N. LATITUDE BY WEIGHTING FUNCTION

	Setlow DNA-damage	Caldwell	R-B meter	Erythema	AΣ21	AΣ9
Total Range	1.44—2.39	1.56—2.84	0.71—1.03	1.37—1.43	1.13—1.43	1.82—2.35
% Spread[a]	66.0	82.1	45.1	29.9	26.5	29.1
Spring Range	2.21—2.39	1.61—2.84	0.73—1.03	1.61—1.78	1.28—1.43	1.93—2.35
% Spread	8.14	76.4	41.1	10.6	11.7	21.8
Summer Range	2.23—2.38	1.63—2.08	0.74—0.88	1.62—1.75	1.28—1.37	1.94—2.15
% Spread	6.73	27.6	18.9	8.0	7.0	10.8
Fall Range	2.18—2.35	1.56--2.36	0.71—0.92	1.58—1.73	1.26—1.36	1.90—2.19
% Spread	7.80	51.3	29.6	9.5	7.9	15.3
Winter Range	1.44—2.33	1.61—2.76	0.73—0.98	1.37—1.72	1.13—1.36	1.82—2.21
% Spread	61.8	71.4	34.2	25.5	20.4	21.4
Sp-Fall Range	2.18—2.39	1.61—2.36	0.71—1.03	1.61—1.78	1.26—1.43	1.90—2.35
% Spread	9.6	46.6	45.1	10.6	13.5	23.7

[a] Max−Min/Min × 100

three different weighting functions (DNA-damage, AΣ21, and Caldwell weighting functions) to obtain percentage reductions in yield as a function of added daily UV-B(BE) doses. The results of the recalculations are presented in Table 5.

The experimentally determined percentage reduction per unit effective dose of UV-B(BE) in Table 5, Column 2, varies because the absolute units of UV-B(BE) dose vary as a result of the somewhat arbitrary parameters of the weighting functions. To obtain a measure of the yield reduction resulting from a 15% ozone layer reduction, which is a quantitative measure of biological effect needed for assessment purposes, one must determine the increase in $(J \cdot m^{-2} \cdot d^{-1})_{eff}$ for the particular latitude and time of year and then multiply it by the dose-response factor. Column 3 shows the current daily UV-B(BE) doses in $(J \cdot m^{-2} \cdot d^{-1})_{eff}$ obtained with the three weighting functions for the summer solstice at 30° N latitude, (a latitude near that of Gainesville, Fla., where the experiments were performed). Column 4 shows the percentage increase in UV-B(BE) resulting from a 15% ozone layer reduction as calculated using the appropriate RAF (1%) values from Table 2 and the Equation, RAF (X%) = RAF (1%) [1/ (1-X%/ 100)]. Column 5 shows the increase in UV-B(BE) in $(J \cdot m^{-2} \cdot d^{-1})_{eff}$ the product of the values in Columns 3 and 4. The percentage reduction in yield for a 15% ozone layer reduction, the product of Columns 5 and 2, is given in Column 6. It may be noted that relative to the results using the DNA-damage weighting function, the percentage increase in UV-B(BE) with the AΣ21 weighting function is 42% lower and with the Caldwell weighting function is 25% lower. However, the biological responses (% yield reduction) for a 15% ozone layer reduction (Column 6) are 56% and 72% *higher* for the AΣ21 and Caldwell weighting functions, respectively, relative to that for the DNA weighting function. This seemingly paradoxical situation arises because the use of those weighting functions, such as AΣ21 and Caldwell, that emphasize the longer UV-B wavelengths — the ones that do not change as greatly with ozone layer reduction — leads to large current and increased UV-B(BE) doses in both relative and absolute terms (Column 3). Therefore, although use of the DNA-damage weighting function yields about 6 times more biological effectiveness per unit dose than, for example, the use of the Caldwell weighting function, the relative percentage increase in UV-B(BE) from a 15% ozone layer reduction using the latter is more than 11 times the increase using the former thus more than compensating for the lower effectiveness.

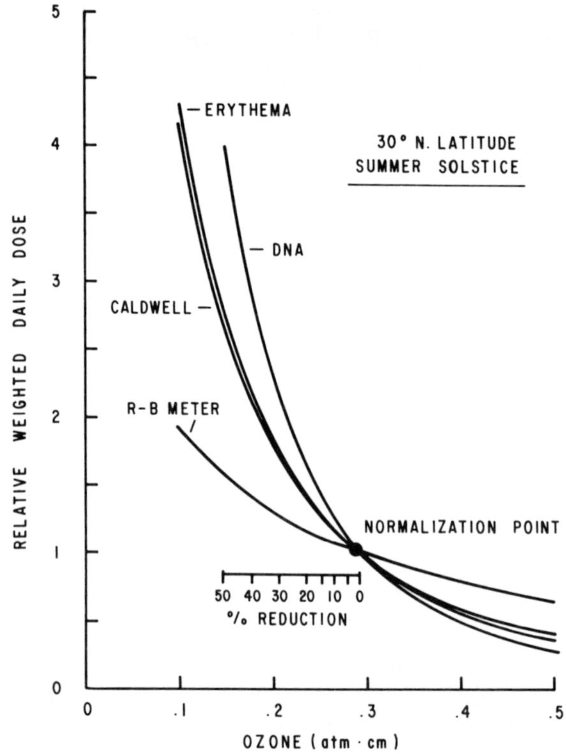

FIGURE 17. Influence of weighting function on relative increase (decrease) in daily UV-B(BE) dose at 30° N. latitude at the summer solstice as a function of ozone layer thickness. The percentage ozone layer reduction from the average for the latitude and season (0.292 atm.cm) is also indicated.

Table 4
ANALYTIC EQUATION AND PARAMETERS FOR CALCULATING THE RADIATION AMPLIFICATION FACTOR FOR DIFFERENT LATITUDES, SEASONS, AND OZONE REDUCTIONS USING THE DNA-DAMAGE WEIGHTING FUNCTION

$$RAF_{DNA}(X\%) = \frac{a_0 + a_1 \sec L + a_2 \sec^2 L}{1 - (k_0 + k_1 \sec L + k_2 \sec^2 L)\,R}$$

where L = ° N. latitude and R = the fractional reduction, $\Delta O_3/O_3$, or X/100.

	Max. error	a_0	a_1	a_2	k_0	k_1	k_2
Yearly average (0—60° N)	1.4%	1.409	1.110	−0.326	1.297	0.103	0
Spring (0—60° N)	2.2%	0.685	2.248	−0.736	1.263	0.152	0
Summer (0—60° N)	0.9%	1.652	0.742	−0.197	1.301	0.095	0
Fall (0—60° N)	1.3%	1.139	1.495	−0.465	1.257	0.131	0
Winter (0—50° N)	1.4%	−1.552	6.508	−2.751	−0.513	3.062	−1.159

The salient point of the example given above is that the interaction of the biological response vs. UV-B(BE) dose and the RAF for a particular season and location precludes the use of the RAF alone in assessing biological impacts of ozone layer reduction. Moreover, a simple projection of results obtained at one location to other latitudes is also precluded. This situation emphasizes the need for studies to determine the appropriate weighting function under fairly natural conditions and, moreover, the

Table 5
COMPARISON OF PERCENTAGE REDUCTIONS IN TOMATO YIELD FOR A 15% OZONE LAYER REDUCTION USING THREE WEIGHTING FUNCTIONS

(1) Weighting function	(2) % Yield reduction per $(J \cdot m^{-2} \cdot d^{-1})_{eff}^{a}$	Rel.	(3) Calculated current UV-B(BE) dose $(J \cdot m^{-2} \cdot d^{-1})_{eff}^{b}$	Rel.	(4) % Increase in UV-B(BE) for 15% O_3 red.	Rel.	(5) Increased UV-B(BE) dose with 15% O_3 red. $(J \cdot m^{-2} \cdot d^{-1})_{eff}$	Rel.	(6) % Yield reduction for 15% O_3 reduction	Rel.
DNA-damage	0.33	(1.00)	141	(1.00)	40	(1.00)	56.5	(1.00)	18.6	(1.00)
AΣ21	0.16	(0.48)	791	(5.61)	23	(0.58)	181.5	(3.21)	29.0	(1.56)
Caldwell	0.05	(0.15)	2099	(14.9)	30	(0.75)	637	(11.3)	31.9	(1.72)

[a] The ratio of the (slope/intercept) × 100 taken from the linear regression of experimentally determined yield of Walter tomato (advanced mature) in grams *vs.* added daily dose of UV-B(BE) determined by weighting the output of filtered fluorescent sunlamps with the weighting functions listed in Column (1). Data of Biggs et al.[36]

[b] For summer solstice, 30° N latitude, O_3 = 0.292, GCS model.

need to understand the biological response to UV-B(BE) doses relative to zero UV-B(BE) dose — not relative to present background UV-B(BE) dose at some particular location. Of course, there will still be uncertainties in extrapolating results determined relative to zero UV-B(BE) dose from one latitude to another because effects of other variables (e.g., visible light) may not be fully understood, but comparing the experimentally determined responses to UV-B(BE) dose on an absolute $(J \cdot m^{-2} \cdot d^{-1})_{eff}$ basis to UV-B(BE) doses at given latitudes and seasons should involve less uncertainty than the current practice of determining responses on an enhanced (relative) UV-B(BE) basis. In any event, the example demonstrates that a great deal of additional work is required before an accurate quantitative assessment of the potential impacts of ozone layer reduction can be made.

X. UNCERTAINTIES ARISING FROM EXPERIMENTAL CIRCUMSTANCES

In addition to comments in Section IX, other difficulties arise in simulating solar irradiances adequately for assessment of the biospheric impact of ozone layer reduction. As pointed out earlier, most investigators use fluorescent sunlamps filtered with cellulose acetate or cellulose triacetate (Kodacel) to remove the shortest wavelengths. As shown in Figure 18, the spectral qualities of such a source do not match well with any particular solar irradiance (under any condition of zenith angle or ozone level). There is too little irradiance above ∿300 nm and too much below for ozone layer thicknesses greater than 0.1 atm.cm, which represents an extremely depleted ozone layer thickness (∿0.3 is the average for mid-latitudes).

To circumvent the mismatch, dosimetry is performed using weighted doses on the assumption that a weighted unit daily dose from the filtered sunlamps is equivalent to a weighted unit daily dose from the sun under specified conditions of ozone layer thickness, latitude, and time of the year. The validity of extrapolating responses to sunlamp UV-B(BE) to expected responses to solar UV-B(BE) is thus strongly dependent on the validity of the weighting function for that response. An example of the pitfalls engendered by the mismatch may help clarify the nature of the problem.

During the early CIAP studies, few investigators had spectroradiometers to measure the spectral irradiances and doses under their particular experimental conditions, so

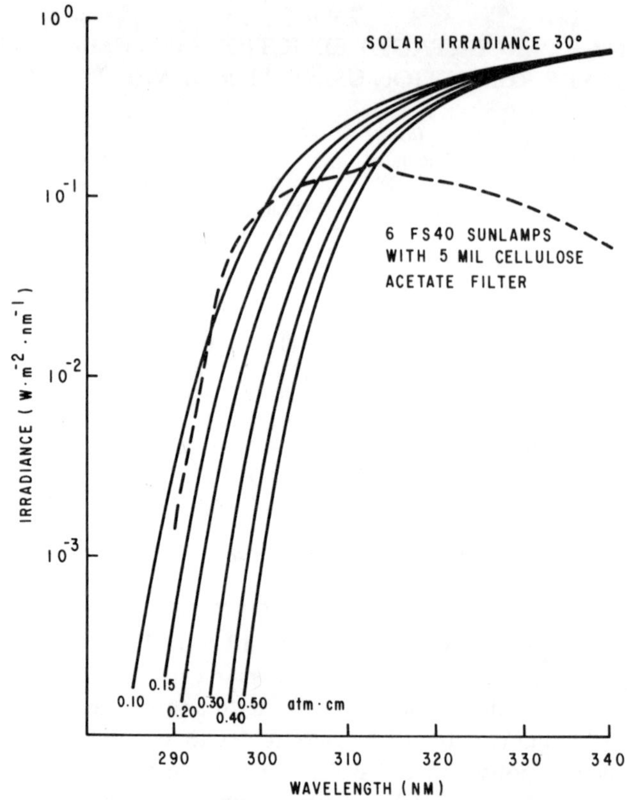

FIGURE 18. Spectral irradiance at 16 cm from six Westinghouse FS-40 sunlamps filtered with 5 mil cellulose acetate superimposed on calculated solar spectral irradiances at θ = 30° with the indicated ozone layer thickness.

many expressed their dosimetry in counts registered on an R-B Meter, or in SU. Although they recognized that the R-B meter response spectrum was designed to simulate the human erythema spectral response and not the spectral response for other biological effects, it was believed naively that the differences would be small on a relative basis and one SU from a filtered fluorescent sunlamp would be fairly equivalent to one SU from the sun. However, *if the DNA-damage action spectrum is the appropriate one to use* for other biological effects, one SU from a filtered fluorescent sunlamp contains about five times the amount of DNA-damaging UV as one SU from the summer sun at noon at 30°N latitude with 0.292 atm.cm of ozone. The discrepancy is even higher at higher latitudes.

Not only is the DNA-damaging dose from filtered fluorescent sunlamps per SU about five times higher than that from the sun, but the instantaneous dose-rate is also about five times higher than that from the sun at noon. The instantaneous dose-rate is an important parameter, because of biological repair processes. Nachtwey[49] has demonstrated a strong dependence of the killing effectiveness of filtered fluorescent sunlamp radiation on the dose rate (Figure 19). It appears as though high-dose rate radiation overwhelms repair processes. When the duration of exposure required to kill 50% or 90% of a model alga is plotted against different dose rates expressed $SU.h^{-1}$ (Figure 20) the effect is even more striking. If the DNA-damage weighting function is the appropriate one to use, almost all of the dose rates studied are greatly in excess of that to be expected from the sun even with extreme ozone layer reductions. Only the points

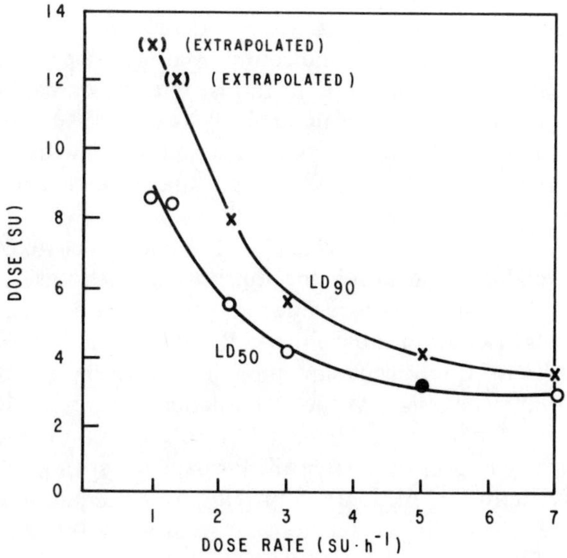

FIGURE 19. UV-B(BE) doses required to kill 50% (LD_{50}) and 90% (LD_{90}) of populations of *Chlamydomonas reinhardi* irradiated at different dose-rates. ($SU.h^{-1}$ = Sunburn Units per hour = 400 counts on a Robertson-Berger Meter.)

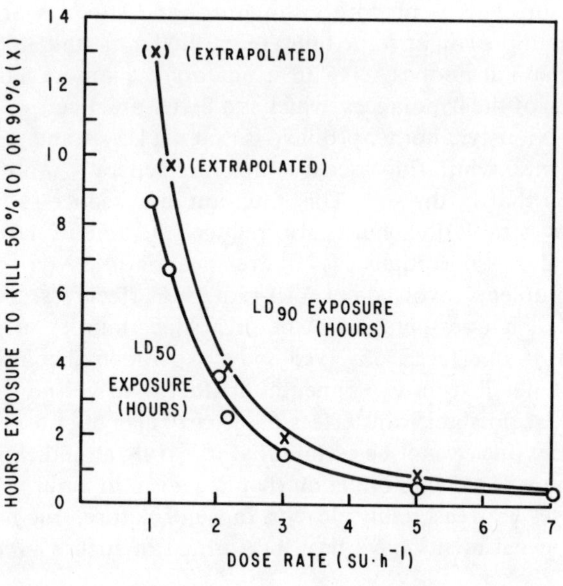

FIGURE 20. Duration of exposure required to kill 50% (LD_{50}) or 90% (LD_{90}) of *Chlamydomonas reinhardi* when irradiated at different dose-rates (data points calculated from values shown in Figure 19.)

at 1 $SU{\cdot}h^{-1}$ are close to current noon-time irradiances on a DNA-damaging dose-rate basis. In this regard it may be noted that the duration required to kill 90% has reached a full daylight period—this duration at a noon-time UV-B(BE) dose rate is not possible. If even lower dose rates had been tested, the duration of exposure required to kill

90% might be expected to approach many days. If the irradiation is protracted over many days, a different dose response indicating even lower sensitivity might be seen because of dark repair processes. Thus in terms of DNA-damaging UV doses, the apparent sensitivity of the algae is invalidated. Thus reports of biological effects resulting from a few SU from fluorescent sunlamps and the conclusions that organisms can be killed by a few hours of summer sunlight under current irradiances must be reevaluated.

In the foregoing statements, a qualifying clause has been repeatedly emphasized: "if the DNA-damage weighting function is the appropriate one to use..." The reason for such a qualification rests on the observation in several studies that some organisms exposed to actual solar radiation show either a R-B-meter-based dosage response similar to that obtained with fluorescent sunlamps or even an increased sensitivity on an SU basis.[50] This finding suggests that the DNA-damage weighting function is not the appropriate one to use for some unicellular organisms and that the R-B meter spectral response is more nearly appropriate after all. Perhaps the strong response of the R-B meter to short wavelength UV-A radiation provides weighting to the influence of UV-A on repair processes. Nonetheless, the above discussion of the DNA-damage dosage response vs. the R-B meter dosage response again emphasizes the importance of the weighting function in the analysis of experimental results with simulated solar UV-B and in the extension of the results to an assessment of what might happen in nature with a reduced ozone layer.

An additional complication in experimental studies is that, in most situations, the sunlamps are turned on for from 4 to 6 hours to obtain a given daily dose comparable to that from the sun. This is obviously unnatural and subjects organisms to much higher intensities than the sun at some times (e.g., 9:00 a.m. and 3:00 p.m.) and lower intensities than the sun at noon. (A few investigators use a more natural stepping upward and downward of the irradiances, which is a better practice.)

As mentioned previously, another problem is that the UV-A and visible components, even with supplemental white fluorescent lamps, are generally more than an order of magnitude less than that in the sun. This situation may reduce some harmful synergisms involving UV-A radiation, but it also reduces the amount of photoreactivating light. In any event, the total irradiance is not even near natural.

This catalog of problems involved in UV-B-radiation effects research and assessment is not exhaustive but it does emphasize the great uncertainties in projections based upon such research. If an effect is observed under experimental conditions, one cannot be certain that it is not just an experimental artifact that will not happen in nature. On the other hand, if no significant effect on a particular organism is found at some high UV-B(BE) dose, one cannot be certain that it is truly an indication that increased UV-B(BE) radiation will have no effect on that organism in nature. Unless the experimental circumstances are reasonably close to those of nature, one may always doubt. To date, very few experiments have been performed in such a manner as to relieve such doubts.

XI. POSSIBLE MEANS TO REDUCE UNCERTAINTIES

Some of the problems discussed above might be avoided if better solar simulators which possess high UV-A and visible light irradiances are used. High intensity solar simulators exist but are very expensive and require considerable care and maintenance. Moreover, they may not be readily varied in intensity and spectral output. For broad band action spectroscopy a large engineering and development effort may be required. Perhaps a national facility for research with such simulators could be established.

The sun at most places near sea level is not a reliable alternative source of high UV-B, UV-A and visible light. The irradiance in the UV-B especially is insufficient to test the effects of enhanced UV-B. Perhaps a suitably enclosed greenhouse laboratory on a mountain top at a low latitude in a usually clear area — a site such as that preferred for a meteorological or astronomical observatory (e.g., Mauna Loa or Haleakala, Hawaii) — may provide a suitable site for environmental photobiological research. Again, this represents a kind of national facility. In either case (accurate solar simulator or reliably sunny site with a high UV-B irradiance) good environmental regulation to eliminate uncontrolled influences or to test the influence of other stresses (e.g., temperature, moisture, etc.) will be required. In a greenhouse situation, temperature control is particularly difficult to achieve but is very necessary. (This also recommends a cool mountaintop site.)

To reduce uncertainties for assessment, dosimetry must be spectroradiometric and, if the sun is employed, continuous. Ozone column measurements would also be required. The greatest difficulty may be in the shaping of the solar irradiances to simulate ozone levels and daily dose-rate changes applicable to other latitudes. It is likely that a perfect simulation will not be possible; simplifying assumptions will still be required. Nonetheless, the employment of the mountain top, low latitude, UV-B-transparent, greenhouse approach coupled with adequate spectroradiometric measurements will eliminate many of the problems discussed above and thus decrease the uncertainties surrounding the assessment of the biospheric consequences of ozone layer reduction.

REFERENCES

1. Meyer, A. L. and Seitz, E. O., *Ultraviolette Strahlen ihre Erzeugung, Messung und Anwendung in Medizin, Biologie und Technik,* de Gruyter, Berlin, 1942.
2. Machta, L., Cotton, G., and Hass, W., Hourly tabulation erythemal spectrum solar ultraviolet radiation and related meteorological data. Supplement to NOAA Final Report to CIAP. Interagency Agreement (with DOT) DOT-AS-20082, NOAA, Washington, D.C., 1975.
3. Nachtwey, D. S., Linking photobiological studies at 254 nm with UV-B, in CIAP Monograph 5, Impacts of Climatic Change on the Biosphere, Part 1, Ultraviolet Radiation Effects, Nachtwey, D. S., Caldwell, M. M., and Biggs, R. H., Eds., DOT-TST-75-55, U.S. Department of Transportation, Washington, D.C., 1975, 3-50.
4. Giese, A. C., *Living with Our Sun's Ultraviolet Rays,* Plenum Press, New York, 1976.
5. Ley, R. D., Sedita, B. A., and Grube, D. D., Absence of photoreactivation of pyrimidine dimers in the epidermis of hairless mice following exposures to ultraviolet light, *Photochem. Photobiol.,* 27, 483, 1978.
6. Regan, J. D., Carrier, W. L., Smith, D. P., unpublished data, 1979.
7. Harm, W., Biological determination of the germicidal activity of sunlight, *Radiat. Res.,* 40, 63, 1969.
8. Carrier, W. L., Smith, D. P., and Regan, J. D., Ultraviolet light dose response of pyrimidine dimer excision in human cells, *Abstr. Am. Soc. Photobiol. Meeting,* WAM-CIO, American Society of Photobiology, Bethesda, M.D., 1979.
9. Nachtwey, D. S. and Giese, A. C., Effects of ultraviolet light irradiation and heat-shocks on cell division in synchronized *Tetrahymena, Exp. Cell. Res.,* 50, 167, 1968.
10. Hodge, F. A. and Nachtwey, D. S., X-ray-induced delay of cell division in synchronized *Tetrahymena pyriformis, Radiat. Res.,* 52, 603, 1972.
11. Rocha, L. R. and Nachtwey, D. S., unpublished data, 1974.
12. Calkins, J., Effects of real and simulated solar UV-B in a variety of aquatic microorganisms—possible implications of elevated UV irradiance, in CIAP Monograph 5, Impacts of Climatic Change on the Biosphere, Part 1, Ultraviolet Radiation Effects, Nachtwey, D. S., Caldwell, M. M., and Biggs, R. H., Eds., DOT-TST-75-55, U.S. Department of Transportation, Washington, DC, 1975, 5.
13. McPhail, L. and Nachtwey, D. S., unpublished data, 1974.
14. NAS Committee on Impacts of Stratospheric Change, Panel on Stratospheric Chemistry and Transport, *Stratospheric Ozone Depletion by Halocarbons: Chemistry and Transport,* National Academy of Sciences, Washington, D.C., 1979.

15. NAS Committee on Impacts of Stratospheric Change, *Protection Against Depletion of Stratospheric Ozone by Chlorofluorocarbons,* National Academy of Sciences, Washington, D.C., 1979.
16. Nachtwey, D. S., Caldwell, M. M., and Biggs, R. H., Eds., CIAP Monograph 5, Impacts of Climatic Change on the Biosphere, Part. 1 Ultraviolet Radiation Effects, DOT-TST-75-55, U.S. Department of Transportation, Washington, D.C., 1975.
17. Howland, G. P., unpublished data, 1977.
18. Conrad, M., Biological Adaptability: the statistical state model *Bioscience,* 26, 319, 1976.
19. Walter, H., *Vegetation of the Earth in Relation to Climate and the Eco-physiological Conditions,* Springer-Verlag, New York, 1973.
20. Fox, F. M. and Caldwell, M. M., Competitive interaction in plant populations exposed to supplementary ultraviolet-B radiation, *Oecologia,* 36, 173, 1978.
21. Hass, W., Cotton, G., and Machta, L., unpublished final report for NASA Order T-669G, 1979.
22. Krizek, D. T., Influence of ultraviolet radiation on germination and early seedling growth, *Physiol. Plant.,* 34, 182, 1975.
23. Damkaer, D. M., Heron, G. A., and Dey, D. B., unpublished EPA-BACER report, 1978.
24. Hunter, J., Taylor, J., and Moser, H. G., unpublished EPA-BACER reports 1977, 1978.
25. Green, A. E. S., Sawada, T., and Shettle, E. P., The middle ultraviolet reaching the ground, *Photochem. Photobiol.,* 19, 251, 1974.
26. Bener, P., Approximate Values of Intensity of Natural Ultraviolet Radiation for Different Amounts of Atmospheric Ozone, Technical Report, European Research Office, U.S. Army, London, Contract No. DAJA37-68-C-1017, 1972.
27. Shettle, E. P. and Green, A. E. S., Multiple scattering calculation of the middle ultraviolet reaching the ground, *Appl. Opt.,* 13, 1567, 1974.
28. Green, A. E. S., Mo, T., and Miller, J. H., A study of solar erythema radiation doses, *Photochem. Photobiol.,* 20, 473, 1974.
29. Green, A. E. S., Cross, K. R., and Smith, L. A., Improved analytic characterization of ultraviolet skylight, *Photochem. Photobiol.,* 31, 59, 1980.
30. Halpern, P., Dave, J. V., and Braslav, N., Sea-level solar radiation in the biologically active spectrum, *Science,* 186, 1204, 1974.
31. Dave, J. V. and Halpern, P., Effect of changes in ozone amount on the ultraviolet radiation received at sea level of a model atmosphere, *Atmos. Environ.,* 10, 547, 1976.
32. Burt, J. E. and Luther, F. M., Effect of receiver orientation on erythema dose, *Photochem. Photobiol.,* 29, 85, 1979.
33. Caldwell, M. M., Solar ultraviolet radiation and the growth and development of higher plants, in *Photophysiology,* Vol. 6, Giese, A. C., Ed., Academic Press, New York, 1971, 131.
34. Setlow, R. B., The wavelengths in sunlight effective in producing skin cancer: a theoretical analysis, *Proc. Natl. Sci. U.S.A.,* 71, 3363, 1974.
35. Thimijan, R. W., Carns, H. R., and Campbell, L. E., unpublished EPA-BACER report, 1978.
36. Biggs, R. H. and Kossuth, S. V., unpublished EPA-BACER report, 1978.
37. Jones, L. W. and Kok, B., Photoinhibition of chloroplast reactions. I. Kinetics and action spectrum, *Plant Physiol.,* 41, 1037, 1966.
38. Smith, R. C. and Baker, K. S., Stratospheric ozone, middle ultraviolet radiation and C measurements of marine productivity, *Science,* 208, 592, 1980.
39. Forbes, P. D., Davies, R. E., and Urbach, F., unpublished data, 1979.
40. Rothman, R. H. and Setlow, R. B., An action spectrum for cell killing and pyrimidine dimer formation in Chinese hamster V-79 cells, *Photochem. Photobiol.,* 29, 57, 1979.
41. Webb, R. B., Lethal and mutagenic effects of near-ultraviolet radiation, in *Photochemical and Photobiological Reviews,* Vol. 2, Smith, K. C., Ed., 1977, 169.
42. Elkind, M. M., Han, A., and Chang-Liu, C-M., "Sunlight"-induced mammalian cell killing: a comparative study of ultraviolet and near-ultraviolet inactivation, *Photochem. Photobiol.,* 27, 709, 1978.
43. Elkind, M. M. and Han, A., DNA single-strand lesions due to "sunlight" and UV light: a comparison of their induction in Chinese hamster and human cells, and their fate in Chinese hamster cells, *Photochem. Photobiol.,* 27, 717, 1978.
44. Tyrrell, R. M., Webb, R. B., and Brown, M. S., Destruction of photoreactivating enzyme by 365 nm radiation, *Photochem. Photobiol.,* 18, 249, 1973.
45. Tyrrell, R. M. and Webb, R. B., Reduced dimer excision following near ultraviolet (365 nm) radiation, *Mutat. Res.,* 19, 361, 1973.
46. Nachtwey, D. S., Comparative effects of UV-B and 254 nm UV on the alga, *Chlamydomonas reinhardi:* evidence for a qualitative difference, in CIAP Monograph 5, Impacts of Climatic Change on the Biosphere, Part 1, Ultraviolet Radiation Effects, Nachtwey, D. S., Caldwell, M. M., and Biggs, R. H., Eds., DOT-TST-75-55, U.S. Department of Transportation, Washington, D.C., 1975, 3-86.
47. Marquardt, D. W., An algorithm for least-squares estimation of nonlinear parameters, *J. Soc. Indust. Appl. Math.,* 11, 431, 1963.

48. Bevington, P. R., *Data Reduction and Error Analysis for the Physical Sciences,* McGraw-Hill, New York, 1969.
49. Nachtwey, D. S., Potential effects on aquatic ecosystems of increased UV-B radiation, in Proceedings of the Fourth Conference on the Climatic Impact Assessment Program. Hard, T. M., and Broderick, A. J., Eds., DOT-TSC-OST-75-38, U.S. Dept. of Transportation, Washington, DC, 1976, 79.
50. Calkins, J., Barcelo, J. A., Grigsby, P., and Martin, S., Studies of the role of solar ultraviolet radiation in "natural" water purification by aquatic ecosystems, in *Research Report No. 108,* University of Kentucky Water Resources Research Institute, Lexington, KY, 1978.
51. Nachtwey, D. S., Caldwell, M. M., and Biggs, R. H., Eds., CIAP Monograph 5, Impacts of Climatic Change on the Biosphere, Part I, Ultraviolet Radiation Effects, DOT-TST-75-55, U.S. Department of Transportation, Washington, D.C., 1975, 2-28.
52. Hanawalt, P. C. and Smith, K. C., *Molecular Photobiology—Inactivation and Recovery,* Academic Press, New York, 1969.
53. Deering, R. A., *Sci. Am.,* 207, 135, 1962.
54. Hanawalt, P. C., *Endeavor,* 31, 84, 1972.
55. Smith, K. C., *Sunlight and Man,* Universityy of Tokyo Press, Tokyo, 1974, 72

Chapter 5

OZONE CHANGE: CLIMATOLOGICAL EFFECTS

Wei-Chyung Wang

TABLE OF CONTENTS

I. INTRODUCTION

It is well recognized that the thermal structure of the atmosphere of the Earth is influenced by the presence of trace gases. The main radiative effect of the gases is modulation of the solar radiation and thermal emission. The principal gaseous absorbers of solar radiation are water vapor (H_2O) in the troposphere and ozone (O_3) in the stratosphere. Water vapor absorbs primarily in the near-infrared spectral region while O_3 is most effective in the ultraviolet (UV) and visual regions (Figure 1).[1]

In the infrared, H_2O effectively blocks thermal emission from the surface of the Earth except for the "window" region between 7 and 12 μm (800 to 1400 cm^{-1})* where O_3, carbon dioxide (CO_2) and several trace gases with strong absorption bands contribute additional atmospheric opacity (Figure 2).[2] Absorption of outgoing thermal radiation in the atmosphere, followed by re-radiation at local temperature can lead to an increase of the surface temperature, the so-called *greenhouse effect.* In fact, these trace gases, together with clouds, are mainly responsible for the globally averaged temperature of the Earth being 288K, some 35K warmer than the calculated temperature for the planet with its existing albedo. This trapping of infrared radiation in the troposphere reduces outgoing thermal radiation in the stratosphere, producing a lowering of stratospheric temperatures.

Consequently, a change of trace gas concentrations in the atmosphere can affect the thermal structure of the atmosphere. For example, it is generally believed that continued injection of anthropogenic CO_2, largely from fossil fuel combustion, can significantly cool the stratosphere and warm the troposphere, causing a significant impact on the global climate.[3]

In recent years another cause for considerable public concern is the possibility that the amount of O_3 in the atmosphere might be perturbed as a result of human activities.[4-7] Such a perturbation could have serious consequences for mankind. Reducing the amount of O_3, for example, could increase the amount of solar UV radiation reaching the surface of the Earth. These processes and the possible resulting effects are discussed in other chapters. Furthermore, change in the amount of O_3 could change the distribution of solar heating and thermal cooling in the atmosphere and result in a change of the temperature profile. These temperature changes may influence the atmospheric circulation and hence weather and climate.

In this chapter, we focus on the climatic effects due to atmospheric O_3 perturbations. First, the role of O_3 in atmospheric thermal balance is discussed. Next, we present studies of the sensitivity of the atmospheric thermal profile to change in atmospheric O_3, and review the combined climatic effects of atmospheric O_3 perturbations, using chlorofluoromethanes (CFM's) and nitrogen fertilizers as examples of anthropogenic impacts. Finally, we discuss briefly the uncertainties involved in the O_3 perturbation calculations.

II. ROLE OF OZONE IN ATMOSPHERIC THERMAL BALANCE

Ozone is formed by the photolysis of oxygen (O_2) at wavelengths < 240 nm. It is removed by reactions catalyzed by trace species such as odd hydrogen (HO_x), odd nitrogen (NO_x), and odd chlorine (ClX). The concentrations of these species in the atmosphere depend, to a large extent, on the abundances of precursor gases such as H_2O, methane (CH_4), nitrous oxide (N_2O), CFM's (CCl_3F, CCl_2F_2), and chlorocarbons (CH_3Cl, CH_3CCl_3) in the troposphere.

* 10000/wavelength (μm) = wave number (cm^{-1}). Wavenumber is directly related to frequency by the velocity of light. It is a commonly used unit in spectroscopy.

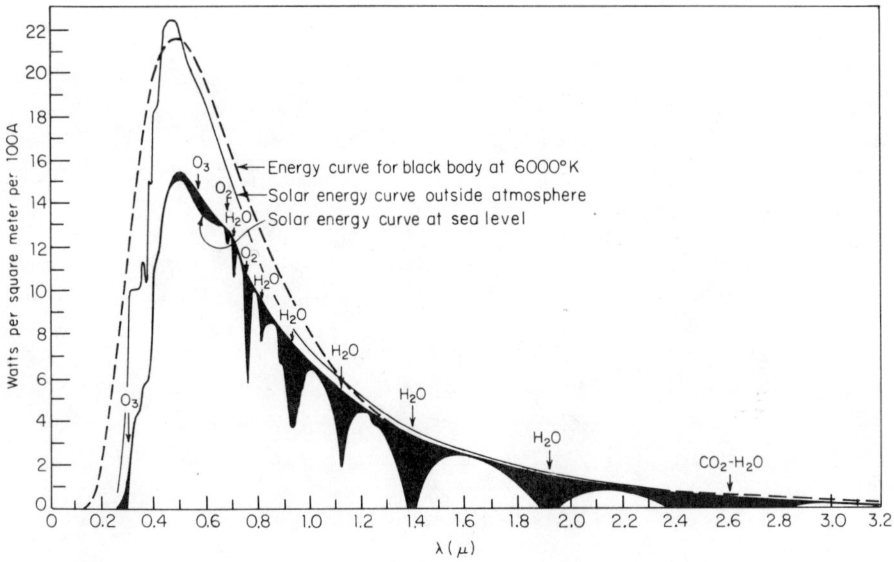

FIGURE 1. Spectral energy curve of solar radiation at sea level and extrapolated outside the atmosphere, as given by Pettit (1951). The darkened areas represent gaseous absorption in the atmosphere.

Most of the O_3 is located in the stratosphere between the tropopause and about 50 km with maximum mixing ratio of a few parts per million by volume at a height between 30 and 35 km. Ozone concentration in the stratosphere is determined largely by the processes of photochemistry and atmospheric transport and varies considerably with season, year, and latitude. Geographically, O_3 is more concentrated towards the poles than at lower latitudes. At mid-latitudes, the natural fluctuations are quite large, for instance, almost 25% between spring and fall. The presence of the small amount of O_3 is important for life on Earth as it prevents essentially all of the solar UV flux of wavelength < 290 nm from reaching the surface, and substantially attenuates the solar UV flux in the wavelength range 290 to 320 nm, processes which are certainly important for the biological cycle.

Ozone is of major importance in maintaining the thermal structure in the stratosphere through its absorption of solar radiation in the UV by the Hartley (200 to 290 nm) and Huggins (290 to 340 nm) bands and in the visual by the Chappuis (500 to 700 nm) band. The Hartley band is a fundamental absorption band in which the absorption cross section is strongly peaked at 255 nm; at altitudes above $\sim$ 45 km, this absorption provides the major heating of the atmosphere. Because of the exponential attenuation of solar radiation at each wavelength, the Hartley band is practically saturated for a nominal amount of O_3, while the visual absorption remains nearly proportional to the O_3 amount. Thus, below $\sim$ 30 km, absorption by the Chappuis band dominates the heating. In fact, it is believed that the stratospheric O_3 is largely responsible for the existence of the tropopause, a nearly isothermal region separating the radiatively equilibrated stratosphere from the more dynamically controlled troposphere. For a moderate change in atmospheric O_3 amount, the changes in UV flux at the surface of the Earth are confined to the spectral region around 300 to 340 nm, which is the long-wavelength edge of the weaker Huggins band.

In the infrared, O_3 has a number of vibration-rotation absorption bands. Stronger bands with maxima at 4.75, 9.57, and 14.2 μm (2105, 1045, and 705 cm^{-1}), and weaker bands at 3.28, 3.57, and 5.75 μm (3050, 2800, and 1740 cm^{-1}), serve to cool the middle

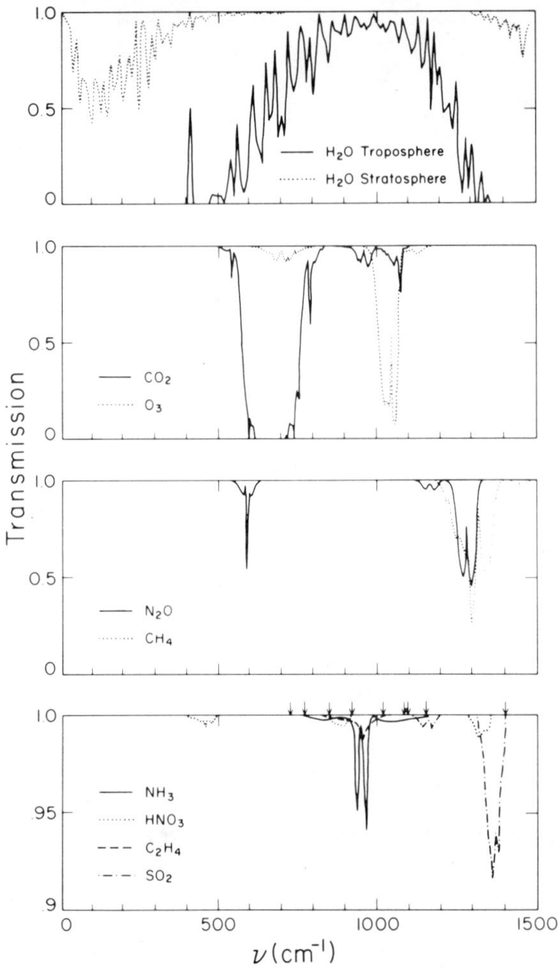

FIGURE 2. Transmission of thermal radiation by atmospheric gases for present day abundances. H_2O effectively blocks emission from the Earth's surface except for the "window" between 600 to 1300 cm^{-1} which is narrowed by the strong CO_2 band. The 9.57 μm band is particularly effective because of its location in the center of the window. The arrows indicate the locations of the CFM and Chlorocarbon bands, which are too weak to be visible. (From Wang, W. C., et al., Greenhouse effects due to man-made perturbations of trace gases, *Science*, 194, 685, 1976. Copyright 1981 by the American Association for the Advancement of Science. With permission.)

and upper stratosphere and to provide a greenhouse effect for the lower stratosphere and the troposphere. In particular, the narrow 9.57 μm (1045 cm^{-1}) absorption band is the most effective because this band lies inside the atmospheric window (7 to 12 μm, 800 to 1400 cm^{-1}) where H_2O is almost transparent (Figure 2). Furthermore, the Planck function at the surface temperature is also peaked in this region. Hence, O_3 can absorb a large quantity of heat emitted from the surface of the Earth; simultaneously, it radiates heat downward to the troposphere and surface and upward into space. Other absorption bands are less effective in causing a greenhouse effect because of the overlap with the strong H_2O absorption bands. Ramanathan and Dickinson,[8] who studied the role of seasonally and latitudinally varying O_3 in the troposphere-surface energy balance, find that on an annual and global basis, O_3 contributes about 20% of the total thermal downward flux from the stratosphere to the troposphere. Thus, it is quite clear that O_3 plays a key role in atmospheric heat balance.

III. SENSITIVITY OF TEMPERATURE PROFILE TO CHANGES IN ATMOSPHERIC OZONE

Past studies[9-11] based on the simple one dimensional radiative-convective (1-D RC) models indicate that the computed temperature profile is sensitive to both the vertical distribution and the column amount of O_3. Manabe and Wetherald[9] studied the sensitivity of temperature profile to O_3 vertical distribution in the 1-D RC model by simply replacing the mid-latitude O_3 profile with the equatorial or polar O_3 distribution, while other model parameters such as solar irradiance are kept the same at mid-latitude values. The equatorial ozone profile, compared to the mid-latitude profile, gives a slightly warmer stratosphere above $\sim$ 22 km, a sharper, colder tropopause, and a surface temperature 1K colder. In contrast, the polar ozone profile, again compared to the mid-latitude profile, gives a nearly isothermal region between 12 and 26 km and a surface temperature 1K warmer.

Based on Manabe and Wetherald's model, Reck[10] has studied the sensitivity of surface temperature change due to O_3 perturbation to changes in the 1-D RC model parameters, including the presence or absence of aerosol layers and changes in surface reflectivity. The results suggest that the magnitude and even the sign of the surface temperature change caused by stratospheric O_3 depletion depend to a significant extent on the aerosol optical properties and the underlying surface reflectivity.

In a different approach, Ramanathan et al.[11] examined the surface temperature sensitivity to changes in total column O_3 and the height of maximum O_3 concentration. In general, they found that the larger the total column O_3 or the lower the height of maximum O_3 concentration, the warmer are the temperatures of troposphere and surface.

Recently, Wang et al.[12] have calculated the separate solar and thermal effects of atmospheric O_3 in determining the surface temperature for a typical mid-latitude atmosphere. To do this, the O_3 radiative contribution to the thermal cooling and solar heating is excluded in computing the thermal structure. Excluding the strong thermal greenhouse effect causes the surface temperature to decrease by 2.6K, while omitting the solar effect warms the surface by only 0.2K. The results suggest that the O_3 thermal effect is far more important than the solar effect in affecting the troposphere and surface temperature.

Of course, all these results are obtained from simple 1-D RC models, with inherent uncertainties resulting from the simplified treatment of some complex physical processes.[7] Nevertheless, these results clearly suggest that O_3 has an important influence on the current climate of the planet. Certainly, a perturbation of the O_3 amount in the atmosphere will perturb the thermal structure and heat balance of the Earth-atmosphere system and eventually could have direct and indirect climatic effects.

Direct climatic effects may follow from the fact that a change of O_3 amount may change the way it modulates the solar and thermal radiation. For example, a reduction of stratospheric O_3 may lower the stratospheric temperature, but will produce the following competing effects on the temperatures of the troposphere and surface. First, more solar UV and visible radiation can reach the troposphere and the surface, thus causing a warming effect. Second, the strength of the greenhouse effect is reduced which, in turn, induces a cooling effect. Consequently, the effects on the troposphere depend ultimately on the difference between the two opposing effects. As it turns out,

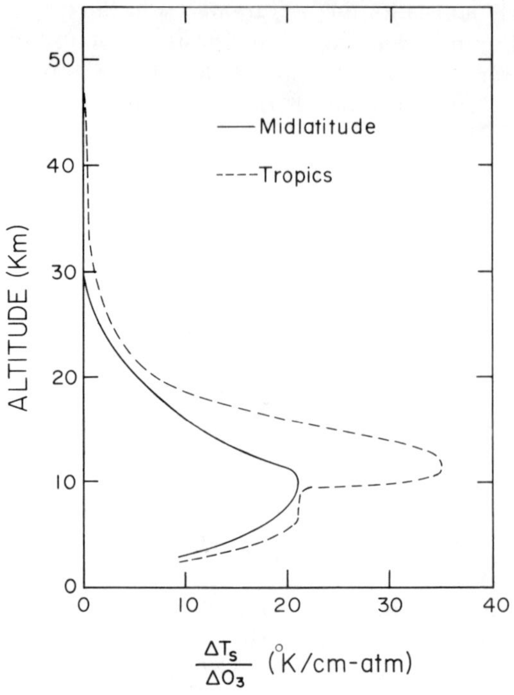

FIGURE 3. Sensitivity of surface temperature change (ΔT_s) to changes in O_3 concentration for midlatitude and tropical atmospheres. (From Wang, W. C., Pinto, J. P., and Yung, Y. L., *J. Atmos. Sci.*, 37, 533, 1980. With permission.)

studies based on the 1-D RC models show a cooling of a few tenths of a degree Kelvin if stratospheric O_3 concentration is uniformly reduced by 20 to 30%.[2,10,11] A more detailed study by Ramanathan and Dickinson[8] who considered the latitudinal and seasonal variation of O_3, confirms the cooling effect, although the effect is found to be smaller because the perturbations in solar and thermal fluxes into the Earth-troposphere system nearly cancel. In addition, they have indicated that the tropospheric climate may be as sensitive to a relatively small vertical redistribution of O_3 as it is to a large uniform reduction of stratospheric O_3. On the other hand, Fishman et al.[13] have studied the sensitivity of tropospheric O_3 perturbation on climate. By varying the tropospheric O_3 amount, they examined the change in the net radiative energy input to the surface-troposphere system and further assumed that such a change could be related to a surface temperature change by using the climate sensitivity parameter (cf. Lian and Cess[14]). Fishman et al.[13] estimate that halving the tropospheric O_3 concentration may cool the surface by 0.5K and that doubling may warm it by 0.9K.

These aspects, in fact, were illustrated in a sensitivity study performed by Wang et al.[12] Figure 3, taken from Wang et al., shows the surface temperature sensitivity as a function of O_3 perturbations at various altitudes for a mid-latitude and a tropical atmosphere. It is quite clear that on a per molecule basis, O_3 changes in the lower stratosphere and upper troposphere are more effective than changes in other regions of the atmosphere in causing a surface temperature change. Furthermore, the results suggest that the magnitude of the surface temperature change also depends on the latitude region, with larger changes found in the tropics. The fact that the tropics have smaller column O_3 amount and higher surface temperature contributes to the larger surface temperature sensitivity. These results further demonstrate that a perturbation in the O_3 profile itself might affect the climate even if the total column amount remains relatively unchanged and, that investigation of climatic effect due to O_3 perturbations require consideration of latitudinal effect (2-D or 3-D models).

Indirect climatic effects may result from changes in atmospheric circulation, especially from changes in phase and amplitude of long planetary waves resulting from changes in the stratospheric thermal structure. Such dynamic changes could themselves have significant feedback to stratospheric trace gas composition with subsequent effect on radiation. However, it is not yet clear whether this will be a positive or negative feedback. The effects are difficult to evaluate since they involve interactions between troposphere and stratosphere.[7]

IV. COMBINED CLIMATIC EFFECTS OF ATMOSPHERIC OZONE PERTURBATION AND ANTHROPOGENIC EMISSIONS

We have discussed the role of O_3 in maintaining the present day climate as well as the sensitivity of the thermal structure to hypothetical changes of atmospheric O_3. However, atmospheric O_3 perturbations associated with various anthropogenic emissions are significantly different.

As discussed earlier, the atmospheric O_3 concentration is determined, to a large extent, on the concentrations of HO_x, NO_x, and Cl_x, which in turn, depend on the abundances of the precursor gases such as N_2O, CH_4, H_2O, and CFM's. The abundance of these trace gases may increase due to the activities of man which have so far been identified, including the use of CFM's, nitrogen fertilizers, nuclear weapons, and high-flying aircraft. Review of the effect on stratospheric O_3 due to nuclear weapons and high-flying aircraft can be found in, for example, studies by Bauer,[15] Chang et al.,[16] and Luther et al.[17] In this section, we review the work* related to the use of CFM's and nitrogen fertilizers.

A. Chlorofluoromethanes

The strong infrared bands of CFM's inside the atmospheric window contribute to greenhouse effects.[2,18] Besides their warming effect, CFM's in the stratosphere are dissociated by UV radiation into free chlorine which may act as a catalyst to reduce O_3. 1979 model calculations[5] estimated that the total atmospheric O_3 column amount could be reduced by $\sim$ 16.5%.

Possible effects of the feedback of stratospheric O_3 depletion on the CFM's climatic effect have recently been examined by Wang et al.[12] In this study, the present day concentrations of CCl_3F and CCl_2F_2 in the atmosphere were assumed to be 0.1 and 0.2 ppb by volume, respectively, and at steady state ($\sim$ year 2100), were calculated to be $\sim$ 0.8 and 2.3 ppb of CCl_3F and CCl_2F_2 if the CFM's were continually released at 1973 rates. Calculations using the one dimensional photochemical model produce the perturbed O_3 profile, as shown in Figure 4 (curve A). Maximum O_3 depletion of $\sim$ 50% is found around 40 km while the reduction is about 10 to 15% between 10 and 30 km. These results are similar to those obtained by other workers.[19] The total

* The work reviewed is based on the 1979 chemistry. Since the publication of the NASA[7] report, there have been major revisions (JPL publication 81-3, 1981) in the absorption cross-section and rate data for several key stratospheric processes which are known to affect the response of atmospheric ozone to the perturbations due to addition of NO_x, N_2O, H_2O, and CFM's. The implications of some of the revisions in rate data for photochemical models are discussed in other chapters. It is clear that the changes in the response of atmospheric ozone associated with anthropogenic activities will have different climatic affect than those calculated previously. Nevertheless, the physical reasonings for climate change due to ozone perturbation will not be affected.

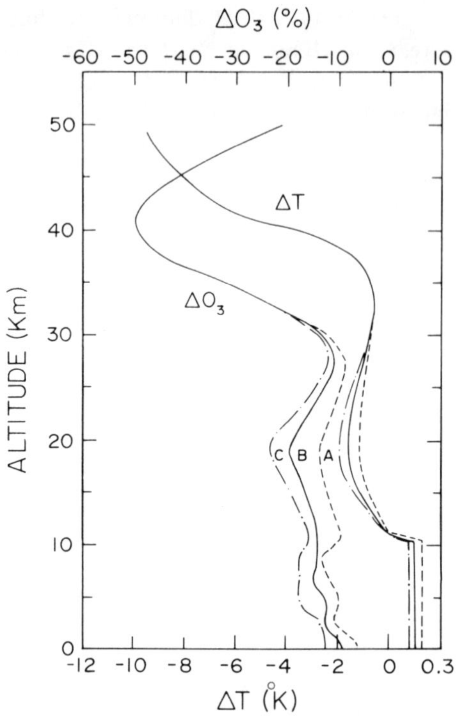

FIGURE 4. Perturbations of the temperature profile (ΔT) and O_3 concentrations (ΔO_3) at steady state due to continued release of CFMs at 1973 rates[12] (Case A). Case B includes O_3 perturbations due to the $ClONO_2$ cycle. Case C includes both the $ClONO_2$ and BrO-ClO cycles.

column O_3 reduction is 15%. These calculations[12] include the catalytic cycle of $ClONO_2$[20] and the effect of bromine[21] for O_3 destruction. Inclusion of these two additional cycles is found to result in significant O_3 reduction in the lower stratosphere and upper troposphere (Figure 4, curve C) while the total column O_3 reduction for this case is 20%. Corresponding changes in the temperature profile (ΔT) for the three cases are also shown in Figure 4. Not surprisingly, the stratospheric temperatures are calculated to decrease significantly. When taking into account both the direct warming effect of CFM's and the cooling effect of O_3 depletion, the net effect on surface temperature is small, as summarized in Table 1. The calculated greenhouse effect associated with the increases of CFM's[2] is significantly reduced if the climatic effect of O_3 depletion (Figure 4, curve C) is included. However, despite the strong negative feedbacks induced by the atmospheric O_3 depletion, the climatic effects due to the CFM's are still potentially large and thus potentially serious. Furthermore, it must be emphasized that the results critically depend on the detailed photochemistry and dynamics in the lower stratosphere, a region where considerable uncertainty exists. Results depend on projected continued emission at 1973 rates. The region of the upper troposphere and lower stratosphere remains the most complicated region to model as well as the most crucial region for tropospheric climate studies. Further studies should address the problem with coupled dynamics, radiation, and photochemistry in this region.

B. Nitrogen Fertilizers

In recent years, there has been serious concern[19] that extensive use of nitrogen fertil-

Table 1
THE CHANGE IN SURFACE TEMPERATURE (ΔT_s) DUE TO THE PRESENCE OF CHLOROFLUOROMETHANES (CFM'S) WITH AND WITHOUT FEEDBACK DUE TO O_3 PERTURBATIONS

Assumptions	ΔT_s(K)
CFM's alone	0.32
CFM's + curve A*	0.21
CFM's + curve B*	0.15
CFM's + curve C*	0.12

* The curves refer to O_3 perturbations in Figure 4.

From Wang, W. C., Pinto, J. P., and Yung, Y. L., *J. Atmos. Sci.*, 37, 533, 1980. With permission.

izers may significantly perturb the nitrogen cycle and thus increase the amounts of N_2O released from the surface into the atmosphere. Such increases could have climatic consequences; N_2O has strong infrared bands (Figure 2) which could contribute to the greenhouse effect.[2,22,23] Doubling the amount of N_2O is calculated to warm the surface by $\sim$ 0.5K, which is somewhat larger than the effect of a tenfold increase in CFM's as discussed earlier.

In addition, N_2O plays an important role in stratospheric chemistry.[19] An increase in N_2O concentration could lead to a corresponding increase in NO_x, with subsequent impacts on atmospheric O_3.

Wang and Sze[24] have examined the coupled N_2O-O_3 climate problem. Figure 5 adapted from Wang and Sze, depicts the interactive one dimensional radiative-photochemical model calculations of the perturbed O_3 profile resulting from doubling the current mixing ratio of atmospheric N_2O (320 ppbv). The total column O_3 amount is increased slightly by 2%, but the vertical O_3 profile is greatly changed with maximum increase and decrease of 19% at 12 and 38 km, respectively. The corresponding change of the temperature profile ΔT calculated from a 1-D RC model is also shown in Figure 5. The stratospheric temperature decrease is much smaller than the one caused by the CFM - induced O_3 depletions as shown in Figure 4. It should be noted, however, that the warming in the lower stratosphere is larger than in the troposphere due to O_3 increases in the lower stratosphere. The surface temperature change with and without the feedback from the O_3 perturbation is summarized in Table 2. The 30% enhancement in the surface temperature change is caused mainly by the increases of O_3 in the lower stratosphere and upper troposphere which is the most sensitive region for O_3 perturbations as discussed earlier.

However, it must be borne in mind that the results strongly depend on the projected increase of the amount of N_2O, for which reliable assessment cannot now be made.[19] Nevertheless, the nitrogen fertilizers deserve further thorough study simply because of its potentially large impacts on global climate.

V. UNCERTAINTIES IN THE OZONE — CLIMATE PROBLEM

The foregoing review of the calculations has clearly shown that O_3 change in the

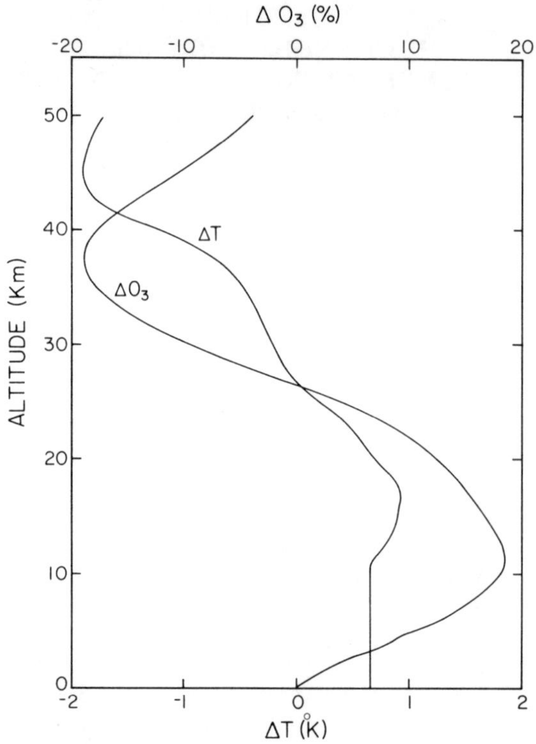

FIGURE 5. Perturbations of the temperature profile (ΔT) and of O_3 concentrations (ΔO_3) due to the increase of N_2O mixing ratio from 320 to 640 ppbv. (Reprinted by permission from Wang, W. C. and Sze, N. D., *Nature (London)*, 286, 589, 1980.)

atmosphere may lead to climate change. On the basis of 1-D RC model calculations, the effects on the troposphere climate depend critically on the perturbation of the O_3 profile, with *the lower stratosphere and the upper troposphere* identified as the most crucial regions.

A. Validation of the 1-D Radiative-Convective Models

The 1-D RC models which reasonably reproduce mean atmospheric temperature profiles play useful roles for climate studies, primarily by estimating the direct global temperature effects of changing atmospheric composition. Studies[25,26] which attempt to verify the 1-D RC models also offer encouraging evidence about the ability of the model. Comparisons with more sophisticated models indicate that different models agree when the same physical processes are taken into account. However, a certain amount of harmonization or "tuning" tends to minimize the differences between the results from the various models,[7] making additional comparisons with observations essential. To this regard, a 1-D RC model has been used to study the temperature perturbations caused by volcanic aerosols produced by the 1963 Mt. Agung eruption.[27] Despite many uncertainties such as the aerosol properties, the observed temperature changes, the model simplifications, and others, good agreement is found between the computed and the observed temperature changes in the stratosphere and the troposphere. It is anticipated that the role of the model is expected to be expanded for several reasons. First, the realism of the models is improving. For example, the positive ice-albedo feedback has been incorporated into the models[26] and can be further im-

Table 2
CHANGE IN SURFACE TEMPERATURE (ΔT_s) DUE TO THE INCREASE OF N_2O MIXING RATIO FROM 320 TO 640 ppbv WITH AND WITHOUT FEEDBACK DUE TO O_3 PERTURBATION

Assumptions	ΔT_s(K)
N_2O alone	0.44
$N_2O + \Delta O_3$ (Figure 5)	0.62

Reprinted by permission from Wang, W. C. and Sze, N. D., *Nature (London)*, 286, 589, 1980.

proved. Second, as more complete multidimensional models come into extensive use, they will generate a need for complementary 1-D models to develop and validate parameterization of the selected phenomenon for the multidimensional models. Third, and most important, as observations of the stratosphere improve, the need increases for studies on the coupling of different processes such as chemistry and radiation. It is generally agreed that studies of simultaneous perturbations and their interactions are necessary to understand atmospheric changes because of the highly nonlinear behavior of coupled processes such as temperature-O_3 coupling[28,29] and CO_2-O_3 coupling.[30-35]

B. Uncertainties in the Ozone-Climate Calculations

It is very difficult to assess quantitatively the 1-D RC model calculations since the models do not include some important features such as the cloud feedback which could be either positive or negative. But it does not necessarily mean that the computed climatic effect will be larger or smaller if the feedback is included (cf. Manabe and Wetherald[36]). In addition, the 1-D RC models can only estimate a globally averaged temperature profile. The fact that O_3 concentration in the atmosphere undergoes seasonal, geographical, and interannual variations implies that ideally one would like to use the more complete three dimensional (3-D) climate models for atmospheric O_3 perturbation studies. However, 3-D models which can include the correct treatment of the coupled processes of chemistry, dynamics, and radiation are in an early stage of development and thus, are not available for use in the near future. Furthermore, even if the 3-D models were capable of making quantitative calculations of the O_3-climate interactions, the cost and complexity would probably limit their use to examination of only the most important feedback processes.[7] In any case, their continued development should be encouraged; meanwhile, 1-D RC models will continually play a key role for climate studies.

ACKNOWLEDGMENTS

This work was done while the author was with Goddard Institute for Space Studies, NASA, New York.

REFERENCES

1. **Pettit, E.,** *The Sun and Stellar Radiation in Astrophysics,* Hynek, J., Ed., McGraw-Hill, New York, 1951, 703.
2. **Wang, W-C, Yung, Y. L., Lacis, A. A., Mo, T., and Hansen, J. E.,** Greenhouse effects due to man-made perturbations of trace gases, *Science,* 194, 685, 1976.
3. National Academy of Sciences, Carbon Dioxide and Climate: A Scientific Assessment, National Academy of Sciences, Washington, D.C., 1979.
4. National Academy of Sciences, Halocarbons: Environmental Effects of Chlorofluoromethane Release, Washington, D.C., 1976.
5. National Academy of Sciences, Stratospheric Ozone Depletion by Halocarbons: Chemistry and Transport, Washington, D.C., 1979.
6. National Aeronautics and Space Administration, Chlorofluoromethanes and the Stratosphere, Ref. Publ. 1010, Hudson, R. D., Ed., Washington, D.C., 1977.
7. National Aeronautics and Space Administration, The Stratosphere: Present and Future, Ref. Publ. 1049, Hudson, R. D. and Reed, E. I., Eds., Washington, D.C., 1979.
8. **Ramanathan, V. and Dickinson, R. E.,** The role of stratospheric ozone in the zonal and seasonal radiation energy balance of the earth-troposphere system, *J. Atmos. Sci.,* 36, 1084, 1979.
9. **Manabe, S. and Wetherald, R. T.,** Thermal equilibrium of the atmosphere with a given distribution of relative humidity, *J. Atmos. Sci.,* 24, 241, 1967.
10. **Reck, R. A.,** Stratospheric ozone effects on temperature, *Science,* 192, 557, 1975.
11. **Ramanathan, V., Callis, L. B., and Boughner, R. E.,** Sensitivity of surface temperature and atmospheric temperature to perturbations in the stratospheric concentration of ozone and nitrogen dioxide, *J. Atmos. Sci.,* 33, 1092, 1976.
12. **Wang, W-C., Pinto, J. P., and Yung, Y. L.,** Climatic effects due to halogenated compounds in the earth's atmosphere, *J. Atmos. Sci.,* 37, 333, 1980.
13. **Fishman, J., Ramanathan, V., Crutzen, P. J., and Liu, S. C.,** Tropospheric ozone and climate, *Nature (London),* 282, 818, 1980.
14. **Lian, M. S. and Cess, R. D.,** Energy balance climate models: a reappraisal of ice-albedo feedback, *J. Atmos. Sci.,* 34, 1058, 1979.
15. **Bauer, E.,** A catalog of perturbing influences on stratospheric ozone, 1955-1975, *J. Geophys. Res.,* 84, 6929, 1979.
16. **Chang, J. S., Duewer, W. H., and Wuebbles, D. J.,** The atmospheric nuclear test of the 50's and 60's: a possible test of ozone depletion theories, *J. Geophys. Res.,* 84, 1755, 1979.
17. **Luther, F. M., Chang, J. S., Duewer, W. H., Penner, J. E., Tarp, R. L., and Wuebbles, D. J.,** Potential Effects of Aircraft Emissions, FAA-EE-79-23, 1979, 234.
18. **Ramanathan, V.,** Greenhouse effect due to chlorofluorocarbons: climatic implications, *Science,* 190, 50, 1975.
19. **Logan, J. A., Prather, M. J., Wofsy, S. C., and McElroy, M. B.,** Atmospheric chemistry: response to human influence, *Philos. Trans. R. Soc. London,* 290, 187, 1979.
20. **Barker, J. L., Chang, J. S., Davenport, H. E., and Golden, D. M.,** Chlorine nitrate photolysis at low pressures, *Chem. Phys. Lett.,* 60, 385, 1979.
21. **Yung, Y. L., Pinto, J. P., Watson, R. T., and Sander, S. P.,** Atmospheric bromine and ozone perturbation in the lower stratosphere, *J. Atmos. Sci.,* 37, 339, 1980.
22. **Yung, Y. L., Wang, W-C., and Lacis, A. A.,** Greenhouse effect due to atmospheric nitrous oxide, *Geophys. Res. Lett.,* 3, 619, 1976.
23. **Donner, L. and Ramanathan, V.,** Methane and nitrous oxide: their effects on the terrestrial climate, *J. Atmos. Sci.,* 37, 119, 1980.
24. **Wang, W-C. and Sze, N. D.,** Coupled effects of atmospheric N_2O and O_3 on the earth's climate, *Nature (London),* 286, 589, 1980.
25. **Coakley, J. A., Jr.,** A study of climate sensitivity using a simple energy balance model, *J. Atmos. Sci.,* 36, 260, 1978.
26. **Wang, W-C. and Stone, P. H.,** Effect of ice-albedo feedback on global sensitivity in a one-dimensional radiative-convective climate model, *J. Atmos. Sci.,* 37, 545, 1980.
27. **Hansen, J. E., Wang, W-C., and Lacis, A. A.,** Mt. Agung eruption provides test of a global climatic perturbation, *Science,* 199, 1065, 1978.
28. **Chandra, S., Butler, D. M., and Stolarski, R. S.,** Effect of temperature coupling on ozone depletion prediction, *Geophys. Res. Lett.,* 5, 199, 1978.
29. **Luther, F. M., Wuebbles, D. J., and Chang, J. S.,** Temperature Feedback in a Stratospheric Model, *J. Geophys. Res.,* 82, 4935, 1977.
30. **Boughner, R. E.,** The effect of increased carbon dioxide concentrations on stratospheric ozone, *J. Geophys. Res.,* 83, 1326, 1978.

31. **Groves, K. S., Mattingly and Tuck, A. F.**, Increased atmospheric carbon dioxide and stratospheric ozone, *Nature (London)*, 273, 711, 1978.
32. **Haigh, J. D. and Pyle, J. A.**, A two-dimensional calculation including atmospheric carbon dioxide and stratospheric ozone, *Nature (London)*, 279, 222, 1979.
33. **Groves, K. S. and Tuck, A. F.**, Simultaneous effects of CO_2 and chlorofluoromethanes on stratospheric ozone, *Nature (London)*, 280, 127, 1979.
34. **Isaksen, I. S. A., Hesstvedt, E., and Stordal, F.**, Influence of stratospheric cooling from CO_2 on the ozone layer, *Nature (London)*, 283, 189, 1980.
35. **Penner, J. E. and Luther, F. M.**, Effect of temperature feedback and hydrostatic adjustment in a stratospheric model, *J. Atmos. Sci.*, 38, 446, 1981.
36. **Manabe, S. and Wetherald, R. T.**, On the distribution of climate change resulting from an increase in CO_2 content of the atmosphere, *J. Atmos. Sci.*, 37, 99, 1980.

EDITORS' NOTE

No accounting of current research on stratospheric ozone would be complete without a discussion of governmental activities. In past years, the Interdepartmental Committee on Atmospheric Sciences (ICAS) performed a valuable function of tabulating and describing concisely the research programs of some 15 Federal agencies. Reorganizations within the Federal Government resulted in the demise of ICAS and the assumption of its activities by the Federal Coordinating Council for Science, Engineering, and Technology (FCCSET). A subunit of FCCSET, the Committee on Atmospheres and Oceans (CAO), published its last resumé of Federal programs for fiscal year 1978 (ICAS-21-FY78). We have tried to find an author for this chapter in the Environmental Protection Agency, the National Science Foundation, CAO, and FCCSET who would have been able to provide the necessary overview of research programs under the aegis of the U.S. and foreign governments. We were unsuccessful. Rather than omit a discussion of this area, we were forced to turn elsewhere.

Dr. B. Peter Block, Chairman of the External Affairs Task Force, Fluorocarbon Program Panel, Chemical Manufacturers Association, is actively engaged in liaison with research programs sponsored by U.S. Government agencies and foreign governments. Dr. Block had already co-authored Chapter 7 on industry research but, nevertheless, volunteered to prepare this chapter when he became aware of our dilemma. For this, we are particularly appreciative.

F.A.B.
R.B.W.

Chapter 6

CURRENT GOVERNMENT RESEARCH RELATED TO OZONE

B. Peter Block

TABLE OF CONTENTS

I. GENERAL

A large fraction of the current research done in the U.S. is supported by U.S. Government agencies either through in-house work or by contract. The missions of some agencies led them to undertake such research, whereas the Clean Air Act Amendments of 1977 (CAAA) assigned some agencies specific responsibilities for research on various aspects of stratospheric ozone protection. Table 1 outlines the responsibilities assigned by the CAAA which, at present, determines the major roles to be played by the various U.S. agencies. The U.S. Environmental Protection Agency (EPA) Administrator has a key position in the CAAA, apparently because he is responsible for deciding on what, if any, regulations are necessary to protect the stratospheric ozone layer. In the discussion that follows, the current activities of each U.S. Government agency in research related to stratospheric ozone will be outlined, agency by agency, starting with the EPA. Funds devoted to pertinent research during U.S. fiscal year 1980 (FY 1980) are summarized in Table 2 for those U.S. agencies that make major commitments to this work. Other U.S. agencies for the most part devoted significantly less than $1 million each to pertinent studies in FY 1980. Comments on activities supported by other countries are presented after the description of U.S.-supported work.

II. UNITED STATES

A. Environmental Protection Agency (EPA)

EPA's primary responsibility is regulatory rather than scientific, but ultimately it is responsible for determining what information is necessary to make sound decisions on the extent of regulation of the environment required to protect the health and welfare of the nation. EPA has supported research on effects of increased UV radiation and on atmospheric science, particularly related to halogenated compounds in the troposphere, although the size of the efforts has not been large.

The research on effects, which started as the Biological and Climatic Effects Research Program (BACER), is currently called the Stratospheric Impacts Research and Assessment Program (SIRA). It is managed through EPA's Office of Research and Development and has taken the form of contracts with other organizations:

- To study the epidemiology of skin cancer
- To determine the effect of increased exposure to UV-B (UV radiation in the 290 to 320 nm region) on selected plants and animals
- To develop standard sources of, and radiometers for, UV-B
- To develop techniques for monitoring exposure to UV-B
- To collect spectroradiometric data in the field

The research on halogenated pollutants includes field measurement of ambient levels of halogenated compounds, laboratory studies of photochemical reactions, determination of transport parameters, and delineation of removal processes. Some of this work is done at EPA's Environmental Sciences Research Laboratory (Research Triangle Park, N.C.), and some is done by contract with outside investigators.

The collection of data and assessment of the state of the science are, by law, contracted to the U.S. National Academy of Sciences. In addition, different groups in EPA have contracted with other organizations for help in assembling and interpreting data. Notable among these organizations have been the Rand Corporation, A. D. Little Inc., SRI International, and the University of Maryland. The CAAA requires the EPA Administrator to report to the U.S. Congress biennially on the status of knowledge of changes in the stratosphere with particular reference to stratospheric ozone.

Table 1
SCIENTIFIC RESPONSIBILITIES ASSIGNED BY THE CLEAN AIR ACT AMENDMENTS OF 1977 (PUBLIC LAW 95-95)

Environmental Protection Agency

- Study substances, practices, processes, and activities that may affect stratospheric ozone
- Study direct or indirect effect of changes in stratospheric ozone
- Study probable causes of changes
- Investigate methods to recover and recycle or prevent the escape of ozone-depleting substances
- Investigate safe substances
- Investigate other methods of regulation
- Enlist cooperation and assistance of universities, industry, other U.S. agencies, and other nations
- Contract with the National Academy of Sciences to study the state of the knowledge and adequacy of research on stratospheric chemistry, effects of ozone depletion, and methods to control ozone-depleting substances
- Chair coordinating committee

National Oceanic and Atmospheric Administration

- Do research and monitor stratosphere for early detection of changes and climatic effects of changes

National Aeronautics and Space Administration

- Continue programs of research, technology, and monitoring of stratosphere for understanding of its physics and chemistry and early detection of changes

National Science Foundation

- Encourage and support research on the stratosphere and effects of changes in it on living organisms and ecosystems

Department of Agriculture

- Study effects of stratospheric changes on animals, crops, and other plant life

Department of Health, Education, and Welfare

- Study effects of stratospheric changes on human health

The EPA was also required to establish a committee to ensure the coordination of stratospheric research pertinent to the CAAA by U.S. Government agencies. The membership of this committee, the Interagency Committee on Stratospheric Ozone Protection (ICSOP), is given in Table 3. Although its primary responsibility is to report to the U.S. Congress, it also has agreed to report to the Federal Coordinating Council for Science, Engineering and Technology (FCCSET) in the Executive Office of the President via the Committee on Atmosphere and Oceans (CAO) to ensure that all U.S. agency research efforts on ozone are coordinated. Subcommittees of ICSOP have been established in the areas of atmospheric research (chaired by the National Oceanic and Atmospheric Administration), biological and ecological effects research (chaired by the National Science Foundation), and health effects research (chaired by the National Institutes of Health) to facilitate this coordination.

B. National Oceanic and Atmospheric Administration (NOAA)

Historically, NOAA and its predecessors have been concerned with stratospheric ozone longer than any other U.S. agency. Its Air Resources Laboratory operates the

Table 2
U.S. GOVERNMENT FUNDING OF RESEARCH IN CONNECTION WITH POSSIBLE STRATOSPHERIC OZONE DEPLETION (MAJOR COMMITMENTS FOR FISCAL YEAR 1980)

	$ million
Environmental Protection Agency	1.41
Stratospheric Impacts Research and Assessment Program	1.26
Administration, etc.	0.15
National Oceanic and Atmospheric Administration	4.91
Field Measurements	2.99
Laboratory Work	0.48
Modeling Studies	1.44
Federal Aviation Administration	2.38
Field Measurements	1.36
Laboratory Work	0.35
Modeling Studies	0.67
National Aeronautics and Space Administration	16.2
Field Measurements	8.1
Laboratory Work	2.4
Modeling Studies	3.2
Other	2.5
National Science Foundation	4.81
Atmospheric Chemistry Program	1.44
National Center for Atmospheric Research	0.24
National Scientific Balloon Facility	2.30
Ecosystems Studies Program	0.48
Other	0.35

Table 3
MEMBERSHIP OF COORDINATING COMMITTEE AS REQUIRED BY THE CLEAN AIR ACT AMENDMENTS

Interagency Committee on Stratospheric Ozone Protection (ICSOP)

Chair: Environmental Protection Agency
Designated Members:
National Oceanic and Atmospheric Administration
National Aeronautics and Space Administration
Federal Aviation Administration
Department of Agriculture
National Cancer Institute
National Institute of Environmental Health Sciences
National Science Foundation
Department of State

U.S. portion of the World Meteorological Organization Total Ozone Network, which includes stations at Caribou, Me.; Wallops Island, Va.; Tallahassee, Fl.; Nashville, Tn.; Bismarck, N.D.; Boulder, Colo.; White Sands, N.M.; Barrow, Ak.; Mauna Loa, Hi.; U.S. Samoa; and Huancayo, Peru, and maintains the world-standard Dobson ozone spectrophotometer. This laboratory also monitors erythemal UV radiation at the ground in collaboration with the Temple University Medical School via a network of Robertson-Berger sunburn meters located at National weather stations and research locations throughout the U.S. In addition, it is developing an improved stratospheric water vapor monitoring network. The data collected and comparable data from throughout the world are analyzed for trends.

The NOAA Aeronomy Laboratory is concerned more with stratospheric research than with monitoring. One of its groups makes *in situ* stratospheric measurements, collecting samples during balloon flights for chlorofluorocarbon (CFC) and N_2O analyses, and flying an instrument for stratospheric water vapor measurement. Another group in this laboratory is concerned with remote measurements, and determines total stratospheric column densities for N_2O and OH using ground-based spectrophotometers. Kinetic studies carried out in the Aeronomy Laboratory have been mainly devoted to stratospheric ion reactions and reactions of $O(^1D)$, HO, HO_2, and ClO radicals. Atmospheric modeling is another activity at the Aeronomy Laboratory pertinent to stratospheric ozone. The NOAA one dimensional (1-D) model is one of the models included in the comparative modeling studies in the U.S.

Atmospheric modeling is also performed at the NOAA Geophysical Fluid Dynamics Laboratory (GFDL). Basic research to improve the understanding of the general circulation of the atmosphere, including the stratosphere, has been under way at GFDL for over 20 years. Since 1970, the work has concentrated on the stratosphere. Tracer transport studies are being made in a global three dimensional (3-D) general circulation model to test simpler transport models, and the 3-D model is being improved in order to permit inclusion of more physical and chemical processes.

Two other NOAA installations, the Satellite Experiment Laboratory and the National Meteorological Center Upper Air Branch, conduct work pertinent to stratospheric ozone. The former is concerned with developing satellite measurements of minor stratospheric species; the latter with analysis of ozone concentrations and distributions in conjunction with atmospheric circulation and thermal fields and with verification of satellite total ozone and ozone mixing ratio data.

Some research on the effects to be expected if the amount of UV-B reaching the Earth's surface increases is also done in NOAA laboratories. NOAA conducts several programs concerned with climate and is the lead agency for the U.S. National Climate Program. Data on weather collected by NOAA are used by GFDL, the Air Resources Laboratory, and the Aeronomy Laboratory in their studies of climatic changes and the factors related to such changes. Furthermore, because NOAA is interested in all aspects of oceans, its Pacific Marine Environmental Laboratory is assessing the effect of increased UV-B on several varieties of marine life.

C. Federal Aviation Administration (FAA)

The prediction that the supersonic transport (SST) would drastically deplete stratospheric ozone led to the involvement of FAA in all aspects of ozone depletion as the lead agency in the CIAP program. Although 1-D models currently predict that fleets of SST's will have little, if any, effect on stratospheric ozone, and may even increase the amount of ozone at certain altitudes, FAA is maintaining an active interest in the area as a matter of prudence via its High Altitude Pollution Program (HAPP).

Because FAA does not maintain laboratories that do research pertinent to the atmospheric sciences in question, HAPP, which is devoted to reducing the uncertainties of the physical effects of aircraft on the environment, is completely a research funding program. It is organized into four areas of work: engines and fuels, laboratory measurements, field measurements, and modeling. Emphasis is on the effect of NO_x and HO_x species on ozone. Simultaneous *in situ* measurements of all NO_x species in the stratosphere, as well as measurement of each species individually, measurement of each HO_x species in the stratosphere, photolysis studies of various compounds such as HO_2NO_2, kinetic studies involving reactions of HO or NO_2, and studies of pressure dependence of selected bimolecular reactions are being supported. Existing data on stratospheric ozone and water are being analyzed, and the modeling efforts are directed toward improving 1-D and 2-D models, making sensitivity studies, studying local variability of trace species, and assessing uncertainties.

D. National Aeronautics and Space Administration (NASA)

The NASA Upper Atmosphere Research Program, which is currently in the Atmospheric Programs Branch of the Office of Space and Terrestrial Applications, was formulated in response to an assignment given to NASA in the FY 1976 NASA Authorization Act. The CAAA also charges NASA with certain responsibilities with respect to research relevant to stratospheric ozone. Because there is concern about the effect on ozone of chlorine compounds in the exhaust from space shuttles, NASA has taken a major role in investigating the chemistry of chlorine compounds in the stratosphere. The studies, which are carried out both in various NASA installations and by NASA contractors, include laboratory investigations, field measurements, satellite missions, and theoretical studies.

Extensive field measurements are made *in situ* from balloon and rocket platforms as well as by remote sensing from ground, aircraft, and satellite platforms. Simultaneous as well as single-species measurements are made in the NO_x, HO_x, and ClX systems. Particular emphasis is being placed on the measurement of stratospheric ClO and the satellite measurement of stratospheric ozone. NASA groups participating in *field measurements* are located at NASA's Ames Research Center, Goddard Space Flight Center, Jet Propulsion Laboratory, Langley Research Center, Lewis Research Center, and Wallops Flight Center. Other field-measurement work is contracted out to academic and contract research organizations.

NASA-supported laboratory work includes study of reaction kinetics and photochemistry and development of instruments and analytical techniques. Most of the latter work is concerned with infrared spectroscopy. NASA groups involved in *laboratory work* are at Ames Research Center, Goddard Space Flight Center, Jet Propulsion Laboratory, and Langley Research Center. A large part of the laboratory work is done by contract with investigators at academic institutions.

The use of satellites as platforms for measurements is gradually being developed. Most of the studies directed toward expanding knowledge of the chemistry of the stratosphere are still in the planning stage and will be included in the experiments run on the Upper Atmosphere Research Satellites (UARS) planned for the mid 1980s. The later Nimbus satellites and the Stratospheric Aerosol and Gas Experiment (SAGE) satellite have produced or are producing essentially continuous ozone monitoring data, and some of them, as well as the Solar Mesosphere Explorer (SME) and Halogen Occultation Experiment (HALOE) satellites, are expected to yield data for species in the sulfur, nitrogen, hydrogen, and chlorine families.

Theoretical studies supported by NASA cover the gamut of modeling from simple 1-D through 3-D models, molecular calculations, data analysis, and assessment studies. NASA *modeling groups* are located at Ames Research Center, Goddard Institute for Space Science, Goddard Space Flight Center, and Langley Research Center. Other kinds of theoretical studies are also performed at most of these NASA locations and also at the Jet Propulsion Laboratory and Wallops Flight Center. Again, a substantial portion of the work is done by contract, primarily with university investigators.

E. National Science Foundation (NSF)

Because much of the work supported by NSF is not mission oriented, it is difficult to relate some of it directly to a specific problem such as that of possible perturbations of stratospheric ozone and their consequences. Most studies pertinent to the stratosphere were funded by the Division of Atmospheric Sciences of the Directorate for Astronomical, Atmospheric, Earth, and Ocean Sciences through its Aeronomy and Meteorology Programs. An Atmospheric Chemistry Program was added in FY 1977. Now most of the basic research on the chemistry and dynamics of the stratosphere is provided through the latter program.

The role of NSF in working towards an understanding of the role of ozone in the stratosphere is long-term, as exemplified by the Global Atmospheric Measurements on Tropospheric Aerosols and Gases (GAMETAG) Program, which has an underlying philosophy that it is necessary to understand the troposphere before a true understanding of its influence on the stratosphere is possible. GAMETAG is an attempt to develop a comprehensive picture of the troposphere based on analysis for key species and their movements. Other studies involve field measurements from ground, aircraft, and balloon platforms for a variety of species, laboratory studies of reactions of possible relevance with several devoted to heterogeneous chemistry, atmospheric transport and dynamics, and modeling.

Four of the independently operated centers funded by NSF are involved in research relevant to the stratospheric ozone problem. Chief among these is the National Center for Atmospheric Research (NCAR) which is operated by the 46-university nonprofit University Corporation for Atmospheric Research. Its long-range objectives include understanding climatic trends and their causes, and determining trends in, and influences on, air quality through study of atmospheric dynamics and chemistry. Work in support of these objectives includes measurement programs based on cryogenic sampling of the stratosphere, collection of samples of the stratosphere on treated filters, and spectroscopic studies. A major activity is the use of 1-D, 2-D, and 3-D models for theoretical studies of the stratosphere.

The other centers supported by NSF serve more as support facilities than as active research groups. The National Scientific Balloon Facility, located at Palestine, Texas, is formally a part of NCAR. It provides launching, tracking and recovery of stratospheric research balloons as well as data acquisition. NCAR also operates a fleet of aircraft that serve as instrument platforms in the troposphere and low stratosphere. The facilities of the Kitt Peak National Observatory and the National Astronomy and Ionosphere Center are occasionally used by scientists for work pertinent to the stratosphere.

The Division of Environmental Biology in the NSF Directorate for Biological, Behavioral, and Social Sciences was given responsibility for research on the effects of changes in the stratospheric ozone layer by CAAA. It is supporting the study of biological responses to changes in solar UV radiation as part of this responsibility.

F. National Bureau of Standards (NBS)

Relevant experimental work at NBS primarily consists of laboratory studies, the results of which can be used in measurement programs or modeling studies. Fundamental spectroscopy, photoabsorption, and chemical kinetics are areas of continuing interest at NBS.

Although NBS is not singled out by CAAA for any particular responsibility in stratospheric ozone chemistry, a number of its scientists are engaged in pertinent research. For the most part, these scientists obtain support for such work from other U.S. agencies or from nongovernment sources. One such pertinent program is to gauge current interlaboratory precision in the analysis of atmospheric samples for CFC-11, CFC-12, and N_2O, and to produce a reference standard for these compounds. Another area is measurement technology in which techniques for measuring critical species by various spectroscopic methods are studied. An improved determination of the solar flux and its variability is under way, and specialized UV sources for use in biological experiments and in instrument calibration are being developed. The NBS Standard Reference Data program provides technical data on thermodynamic properties of gases, infrared, microwave, and UV spectroscopic information, cross sections for molecular collision and ionization processes, and gas phase chemical reaction kinetics. A principal effort involves chemical rate constants, quantum yields, and optical absorption coefficients of atmospheric species. Work is also proceeding on the catalytic photodecomposition of halocarbons.

G. Department of Defense (DOD)

The CAAA does not assign DOD specific responsibilities with respect to stratospheric ozone, but the U.S. Air Force, Army, and Navy all have concerns involving the stratosphere and, consequently, sponsor some pertinent work. Fundamentally, all three services are interested in two things — the effects of natural and man-made phenomena in the upper atmosphere on the performance of military systems and the impact of military operations on the stratosphere.

The U.S. Air Force has the most extensive program of the three services. Its Cambridge Research Laboratories and Geophysical Laboratory are concerned with the measurement of stratospheric species (primarily from balloon platforms), the study of stratospheric dynamics, and atmospheric modeling. It operates a balloon-launching facility at the Holloman Air Force Base which is used not only for its own experiments but also by other investigators. Occasionally, the Air Force Office of Scientific Research (AFOSR) funds work pertinent to the stratosphere, but it does not have an organized program dedicated to stratospheric research.

The U.S. Army supports field measurement of several stratospheric species as well as physical properties of the stratosphere. Both balloon and rocket platforms are used in the work, which is mainly the responsibility of the U.S. Army Atmospheric Sciences Laboratory and Ballistic Research Laboratories. There sometimes are relevant studies among those funded by the Army Research Office (ARO), but ARO makes no formal attempt at a planned program.

The U.S. Navy, through the Naval Research Laboratory, supports field measurements and laboratory studies pertinent to the stratosphere. The field work is concerned with H_2O, CFC's, and ozone; the laboratory work with spectra and reactions of several species and various physical phenomena. The Office of Naval Research, in like manner to AFOSR and ARO, does occasionally support studies that prove to be relevant to some aspect of stratospheric ozone, but it has no continuing program in this area.

H. Department of Energy (DOE)

Another U.S. agency not specifically mentioned in the CAAA but concerned with stratospheric phenomena is the U.S. Department of Energy in continuation of its predecessors' programs. Although its primary concern insofar as the stratosphere is concerned is the monitoring of nuclear debris and modeling atmospheric processes, its field measurement programs and modeling studies contribute to the effort to understand possible perturbations of the stratosphere. Field measurements are made by investigators at DOE's Emission Measurements Laboratory, Los Alamos Scientific Laboratory, and Sandia Laboratories. The bulk of the modeling work is carried out at Lawrence Livermore Laboratory. The DOE studies are pertinent to both NO_x and Cl_x effects.

I. National Cancer Institute (NCI)

Most of the work described up to this point is concerned with stratospheric phenomena. It is, however, also necessary to establish the nature of the consequences of perturbing stratospheric ozone with a concomitant change in intensity of UV-B at the Earth's surface. Skin cancer incidence is greatest principally for fair-skinned people and appears related to their exposure to UV-B. Epidemiological studies managed by NCI are expected to ascertain how much correlation there is between exposure to UV-B and the three kinds of cancer in question — the rarely fatal basal cell and squamous cell carcinomas, and the more dangerous but rarer melanoma. Although the CAAA call for the National Institute of Environmental Health Studies to participate in the study of the effect of increased UV-B on human health, at present the major effort is the work at NCI.

J. Department of Agriculture (DOA)

The charge in the CAAA to the Department of Agriculture is to study the effect of changes in stratospheric ozone on domestic animals, crops, and other plant life. Although current work in DOA has been substantially cut back, much of the earlier effort was carried out in the laboratories of the Agricultural Research Service, but some was done in field stations and by contract with other investigators. The studies included development of adequate instrumentation both for laboratory and field experiments and investigation of the effects of increased UV-B on crop plants and farm animals under laboratory and field conditions. Although much qualitative information was developed, definitive results require long-range programs.

K. Smithsonian Institution

Research on the biological effects of radiation has been the province of the Radiation Biology Laboratory of the Smithsonian Institution for some time, and it is carrying on studies relevant to the CAAA charges although it is not required to participate in ICSOP. It monitors incident UV radiation on a detailed spectral, rather than a total broad-band basis, in order that photobiological effects can be more accurately related to natural radiation. The effect of increased UV-B on plant growth is also being investigated.

III. UNITED KINGDOM

The fact that the environmental effects of SSTs required investigation led the U.K. into a study of possible perturbations of the stratosphere and their effects on man. Thus, the U.K. Department of the Environment was in a good position to implement a timely study of the chlorofluorocarbon/ozone issue. The program has placed major emphasis on improvement of 1-D, 2-D, and 3-D modeling with contracts at Oxford University, the Atomic Energy Research Establishment (AERE) at Harwell, and the U.K. Meteorological Office. Field measurements from aircraft, balloons and satellites, kinetic studies, and related support work are variously supported at the three institutions already mentioned and at the National Physical Laboratory, the University of East Anglia, Cambridge University, and Queen Mary College of the University of London. The effect of climate change, including changes in UV, on plants is being studied.

Position papers, which are based on all the results available to the authors, are issued approximately biennially by the U.K. Department of the Environment.

IV. FRANCE

France is in a position similar to that of England, i.e., it is a producer of both SST's and CFC's. It, therefore, is also actively interested in studies related to perturbation of stratospheric ozone and sponsors work in several fields: dynamics, chemical reactions, monitoring of ozone and solar radiation, measurement of minor atmospheric constituents, statistical evaluation of ozone trends, modeling, and studies of biological effects of increased UV-B. French installations involved in these studies include Establissement D'Etudes et Recherches Meteorologiques, Office National D'Etudes et de Recherches Aerospatiales, and a number of Centres D'Etudes and Centres de Recherches which are usually located at universities. Although the French studies have emphasized nitrogen-oxygen chemistry because of the French interest in SST's, they also include both laboratory and field studies pertinent to chlorine. Ground-based, balloon, aircraft, rocket, and satellite platforms are used in various aspects of the field work. Both 1-D and 3-D modeling are in progress.

V. CANADA

Canada, a producer of CFC's, supports, through its Atmospheric Environment Service, multifaceted balloon experiments for simultaneous measurement of several species and phenomena. The work has been mainly devoted to the NO_x system, but also includes other species. Laboratory work includes studies of reaction rates and photoabsorption cross sections, and 1-D and 2-D modeling are performed. There is considerable effort devoted to monitoring ozone, both via field measurements from the ground, aircraft, balloons and satellites, and via development of measurement techniques. The World Ozone Data Centre, an organization sponsored by the World Meteorological Organization (WMO) that collects the worldwide observations on ozone and publishes *Ozone Data for the World,* is housed in Toronto.

VI. FEDERAL REPUBLIC OF GERMANY

West Germany, which is also a producer of CFC's, supports monitoring and other atmospheric measurement work, laboratory studies, modeling studies, and study of effects should there be stratospheric ozone perturbation. This work is performed at several of the Max Planck Institutes and Kernforschungsanlage Jülich as well as in university laboratories. Of major interest, in addition to monitoring of ozone and UV, have been simultaneous measurement of trace species including radicals in the troposphere and stratosphere, the study of atmospheric reactions of radicals, the photochemistry of CFC's, kinetic studies at stratospheric temperatures and pressures, 1-D, 2-D, and 3-D modeling, and the effect of increasing UV-B on skin cancer and terrestrial and aquatic ecosystems.

VII. OTHER CHLOROFLUOROCARBON-PRODUCING COUNTRIES

In Australia, the principal activity is monitoring ozone, trace species that are possible ozone depleters, and UV, both with ground-based instruments and from balloons. There is also an effort under way to establish the cause of the current increase in skin cancer. The Division of Atmospheric Physics of the Commonwealth Scientific and Industrial Research Organization has the major scientific responsibility.

Italian work relevant to stratospheric ozone is performed at branches of Consiglio Nazionale delle Ricerche and in universities. Measurements in the stratosphere with balloon-borne submillimeter spectrometer, measurements of CFC's in the troposphere, modeling of climatic effects, and a study of tropospheric sinks for CFC's are all in progress.

Although several Japanese companies produce CFC's, there is little government-supported work related to stratospheric ozone problems. Some effort is devoted to the measurement of trace species in the troposphere, and Japanese stations are maintained and are part of the Dobson network.

Relevant research in the Netherlands is limited and is primarily devoted to a study of the dose-response relationship of UV and skin cancer.

Work in the USSR, which is apparently a relatively minor producer of CFC's but also makes SST's, includes monitoring of ozone and UV and 1-D and 2-D modeling with emphasis on surface and high-altitude sources of pollutants and climate. It maintains a network of more than 40 stations for the determination of total ozone. These stations, however, use filter ozonometers rather than Dobson instruments, so that the Russian data are not necessarily compatible with the data from the Dobson network.

VIII. NONCHLOROFLUOROCARBON-PRODUCING COUNTRIES

Some countries that do not produce SST's or CFC's maintain programs relevant to the stratospheric ozone problem, usually because their scientists have historically been interested in the stratosphere or in ozone. One group is at Institut d'Aeronomie Spatiale de Belgique. The Belgian investigators are active in both field measurements, primarily balloon-based, and 1-D and 2-D modeling.

In Denmark, the interest is more on the effects if ozone is depleted; however, ozone and UV are monitored. The epidemiology of skin cancer and the effect of increasing UV on plant growth and development are being investigated.

Norwegian activity involves monitoring ozone for the Dobson network, 1-D and 2-D modeling of the troposphere and stratosphere, and a study of skin cancer with emphasis on malignant melanoma. Sweden also supports modeling studies and studies of the role of chromosomal damage in skin cancer.

One of the most carefully maintained Dobson instruments is kept at the station in Arosa, Switzerland. A continuous series of ozone observations covering more than 50 years has been obtained at this station. Modeling with emphasis on photochemistry is also carried on in Switzerland.

IX. UNITED NATIONS

The Coordinating Committee on the Ozone Layer (CCOL), which was established by the United Nations Environment Programme (UNEP) in 1977, is charged with exercising a broad coordinating and catalytic role aimed at the integration of research efforts relevant to the Earth's ozone layer. It has developed a World Plan of Action on the Ozone Layer and meets semiannually to review current results and consider suggestions for further action. The individual countries and international organizations represented on CCOL include most of the groups supporting pertinent work; hence, the *UNEP Ozone Layer Bulletin* is a good source of information on current government-sponsored research activities in many countries.

Chapter 7

CURRENT INDUSTRY RESEARCH

B. Peter Block and
J. Richard Soulen

TABLE OF CONTENTS

I. GENERAL

In July of 1972 E. I. du Pont de Nemours & Company issued to chlorofluorocarbon manufacturers worldwide an invitation to a "Seminar on the Ecology of Fluorocarbons." Its purpose was to establish a technical program for the investigation of the effects of chlorofluorocarbons on the environment because, as stated in the invitation,

> "Fluorocarbons are intentionally or accidentally vented to the atmosphere worldwide at a rate approaching one billion pounds per year. These compounds may be either accumulating in the atmosphere or returning to the surface, land or sea, in the pure form or as decomposition products. Under any of these alternatives, it is prudent that we investigate any effects which the compounds may produce on plants or animals now or in the future."

Representatives of 15 companies attended the meeting, agreed that such a program was important and established and funded a fluorocarbon research program under the administration of the Chemical Manufacturers Association (CMA).* Thus, in 1972, with no evidence that fluorocarbons could harm the environment, the producers of these chemicals agreed that there was a need for more information and proceeded to act.

The fluorocarbon producers supporting this program (see Table 1) represent almost the total free world production of fluorocarbons. The research is directed by the CMA Fluorocarbon Program Panel (the Technical Panel),** which consists of one member from each supporting company. This Technical Panel meets regularly to review progress on current research, evaluate new proposals, and exchange data and information with contractors, with government agencies, and with other scientists.

Three projects were initiated in 1973:

1. Dr. J. E. Lovelock, University of Reading, use of the electron-capture gas chromatograph for the measurement of several halocarbons in the troposphere and lower stratosphere, particularly over Europe and the Atlantic Ocean[1]
2. Dr. C. Sandorfy, University of Montreal, measurement of the vacuum UV and photoelectron spectra of all industrially important chlorofluorocarbons[2]
3. Dr. O. C. Taylor, University of California at Riverside, use of the electron-capture gas chromatograph for the measurement of F-11 and F-12 in the troposphere over southern California and in the lower stratosphere over New Mexico and Colorado and simulation of atmospheric reactions of chlorofluorocarbons[3]

Publication of the Rowland-Molina hypothesis in 1974[4] identified a potentially serious problem, and the CMA research program was expanded considerably in order to answer the critical question: to what extent will chlorofluorocarbons affect the stratospheric ozone layer?

To strengthen the overall effort to find the answer, the Technical Panel has attempted to coordinate its efforts with others working on the same or related problems such as the supersonic transport (SST) and the space shuttle. All of these concern the U.S. government, and interactions with a number of its agencies have been especially helpful in:

1. Taking advantage of the knowledge and experience that was gained in the Climatic Impact Assessment Program

* Formerly the Manufacturing Chemists Association (MCA). Changed June 1979 to CMA.

** Formerly the MCA Technical Panel on Fluorocarbon Research (TPFR).

Table 1
FLUOROCARBON MANUFACTURERS REPRESENTED ON THE CMA FLUOROCARBON PROJECT PANEL

Akzo Chemie bv (Holland)
Allied Corporation (U.S.)
Asahi Glass Co., Ltd. (Japan)
Ausimont (Montedison Group, Italy)
Australian Fluorine Chemicals Pty. Ltd. (Australia)
CHLOE Chimie (France)
Daikin Kogyo Co., Ltd. (Japan)
Du Pont Canada, Inc. (Canada)
E. I. du Pont de Nemours & Company, Inc. (U.S.)
Essex Chemical Corporation (Racon) (U.S.)
Hoechst AG (West Germany)
Imperial Chemical Industries Limited (England)
I. S. C. Chemicals Ltd. (England)
Kaiser Aluminum & Chemical Corporation (U.S.)
Kali-Chemie Aktiengesellschaft (West Germany)
Mitsui Fluorochemicals Co. Ltd. (Japan)
Pennwalt Corporation (U.S.)
Showa Denko K. K. (Japan)
Ugine Kuhlmann, Produits Chimiques (France)
Union Carbide Corporation (U.S.)*

* Does not currently manufacture chlorofluorocarbons. Supported the CMA program through June, 1977.

2. Coordinating funding of programs addressing the halogen-ozone problem
3. Planning joint experiments with government research groups
4. Helping to set priorities for industry-sponsored research

Annual commitments by the Technical Panel since 1976 have been $1 to 2 million per year, and total expenditures through 1981 will be greater than $11 million.

The Technical Panel operates in a manner quite similar to government agencies. It receives both solicited and unsolicited proposals, evaluates them, and decides which to fund. Because the science is new and difficult, it has engaged a number of consultants to help it evaluate the proposals and the program in general. Each acts individually in his area of expertise rather than as part of a collective advisory body.

The CMA program has been subdivided into several areas overseen by separate task forces of the Technical Panel, the major ones being the Atmospheric Measurements Task Force, the Chemistry Task Force, and the Modeling Task Force. In addition, one person from a member company acts as liaison between each principal investigator and the Technical Panel. At 3- to 6-month intervals the CMA program is summarized cumulatively, and the latest Summary[5] can be obtained from the Program Manager, Fluorocarbon Program, Chemical Manufacturers Association, 1825 Connecticut Avenue, N.W., Washington, D.C. 20009. Investigators supported by the CMA program are encouraged to publish the results of their work. The Summary of the CMA program contains a bibliography of these publications as well as those by scientists from the individual companies.

II. CURRENT CMA PROGRAM

A. Scope

The emphasis of the CMA-administered program is overwhelmingly in the major areas for further work recommended by the groups in the U.S. that have been charged with studying the scientific aspects of the halocarbon-ozone problem and making recommendations for further work, i.e., the Interdepartmental Committee for Atmospheric Sciences of the U.S. government (ICAS)[6] and the Panel on Atmospheric Chemistry of the U.S. National Academy of Sciences (NAS).[7] The details of each project are given in the Summary[5] of the program obtainable from CMA, so only selected areas of greatest importance will be discussed here. At the time of this writing much of the CMA program is attempting to help answer the following key questions:

1. How can current atmospheric measurements that do not agree with the models be explained?
2. Is there a tropospheric sink for chlorofluorocarbons?
3. Is it possible to glean sufficient information from existing data on ozone to detect changes in the ozone layer in time to take corrective action?

B. Atmospheric Measurements Program

The studies funded through CMA for work in atmospheric measurements have been of two general types, (1) laboratory development of methods and instruments, and (2) actual field measurements of various species critical to the chlorofluorocarbon-ozone problem.

It was evident fairly early in the program that simultaneous measurements of ClO and other key species in the stratosphere would provide a test of the completeness of the chemistry in the model. A vacuum UV resonance fluorescence method for measuring ClO developed under an early CMA contract at the University of Michigan[8] has been used by Dr. J. G. Anderson (Harvard University) in balloon-based measurements supported by the National Aeronautics and Space Administration (NASA). The ClO concentrations observed have shown wide variations in the stratospheric concentration of this important species. The Technical Panel has funded a number of other projects on ClO detection and measurement, including a program begun in 1977 for the ground-based measurement of ClO by microwave spectroscopy.[9] If this technique has sufficient sensitivity, observations of the concentration of ClO will be possible over much longer time periods. In parallel with the measurement effort have been laboratory studies aimed at the development of ClO spectra.[10-13]

Another key measurement to test the correctness of the predicted ozone reductions is the determination of total chlorine in the stratosphere. Several attempts to develop reliable procedures for this measurement have been or are being supported by the CMA program. They include neutron activation analysis following total absorption on a medium such as charcoal,[14] infrared absorption spectrophotometry,[15] resonance fluorescence techniques,[16] vacuum UV emission,[17] and flame excitation.[18] It is interesting that some of the measurements of stratospheric ClO alone have yielded quantities much higher than the total amount of chlorine-containing species expected in the stratosphere. The determination of total chlorine in the stratosphere continues to be an area of great interest for which reliable results are still needed.

The stratospheric measurements program supported by CMA is designed to permit considerable flexibility, so that searches for species that become of interest can be instituted as quickly as possible. Almost as soon as speculation on the role of $ClONO_2$ began, studies directed at its determination were started,[19-23] leading to an estimate of 2 ppb of $ClONO_2$ at 26 km.[24] This work is continuing. Studies have also been supported in order to measure the concentration of HO_2 (electron spin resonance on a cryogenic sample),[25] search for HO_2NO_2 (infrared),[23,24] measure the concentration of OH in the stratosphere,[26,27] and search for HOCl.[23,24]

Other species of importance in assessing the degree to which chlorofluorocarbons affect stratospheric ozone are HCl and HF. These compounds have been measured by infrared techniques during CMA-sponsored balloon flights and ground-based monitoring programs.[28-30] Although there have been many HCl measurements reported, the critical altitude region above about 28 km where there appears to be an unexplained decrease in HCl mixing ratio has not been adequately studied. One goal of the CMA measurement program is to measure HCl in this region more fully; another is to collect adequate data on HF and the ratio HF/HCl.

A necessary part of the support of the spectroscopic work is the development of reference spectra with resolution comparable to that of the spectra made in the field. The CMA program includes studies directed toward this end in which the instruments that are flown or sister instruments are used to generate reference spectra.[20,31,32]

The development of an accurate materials balance for chlorofluorocarbons 11 and 12, which, in theory, could be used to determine their atmospheric lifetimes, has been a particular concern of the CMA program. Atmospheric concentrations have been determined by extensive tropospheric and stratospheric measurements, both gas chromatographic[1,3,33-38] and spectroscopic,[21,39,40] *in situ* and with grab samples. The low concentrations present have required considerable effort to ensure that the techniques used are valid. Interlaboratory comparisons[41] and development of primary[42,43] and reference[44] standards continue to be supported. In parallel with determination of the amounts of chlorofluorocarbons 11 and 12 present, the compilation of complete worldwide production and release data has been undertaken by CMA.[45] The data are brought up to date annually. Careful analysis of the data from these and other measurement programs by a CMA modeling contractor has shown that the uncertainties in the measurements allow a wide range of tropospheric lifetimes, from 10 to an infinite number of years.[73] Thus the extent to which these fluorocarbons are destroyed in the troposphere, assumed to be zero in the model calculations of ozone depletion, cannot be determined by this approach. The shorter lifetimes allowed by this analysis would profoundly reduce the calculated ozone depletion.

The CMA Technical Panel, therefore, has undertaken support of a major program to gather enough data of sufficient precision to determine via a trend analysis the tropospheric lifetimes of chlorofluorocarbons 11 and 12. Automated stations have been set up in four locations — one in Adrigole, Ireland, in the north temperate zone,[46] one in Barbados, in the north tropical zone,[46] one in American Samoa, in the south tropical zone,[47] and one in Tasmania, in the south temperate zone[47] — to measure chlorofluorocarbons 11, 12, and 113, methyl chloroform, carbon tetrachloride, and nitrous oxide at hourly intervals. Methyl chloride will be measured periodically by the principal investigators during their visits to the stations.* The fourth station began operating in mid 1978, and the first year's data are now being prepared for analysis.[48] The longer the lifetime of a chlorofluorocarbon, the longer it will take to determine it. Preliminary calculations indicate that trend analysis should detect a 10-year lifetime (which would show that most of a chlorofluorocarbon does not reach the stratosphere) in 3 years, a 15-year lifetime in 4 years, and a 20-year lifetime in 5 years. Many aspects of this study are at the limit of present knowledge, particularly regarding the routine operation of sophisticated analytical equipment in remote locations. Whatever the outcome, the experiment should be a model for future studies of a similar type.

C. Chemistry Program

The chemistry program supported by the CMA Technical Panel is primarily concentrated in three areas — the study of new reactions, the determination of rate constants,

* Measurement studies supported by CMA have shown that the oceans and slash burning of tropical vegetation are significant sources of methyl chloride,[36,49] which has been found in the stratosphere in significant amounts[38] in spite of its reactivity with OH.

and the measurement of absorption cross sections. One reaction that could be of great importance is the heterogeneous decomposition of halomethanes on sand and silicates. That such a reaction might take place was suggested by tropospheric measurements showing depressed concentrations of carbon tetrachloride near the Sahara desert[35] and later confirmed in the laboratory. Because calculations demonstrated that the effect could be significant as a sink for chlorofluorocarbons 11 and 12,[50,51] more detailed laboratory investigations are in progress.[52,53] The real-world significance of this sink must also be established. Other reactions that have been studied as potential sinks for chlorofluoromethanes are their reactions with atmospheric ions;[54,55] results have so far been equivocal.

Some reactions that could affect the chlorine-ozone cycle have also been investigated. Chief among them are the reaction of Cl or ClO with sulfate-type aerosols[56,57] and the H_2SO_4 catalyzed hydrolysis of $ClONO_2$.[58] Although these reactions do occur, they do not appear to be significant.

The rates for a number of key reactions have been measured or redetermined more precisely by investigators in the CMA chemistry program. Among these are the following: $ClO + NO_2 + M \rightarrow ClONO_2 + M$;[59] $Cl + HO_2 \rightarrow HCl + O_2$;[59] $ClO + HO_2 \rightarrow HOCl + O_2$;[60,61] $HO_2 + NO \rightarrow NO_2 + HO$;[60] $HO_2 + O_3 \rightarrow HO + 2O_2$;[60] $HO_2 + NO_2 + M \rightarrow HO_2NO_2 + M$;[60] $HO_2 + O \rightarrow HO + O_2$;[62] and $HO_2 + HO \rightarrow H_2O + O_2$.[62] Some of the values found led to dramatic changes in calculated ozone depletions, which underscores the point that models are at best only indicative and that the validity of their results cannot be accepted without reservation.

Other reactions for which the rates have been measured include: $O(^1D)$ + various chlorofluorocarbons;[34] OH + various chlorofluorocarbons;[34] $O(^1D) + H_2O$;[63] $Cl + SO_2 + M$;[56] $ClONO_2$ + NO, NO_2, O_3, HCl, O;[59] $ClONO_2$ + Cl, O;[64] $ClO + O_3$;[59] $ClO + SO_2$;[65] $OClO + O_3$;[59] $ClO + H_2O_2$;[59] $ClO + SO_2 + O_2$;[59] ClOCl + Cl, O, N, ClO;[66] $HO_2 + HO_2$;[66] $Cl + H_2O_2$; and[59] Cl + OCS.[65] Most of these latter reactions do not appear to have an important effect on the chlorine-ozone cycle, but some work on them is still in progress.

Absorption cross sections and photolysis rates have been or are being measured for chlorofluorocarbons,[2,67,68] HOCl,[61,68,69] SO_2Cl_2,[56,65] $HOSO_2Cl$,[56] $SOCl_2$,[70] $HClO_4$,[70] CCl_3NO,[70] CF_3NOCl,[70] $ClONO_2$,[71] OClO,[61] ClOO,[61] OClOO,[61] and HO_2NO_2.[72] At present the UV cross section for HOCl is a matter of some controversy. If HOCl proves to be fairly stable in the stratosphere, it will provide another sink for chlorine and reduce the calculated depletion of ozone by Cl.

Higher chlorine oxides are important not only as sinks for chlorine, but also because their subsequent degradation may generate oxygen atoms and hence provide an ozone *augmentation* process.

D. Modeling Program[50,51,73,74-76]

The CMA Technical Panel has supported modeling work to permit prompt assessment of the changes in predicted ozone depletion resulting from revised rate data, the introduction of new species or reactions, and changes in concentrations indicated by new measurements. The Technical Panel has also used modeling in evaluating certain proposals it has received and the merit of various suggestions and speculations that have appeared in the literature or been publicized from nonscientific platforms.

Some of the achievements of the CMA modeling program include recognition that a tropospheric lifetime as short as 10 years is not inconsistent with atmospheric measurements of chlorofluorocarbons 11 and 12,[73] that meteorological conditions must be considered in evaluating atmospheric measurements,[76] that desert sands could provide a significant sink for chlorofluorocarbons,[51,76] that current one dimensional models have important shortcomings that impair their usefulness,[73,74] that coupling of the various cycles involving O_3 can markedly affect modeling predictions,[51,73,74] that current models do not yield profiles of species in the stratosphere in agreement with observed profiles,[73] that certain species if present in the stratosphere will markedly alter current predictions of ozone depletion,[74] and that any climatic effect due to chlorofluorocarbons is extremely uncertain and must be considered in parallel with a significant number of other factors that also could induce climatic change.[76]

E. Other Programs

The CMA Technical Panel believes that the final answer to the ozone-depletion problem must come from observations of what is actually happening to the ozone layer in the stratosphere. Although the CMA program does not support routine monitoring of the ozone layer (currently a cooperative international effort), it has undertaken support of analysis of extant data to determine what trends are detectable.[77,78] It is anticipated that further analysis of the ozone measurements and refinement of time series analysis techniques will make it possible to test the correctness of calculated ozone depletions and to recognize changes in the ozone layer in time to permit corrective action to be taken.

Because most ozone monitoring is done with Dobson spectrophotometers, the CMA is supporting a study of the effect of aerosol scattering on ozone measurements made with them.[79]

The monitoring of solar UV radiation at ozone measuring sites, which was initiated by the Climatic Impact Assessment Program, was supported by CMA on an interim basis[80] until the government was able to resume sponsorship in order that collection of these important data could continue without interruption.

III. OTHER INDUSTRY SUPPORTED STUDIES

In addition to the work supported by the chlorofluorocarbon industry as a whole via the CMA Fluorocarbon Program Panel, there are studies underway in the laboratories of individual member companies who have scientists able to make significant contributions to the resolution of the chlorofluorocarbon-ozone problem and in university laboratories with support from a single member company. Scientists in the Du Pont laboratories have made contributions in modeling and chemical kinetic studies, and statisticians at Allied Corporation have pioneered in the application of time series analysis techniques to identify abnormal ozone trends.

Other industry-sponsored work has been done at Hoechst (in large part analytical), Ugine Kuhlmann (primarily developing a method for total Cl), and ICI (support of various academic investigators).

IV. ADDENDUM

A. Atmospheric Measurements Program

Ground-based measurement of stratospheric ClO has yielded column densities that are apparently in good agreement with Anderson's *in situ* resonance fluorescence measurements.[9] The Technical Panel is co-funding an airborne millimeter-wave experiment that it is hoped will help resolve the differences.[81] It is also co-funding ground-based measurements of HCl, HF, O_3, and CH_2O in southern France in an effort to study the stratospheric variability of these species[82] and supporting additional work on interpretation of and/or collection of stratospheric spectra.[83,84]

The tropospheric concentration of chlorofluorocarbon 21, which is not manufactured for any purpose, is being determined.[85] Should significant quantities be found, it will be necessary to establish the source, which will presumably be a commercially produced chlorofluorocarbon that undergoes transformation in the troposphere. Other projects that bear on the tropospheric lifetime of chlorofluorocarbons are the study of the rate of release of chlorofluorocarbon 11 from rigid foams[86] and the measurement of chlorofluorocarbon 22 in the troposphere.[87]

B. Chemistry Program

The possibility that HO reacts with ClO to yield HCl + O_2 as well as HO_2 + Cl is being studied[88,89] because any significant branching in that direction would have a substantial effect on the calculated ozone depletion. Other investigators have been funded for work on reactions of Cl and HO[90] and the photolysis of HOCl.[91] The use of far-infrared laser electronic resonance and of tunable F-center laser spectroscopy for the detection of small molecules is being investigated.[92]

C. Other Programs

Theoretical models for ozone trends and their predictions are being checked against observed ozone as a function of time and position,[93] and the effect of uncertainties on the benefit-cost of chlorofluorocarbon control is being assessed.[94] A 2-D model is being used in conjunction with a 1-D model for the comparison of observed with calculated distributions of trace species and the study of simultaneous perturbations.[95]

REFERENCES*

1. **Lovelock, J. E.,** Measurement of fluorocarbons in the atmosphere, FC Project No. 73-1.
2. **Sandorfy, C.,** Investigation of spectroscopy of and photochemical changes in fluorocarbons, FC Project No. 73-2.
3. **Taylor, O. C.,** Monitoring of fluorocarbons in the atmosphere and simulation of atmospheric reactions of fluorocarbons, FC Project No. 73-3.
4. **Molina, M. J. and Rowland, F. S.,** *Nature (London),* 249, 810, 1974.
5. **Block, B. P., Magid, H., and Ward, R. B.,** Summary, Research Program on Effect of Chlorofluorocarbons on the Atmosphere, Revision No. 15, Chemical Manufacturers Association, Washington, D.C., June 15, 1981.
6. The Possible Impact of Fluorocarbons and Halocarbons on Ozone, ICAS 18a-FY 1975, National Technical Information Service, Springfield, Va., May 1975.
7. Interim Report of the Panel on Atmospheric Chemistry, Climatic Impact Committee, Assembly of Mathematical and Physical Sciences, National Research Council, National Academy of Sciences, Washington, D.C., July 1975; Halocarbons: Effects on Stratospheric Ozone, Panel on Atmospheric Chemistry, Committee on Impacts of Stratospheric Change, Idem., September 1976.
8. Stedman, D. H., Laboratory investigation of the feasibility of measuring ClO in the atmosphere by the chemical conversion-resonance fluorescence detection method, FC Project No. 74-7.
9. **Solomon, P. M.,** Millimeter-wave observations of chlorofluoromethane byproducts in the stratosphere, FC Project No's. 76-130 and 77-225.
10. **Davis, D. D.,** Laboratory determination of sensitivity of laser-induced fluorescence for the detection of ClO under atmospheric conditions, FC Project No. 74-10.
11. **Davis, D. D.,** Laboratory measurement of spectroscopic absorption cross sections of ClO, FC Project No. 75-73.
12. **Nicholls, R. W.,** Laboratory and theoretical studies of the ultraviolet and visible spectra of ClO, FC Project No's. 75-11, 75-11 II, and 75-30b.

* Chemical Manufacturers Association Project Numbers are given in a number of the references. More information on each project is given in the summary of the CMA Fluorocarbon Program research program that is published periodically and can be obtained from the Program Manager, Fluorocarbon Program, Chemical Manufacturers Association, Washington, D.C. The summary also lists publications in the technical literature as they appear.

13. **Coxon, J. A.**, The $A^2\pi_i \leftarrow X^2\pi_i$ band system of ClO: absolute absorption cross sections at high resolution for bands of stratospheric interest, FC Project No. 78-255.
14. **Eggleton, A. E. J.**, Total chlorine measurements in the troposphere and stratosphere, FC Project No. 76-116.
15. **Buijs, H. L.**, Measurement of halogen compounds for determination of total chlorine and total fluorine in the stratosphere using long-path interferometric spectroscopy, FC Project No. 77-168.
16. **Young, R. A.**, Development of an instrument to measure O, ClO, O_3, and total Cl in the stratosphere, FC Project No's. 75-50 and 75-86.
17. **Howard, C. J. and Birks, J. W.**, Development of technique for measuring total chlorine content of air, FC Project No. 77-222.
18. **Skogerboe, R. K.**, Development of a measurement system for the determination of total chlorine in air, FC Project No. 77-206.
19. **Girard, A.**, Measurement of HCl, HF, ClO, etc., in the stratosphere by high resolution infrared spectroscopy, FC Project No. 75-88.
20. **Murcray, D. G.**, Laboratory measurement of high resolution infrared spectra of chlorine-containing molecules of stratospheric interest, FC Project No. 75-92.
21. **Bonetti, A., Carli, B., and Harries, J. E.**, Submillimeter-infrared balloon experiment, FC Project No. 76-137.
22. **Buijs, H. L.**, Continuation of FC Project No. 75-98, FC Project No. 77-156.
23. **Murcray, D. G.**, Detection of selected molecules by ground-based solar spectroscopy, FC Project No. 77-228.
24. **Murcray, D. G.**, Continuation of FC Project No. 75-13, FC Project No. 76-101.
25. **Ehhalt, D. H.**, Electron spin resonance detection of stratospheric radicals, FC Project No. 76-145. CMA funding augmented a program funded by the Federal Republic of Germany.
26. **Murcray, D. G.**, Stratospheric Measurement of ClO and OH, FC Project No. 76-135.
27. **Davis, D. D.**, Measurement of OH in the stratosphere by laser induced fluorescence, FC Project No. 75-87.
28. **Buijs, H. L.**, Construction of and flying a Fourier-transform spectrophotometer, FC Project No. 75-90.
29. **Buijs, H. L.**, Measurement of HCl and HF in the stratosphere by Fourier-transform spectroscopy, FC Project No. 75-98.
30. **Zander, R.**, Ground-based infrared measurements, FC Project Nos. 76-141 and 78-232.
31. **Murcray, D. G.**, Continuation of FC Project No. 75-92, FC Project No. 77-152.
32. **Buijs, H. L.**, Laboratory measurement of infrared spectra of selected stable molecules, FC Project No. 77-221.
33. **Lovelock, J. E.**, Continuation of FC Project No. 73-1, FC Project No. 74-3.
34. **Pitts, J. N., Jr. and Taylor, O. C.**, Continuation of FC Project No. 73-3, FC Project No. 74-2.
35. **Lovelock, J. E.**, Exploration of unidentified factors in the fluorocarbon-ozone problem, FC Project No. 75-67.
36. **Lovelock, J. E.**, Continuation of FC Project No. 75-67, FC Project No. 77-144.
37. **Rasmussen, R. A.**, Measurement of fluorocarbons and related chlorocarbons in the stratosphere and upper troposphere, FC Project No. 75-2.
38. **Rasmussen, R. A.**, Continuation of FC Project No. 75-2, FC Project No. 75-59.
39. **Murcray, D. G.**, Measurement of stratospheric distribution of fluorocarbons and related species by infrared absorption spectroscopy, FC Project No. 75-13.
40. **Ridley, B. A.**, Measurement of fluorocarbons and related chlorocarbons in the stratosphere by collection and analysis, FC Project No. 76-102.
41. **Rasmussen, R. A.**, Interlaboratory comparisons of fluorocarbon measurements, FC Project No's. 76-142 and 78-247.
42. **Stedman, D. H.**, Absolute calibration of fluorocarbon measurements, FC Project No. 76-132.
43. **Lovelock, J. E.**, Development of primary fluorocarbon standards, FC Project No. 78-226.
44. **Lovelock, J. E.**, The electron capture detector as a reference standard in the analysis of atmospheric halocarbons, FC Project No. 76-120.
45. **Alexander Grant & Company**, Environmental analysis of fluorocarbons FC-11 and FC-12, annual reports to Chemical Manufacturers Association starting in 1976, Washington, D.C.
46. **Lovelock, J. E. and Simmonds, P. G.**, Operation of Stations at Adrigole and Barbados for atmospheric lifetime experiment, FC Project No's. 77-193 and 78-243.
47. **Rasmussen, R. A.**, Operation of stations in American Samoa and Tasmania for atmospheric lifetime experiment, FC Project No's. 77-201 and 78-248.
48. **Cunnold, D. M., Alyea, F. N., and Prinn, R. G.**, Coordination and analysis of data for atmospheric lifetime experiment, FC Project No's. 77-213 and 78-251.
49. **Rasmussen, R. A.**, Measurement of the concentration of methyl chloride in air in Kenya, FC Project No. 77-181.

50. **Cunnold, D. M., Alyea, F. N., and Prinn, R. G.**, Meteorological and multi-dimensional modeling considerations relating to atmospheric effects of halocarbons, FC Project No. 76-122.
51. **Sze, N. D.**, Theoretical models of stratospheric chemistry, perturbations, and trace gas measurements, FC Project No. 77-173.
52. **Ausloos, P.**, Follow-up for photodecomposition of chloromethanes absorbed on silica surfaces, FC Project No's. 77-186 and 78-254.
53. **Korte, F.**, Photodegradation of chlorofluoromethanes in the troposphere, FC Project No. 77-194.
54. **Campbell, M. J.**, Investigation of the destruction of chlorofluoromethanes by naturally occurring ions, FC Project No. 75-53.
55. **Mohnen, V. A.**, Investigation of ion-molecule reactions involving chlorofluorocarbons, FC Project No. 75-64.
56. **Kaufman, M.**, Studies of compounds of sulfur, oxygen, and chlorine, FC Project No. 76-126.
57. **Martin, L. R.**, Laboratory investigation of the heterogeneous interaction of Cl and ClO with H_2SO_4, FC Project No's. 75-81 and 75-81 II.
58. **Birks, J. W.**, Studies of heterogeneous reactions, FC Project No. 76-117B.
59. **Birks, J. W.**, Continuation of FC Project No. 75-1, FC Project No. 76-117A.
60. **Howard, C. J.**, Laser magnetic resonance study of HO_2 chemistry, FC Project No. 76-100.
61. **Birks, J. W.**, Studies of homogeneous and heterogeneous reactions of importance in the stratosphere, FC Project No's. 77-192 and 78-244.
62. **Thrush, B. A.**, Studies of reactions of HO_2 by laser magnetic resonance, FC Project No's. 75-58 and 75-58 II.
63. **Zellner, R.**, Experimental investigation of the branching ratio in the $O(^1D) + H_2O$ reaction, FC Project No. 75-195.
64. **Wiesenfield, J.**, Continuation of FC Project No. 76-128, FC Project No. 77-220.
65. **Kaufman, M.**, Continuation of FC Project No. 76-126, FC Project No. 77-197.
66. **Birks, J. W.**, Measurement of reaction rates relevant to the fluorocarbon-ozone problem, FC Project No. 75-1.
67. **Stuhl, F.**, Determination of the photodissociation process and absorption cross section of FC-11 and 12 in the near UV, FC Project No. 77-170.
68. **Timmons, R. B.**, Photochemical and chemical kinetic measurements of stratospheric importance with respect to the fluorocarbon issue, FC Project No's. 76-129, 77-214, and 78-258.
69. **Knauth, H. D.**, Laboratory study of the UV and IR spectra of HOCl, $HOONO_2$, and $HClO_4$ in the temperature range of the stratosphere, FC Project No's. 77-171 and 77-224.
70. **Takacs, G. A.**, Photoabsorption cross sections for compounds of atmospheric interest, FC Project No. 77-196.
71. **Wiesenfeld, J.**, Photochemistry of small chlorinated molecules, FC Project No. 76-128.
72. **Pitts, J. N., Jr.**, Atmospheric chemistry of peroxynitric acid, FC Project No. 77-190.
73. **Sze, N. D.**, Modeling of the fluorocarbon-ozone system, FC Project No's. 75-32 and 76-115.
74. **Sze, N. D.**, Development and implementation of a simplified multi-dimensional model for stratospheric chemistry perturbations, radiation feedback, and trace gas measurements, FC Project No. 78-234.
75. **Cunnold, D. M., Alyea, F. N., and Prinn, R. G.**, Critique of models used to estimate chlorofluorocarbon effects on ozone, FC Project No. 75-24.
76. **Cunnold, D. M., Alyea, F. N., and Prinn, R. G.**, Continuation of FC Project No. 76-122, FC Project No's. 77-199 and 78-252.
77. **Pagano, M. and Parzen, E.**, Total world ozone: statistical analysis, FC Project No. 76-106.
78. **Tiao, G. C. and Box, G. E. P.**, Statistical analysis of worldwide stratospheric ozone data for detection of trend, FC Project No. 78-250.
79. **Moe, K.**, Effect of aerosol scattering on ozone measurements with the Dobson spectrophotometer, FC Project No. 78-235.
80. **Berger, D.**, Continuation of program for ground level monitoring of ultraviolet solar radiation, FC Project No. 75-62.
81. **Beckman, J. E.**, Determination of the rotational emission due to ClO using a cooled superheterodyne receiver, FC Project No. 79-282.
82. **Jouve, P.**, Measurement of the vertical distribution of HCl, O_3, HCHO. Measurement of the ratio of HF/HCl, FC Project No. 79-290.
83. **Goldman, A. and Barbe, A.**, Collaborative studies on atmospheric spectroscopy, FC Project No. 80-322.
84. **Traub, W. A. and Chance, K. V.**, Far-infrared laboratory spectroscopy of halogen-containing molecules, FC Project No. 80-318.
85. **Bruner, F.**, Determination of F-21 and other halocarbons in the troposphere, FC Project No. 78-256.

86. **Shamel, R. E.**, Analysis of release of FC-11 from rigid plastic foam products in the U.S., FC Project No. 79-275.
87. **Rasmussen, R. A.**, F-22 Measurements in the atmosphere, FC Project No. 80-308.
88. **Donovan, R. J.**, Reaction of ClO with HO: a potential sink for ClO_x, FC Project No. 79-286.
89. **Ravishankara, A. R.**, A study of the reaction of HO with ClO, FC Project No. 80-295.
90. **Kurylo, M. J.**, Rates of reaction of Cl atoms with the primary products of alkane photooxidation, FC Project No. 78-233 and reactions within the HO_x cycle, FC Project No. 80-307.
91. **Phillips, L. F.**, Determination of atomic oxygen yields in the photolysis of HOCl and ClOO, FC Project No. 78-241.
92. **Saykally, R. J.**, Laser electronic resonance and F-center laser spectroscopy, FC Project No. 80-300.
93. **Watson, G. S.**, Statistical investigation of the CFM problem, FC Project No. 78-257.
94. **Bailey, M. J.**, Uncertainties and benefit-cost analysis of CFC control, FC Project No. 80-317.
95. **Brasseur, G.**, Modeling of the CFC effect on the ozone layer, FC Project No. 80-320.

Section 3: Public Policy

Chapter 8

THE FORMULATION OF FEDERAL POLICY*

Carroll Leslie Bastian

TABLE OF CONTENTS

* This chapter expresses the views of the author and does not necessarily reflect those of her former agencies, the National Science Foundation and the Council on Environmental Quality; her present agency, the Environmental Protection Agency; or the Interagency Task Force on Inadvertent Modification of the Stratosphere, which she co-chaired.

I. A MODEL FOR "SUCCESSFUL" POLICYMAKING?

On March 17, 1978, small headlines in newspapers throughout the country announced a U.S. ban on most aerosol products containing chlorofluorocarbon propellants, but this news came as no great surprise to most consumers and caused no shock waves in the fluorocarbon or aerosol industries. The regulatory announcement was the culmination of 3½ years of intensive study and preparation that provided an unusual (if not unique) example of Federal interagency coordination on an environmental regulatory question, coupled with a high degree of communication and cooperation among legislators, regulators, scientists, industry, and other interested groups. This experience is educational for what it indicates of the possibilities and limitations of such an approach to the formulation of Federal policy; however, the policymaking process that operated for fluorocarbons may be a difficult precedent to follow under somewhat different circumstances.

This account of Federal policymaking on stratospheric ozone protection is centered around fluorocarbons more than other stratospheric modifiers. Federal and non-Federal activities pertaining to high altitude aircraft are covered elsewhere in this volume. This chapter will trace events, principally within the Executive Department, from the surfacing of the Rowland-Molina theory in 1974 through to the initiation of rulemaking in the late fall of 1976, plus a brief summary of events before and since that time. The interactions among Federal officials and scientists, environmentalists, affected industries, legislative bodies, the general public, and other nations and international organizations will be noted as they affected the Federal policymaking process.

The account that follows is concerned with evaluating the effectiveness of the decisionmaking process. The account is admittedly somewhat subjective and personal, although factually accurate. The bias of the author is that the experience was on the whole "successful", principally because an articulated framework was provided for the decision dialogue among all interested parties. While the problem was one of great scientific complexity, the economic, social, and legal ramifications were equally challenging. For the most part, the appropriate roles of the scientist and the policymaker were observed: the former to identify and assess information, the latter to select and implement appropriate actions. As Dr. Karim Ahmed (Natural Resources Defense Council, NRDC) once expressed it in a public meeting, the fluorocarbon regulatory question was a matter of public policy and judgment, not of science. Scientists can indicate public policy alternatives, but public agencies must make choices.[1] The dialogue among interested groups was for the most part open and forthright, resulting in a reasonable and clear decision. The timeliness of the decision surprised industry, environmental groups, the press, and the government itself.

Several factors contributed greatly to this success, some of which were unique to the fluorocarbon-ozone issue. The unusual nature and scope of the problem — seemingly more appropriate for science fiction than science — immediately captured the imagination of scientists, policymakers, and the public. Awareness of the potential threat developed suddenly and seemed to demand a rapid response, unlike more typical environmental problems that are recognized to be serious only after a period of years or decades. Within less than a year of its reported discovery, the fluorocarbon-ozone problem received attention by high levels of government before it could be smothered by bureaucratic deliberation. A vehicle was developed within government for interested and knowledgeable scientists to be involved in the decision process. Industry scientists were for the most part responsive and responsible rather than obstructionist. They became significant participants in the dialogue with government in addition to the legal and public relations officials of their companies, whose explanations were more likely to distort the scientific findings. Scientists from inside and outside government were aware of the important social implications of this scientific problem and were willing to talk publicly in order to bring the scientific debate into the open. They also maintained a close communications network among themselves, all over the world, and even modified their traditional peer review practices to enable a speedy comparison of results and informal dissemination prior to publication.

Such practices led to a higher degree of scientific accuracy in the content of the policy decisions than is often the case and to a greater responsiveness by the scientists in answering policy-relevant questions. A pitfall of this approach was that media coverage of preliminary results led to wide misunderstanding by the public as to what was certain and what was uncertain. To some extent, this undermined the credibility of the scientists. But because the issue was a matter of public policy as much as of science and deserving of wide public attention, it is difficult to imagine how the situation could have been improved by attempts to stifle discussion.

Some persons would add that the success of this policymaking exercise was largely the result of the "ease" of the decision because of the nonessential nature of aerosol products and the fact that spray cans were already under attack by consumer and environmental groups for a variety of reasons unrelated to the ozone layer. The decision did not seem particularly easy at the time, however. The fluorocarbon, aerosol, and related industries were large and pervasive in their effects in the consumer market. In late 1974 and early 1975 the idea of the government considering tackling the aerosol problem scared most government officials, to say nothing of the implications of fluorocarbon uses in air conditioning and refrigeration. At the time there were few precedents for regulatory action against chemicals for environmental reasons — DDT, vinyl chloride, mercury, and a few others; but none were at all analogous to the fluorocarbon situation. Never had the government been faced with such a seemingly innocuous substance causing such indirect, remote effects which would occur decades in the future, but which were not directly demonstrable at the time when controls needed to be initiated.

Federal "policy" often is not deliberately made, but just happens. The fact that any decision was made at all in this situation is somewhat miraculous to anyone familiar with Washington intergovernmental politics. The experience was not completely successful, however. The development and coordination of the government-sponsored research programs did not entirely escape the bureaucratic clutches of special interests nor the obscure politics of inflation fighting, particularly in the case of long-term research on the effects of stratospheric modification. The difficult questions associated with fluorocarbons as solvents, refrigerants, and blowing agents in the production of foam have yet to be faced, and no one foresees easy or speedy solutions. Guidelines have not been developed as to how the stratosphere can be protected from the cumulative assaults of the variety of potential ozone-destroying substances that might be discovered or released in greater quantities in the future. The effect of regulatory actions on aerosols taken by the U.S. and some other nations to date may soon be offset by growth in nonaerosol uses worldwide or by increased use of fluorocarbons in other countries. So far the results from international diplomacy have been disappointingly slow, although the pace has increased recently. Finally, as Mr. Kenneth L. Johnson, formerly with the Environmental Protection Agency (EPA), points out in a case study of the fluorocarbon controversy, the decision to regulate fluorocarbons was made on a completely *ad hoc* basis and not as part of any enlightened overall strategy for dealing with chemicals posing a threat to man or the environment.[2] Although the desire to control fluorocarbons provided an argument for passage of the Toxic Substances Control Act, it did not provide a blueprint for evaluating the potential threats of the other tens of thousands of substances of less dramatic interest.

An examination of the interactions that led to the formulation of Federal policy and to regulatory actions follows: the principal events are charted in Figure 1.

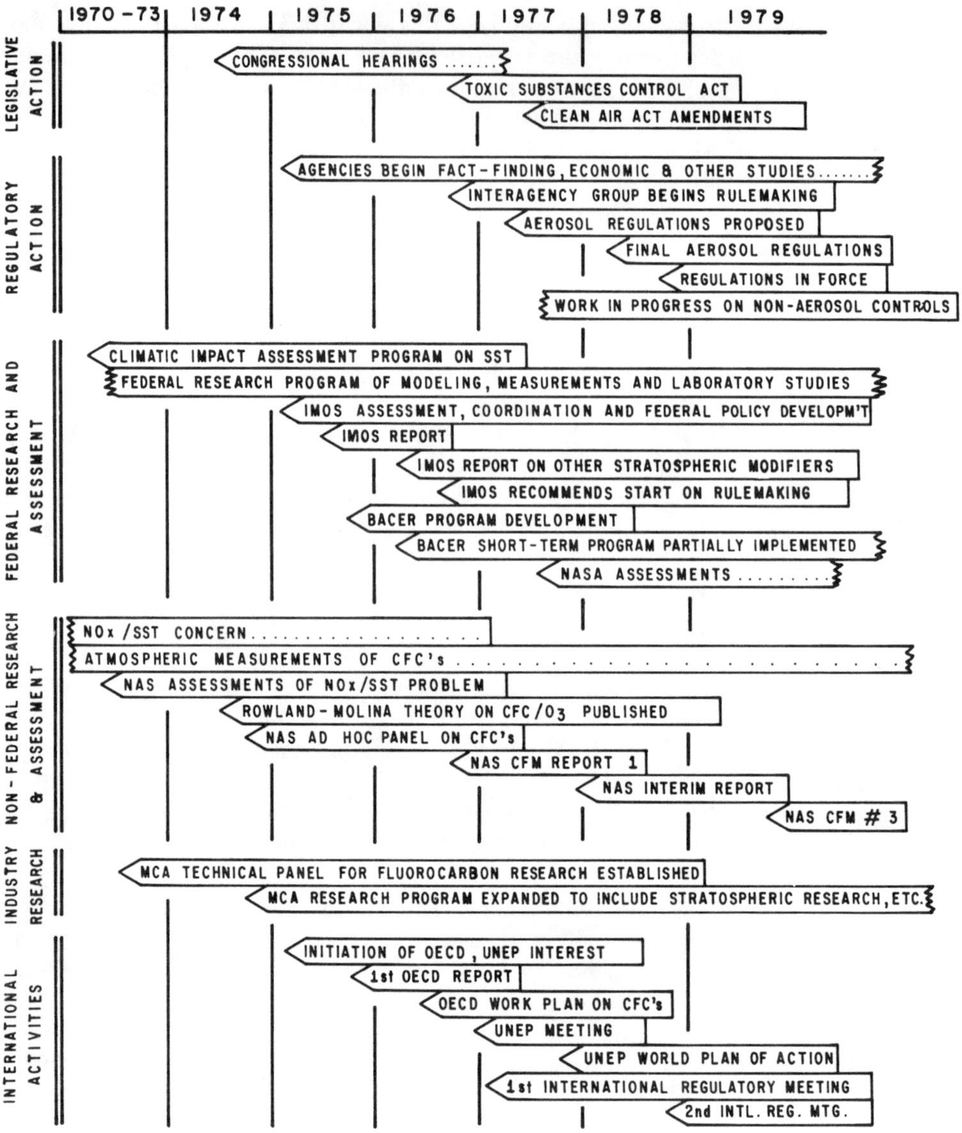

FIGURE 1. Major events relating to the formulation of Federal policy on chlorofluorocarbons.

II. LAYING THE GROUNDWORK: INTEREST IN THE STRATOSPHERE, 1970 TO 1974

The rapid governmental assessment and response during 1974 to 1976 related to fluorocarbons would not have been possible without the earlier growth of a corps of scientists and policymakers who knew and cared about the stratosphere and its potential influence on life on Earth. Interest in this previously neglected area had grown during the debate on commercial development of the supersonic transport (SST). An early concern about possible reduction of ozone by water vapor from the SST exhaust was largely dismissed, but new attention focused on the oxides of nitrogen (NO_x) in 1970 to 1971 with the work of Dr. Paul Crutzen, at that time working in Sweden but later serving as director of the Upper Atmospheric Project at the National Center for Atmospheric Research (NCAR), and Dr. Harold S. Johnston, a chemist at the University of California, Berkeley.[3-5] The NO_x issue was explored further as part of the $22 million Climatic Impact Assessment Program (CIAP) and the parallel study by the National Academy of Sciences (NAS), both taking place between 1971 and 1974, with reports released early in 1975.[6,7] Similar studies were undertaken in Great Britain and France.[8,9] These studies and the attendant policy formulation processes are described in Chapter 1. They provided (1) the interdisciplinary approach necessary for study of stratospheric ozone protection, (2) an international network for comparison of theoretical calculations, atmospheric measurement, monitoring, and laboratory experiments,* and (3) increased public awareness of the relationship between the upper atmosphere and life on Earth. Within the U.S., CIAP was an example of an innovative approach to problem-solving across agency lines: Congress designated the Department of Transportation (DOT) as "lead agency" for the research, with "pass-through" funds going directly to DOT for subsequent transfer in smaller pieces to the several other agencies with relevant expertise.

Most of the atmospheric research under CIAP concerning the oxides of nitrogen later became applicable to the oxides of chlorine (Cl/ClO) problem of fluorocarbons, especially that on the effects of increased UV radiation. The possibility of a stratospheric threat from chlorine was recognized early, but initially was of little concern (except beginning in 1972 and 1973 in reference to HCl from the Space Shuttle), because large, manmade sources of stratospheric chlorine were unknown.

In the meantime, however, in 1971, British scientist Dr. James E. Lovelock reported his 1970 detection of $CFCl_3$ over Ireland. These and later measurements at various locations and altitudes indicated a large buildup of fluorocarbons in the lower atmosphere that led to questions about the eventual fate and effects of these substances.[10,11] The fluorocarbon industry then entered the research arena and in July 1972, created the Technical Panel for Fluorocarbon Research under the auspices of the Manufacturing Chemists Association (MCA)** as an international coordinating body for sponsoring scientific research on fluorocarbon fate and effects.[12] The industry was able to redirect this effort quickly when the possible stratospheric problem of fluorocarbons was identified in 1974 by the January preprint and June publication of the work of Professor F. S. ("Sherry") Rowland and Dr. Mario J. Molina (University of California, Irvine) on photodissociation of CFC-11 and CFC-12 following upward transport to the stratosphere, and subsequent catalytic destruction of ozone by the resulting Cl and ClO.[13] Soon Dr. Ralph J. Cicerone (University of Michigan) contributed a timetable for the expected ozone reductions, and his group at the University of Michigan and other investigators developed confirming calculations and measurements that added to the credibility of the new theory.[14-16]

III. FIRST ASSESSMENTS OF THE FLUOROCARBON PROBLEM, JUNE 1974 TO JUNE 1975

Between June and December 1974 interest in the fluorocarbon theory of ozone destruction was centered primarily in the scientific community, as the startling calculations were presented and discussed at various scientific meetings. The first flurry of

* See, for example, the testimony of Dr. Robert Murgatroyd of the U.K. Meteorological Office and Chairman of COMESA (Committee on Meteorological Effects of Stratospheric Aircraft) concerning the good relations established among scientists of the different countries during this period in Reference 68, pp. 119-120.

** Subsequently the Fluorocarbon Program Panel (FPP) of the Chemical Manufacturers Association (CMA).

wider attention followed the meeting of the American Chemical Society in Atlantic City, New Jersey in September. One account reported that the media did not take the theory seriously until *The New York Times* began covering it in September.[17] By November 1974 a *New York Times* editorial went so far as to ask for action, not complacency: "It is comparable to a child skipping through a minefield on the theory that he won't necessarily step on a mine, and if he does it won't necessarily prove fatal."[18]

The initial public reaction was largely polarized into two extremes. Many thought that prediction of catastrophic effects from minute quantities of gas released at great distances from the stratosphere was an absurd hypothesis not worthy of attention. People had difficulty grasping the concept that something which did not yet have any measurable effect could be so dangerous in the distant future, or the concept that small manmade changes in average ozone concentration could be significant even if less than natural fluctuations. One outraged citizen called the theory "medievally ridiculous," a "back-to-the-mud-hut movement" of "antiindustrial radicals."[19] Others, including many eminent scientists, recognized the theoretical capability of fluorocarbons to affect ozone but had faith in Mother Nature's ability to self-correct the situation through one of the many atmospheric mechanisms that man does not fully understand. The other extreme reaction came from those who immediately predicted doomsday or an ice age or some other catastrophe if all fluorocarbons were not banned immediately.[17,20-22]

The National Academy of Sciences became involved at an early stage since its Climatic Impact Committee formed for the SST study was already interested in the stratosphere. On October 26, 1974, an *ad hoc* NAS panel chaired by Dr. Don Hunten of Kitt Peak National Observatory met and concluded that (1) the fluorocarbon-ozone problem was serious, (2) production of fluorocarbons would need to be curtailed in the future unless something major was missing from current understanding, and (3) the matter should immediately receive more detailed attention by the NAS in a 1-year study. The NAS began initial planning for a study and sought Federal funds (see Chapter 9).

Shortly thereafter, on November 20, 1974, the Natural Resources Defense Council formally petitioned the Consumer Product Safety Commission (CPSC) to ban aerosol products using fluorocarbon propellants. In the spring of 1975, NRDC filed suit because CPSC had failed to act on its earlier petition.

As could be expected, the first real attention to the problem within the Federal government came within the scientific agencies, particularly those sponsoring atmospheric research. This attention was initially focused in the Interdepartmental Committee for Atmospheric Sciences (ICAS), formed in 1959 as a subcommittee of the Federal Council for Science and Technology (FCST). In the fall and early winter of 1974, ICAS briefed Dr. H. Guyford Stever, Chairman of the FCST as well as Science Adviser to the President and Director of the National Science Foundation. At a special December 1974 meeting, ICAS charged its standing Panel on Inadvertent Modification of Weather and Climate, chaired by Dr. Lester Machta of the National Oceanic and Atmospheric Administration (NOAA), with continuing to follow the issue and preparing a full report for ICAS by the next spring. Within ICAS there also arose the beginnings of an awareness of several complex policy issues not strictly within the bounds of atmospheric research concerns: What is the value of the affected industries and how much would they be affected by regulatory measures? How much time do we have, or should we take, to make a decision about what to do? What is the relationship between expected research results to be sought and the decisions that need to be made?

The Congress then stepped into the act, as several versions of "ozone protection" bills were introduced, most providing for some type of regulation preceded by mandated studies. The first Congressional hearings were held on December 11 to 12, 1974, by Congressman Paul Rogers, Chairman of the House Subcommittee on Public Health and the Environment (of the Committee on Interstate and Foreign Commerce).[23] These hearings, called in one account "...the premiere performance of the Incredible Stratospheric Traveling Road Show and Debating Society,"[24] led to the coalescing of governmental concern, while providing the first comprehensive overview of current scientific knowledge and policy opinions of various key groups. The hearings seemed to impress the Congressional participants, press, government officials, and other observers with the potential seriousness of the problem. Congressman Richardson Preyer summarized his reaction:

> Mr. Chairman, when I came over today I thought we would have a hearing that would be a standard conflict between the environmentalists and industry and I can only say I am stunned by what we have heard here.
>
> There has not been inflammatory rhetoric or alarmist language but here we have some of the most distinguished scientists in America telling us about the problem.
>
> I think these could be the most important hearings the committee has ever had. It looks like all of us laymen in this country have got to learn a new subject now.[23]

At the hearings, the general impression from government witnesses was one of a somewhat piecemeal, *ad hoc* approach to the research and assessment. Some interagency cooperation was reported, such as the efforts of ICAS and a November 13 meeting at NOAA of governmental, industry, and academic scientists to describe research needs. Chairman Rogers repeatedly asked government witnesses for their plans for the formulation of future government policy. He finally requested the identification of a lead agency, a schedule for regular meetings and monitoring of events, and a report on interagency coordination.

The testimony of witnesses from the fluorocarbon industry contained both scientific and production information, as well as a program of planned research and a proposed policy approach: three more years for research and assessment by NAS, government, and industry before making a regulatory decision. Some nonindustry scientists were suggesting a similar timetable, such as Dr. Michael McElroy of Harvard, who asked for 1 to 3 years for further study, and NOAA's Machta, who suggested establishing restrictions after a definite study period such as 3 years. Rowland, however, began to call for aerosol restrictions immediately or after a period of study not to exceed 1 year; Cicerone favored an immediate aerosol ban. NRDC supported study by NAS for 6 months and regulations to be in effect within 9 months. In the end, regulations were not promulgated until 3½ years later which did not take effect for yet another 7 months after that (although in the meantime industry and some other scientists asked for additional study time).

Two Federal agencies that followed the December 1974 hearings with particular interest were the National Science Foundation (more correctly, Stever and his staff office in his role as the President's Science Adviser, the Science and Technology Policy Office) and the Council on Environmental Quality (CEQ) in the Executive Office of the President. The problem was, after all, both a scientific and an environmental one. Stever was concerned that "good science" enter into any decisionmaking to be done. He realized that more was at issue than the atmospheric aspects that the Federal Council for Science and Technology, through ICAS, was prepared to address. CEQ, led by Chairman Russell Peterson,* was concerned that any issue of such potential far-reaching global environmental impact be faced in a responsible and timely manner, although at this stage CEQ hoped that the matter was a nonproblem. (The responsible CEQ staffer, Dr. Warren Muir, Senior Staff Member for Environmental Health, later re-

* Peterson, an ardent environmentalist, was interestingly enough for over 25 years a chemist with Du Pont, the world's largest fluorocarbon producer.

vealed that he had relished the assignment of sorting out the fluorocarbon question as an opportunity to earn a "white hat" from industry by debunking the fluorocarbon theory, since he already had a "black hat" because of his position on issues such as PCBs, asbestos, lead, and toxic substances legislation. He soon came to realize, however, that the fluorocarbon problem was not going to disappear.) CEQ is charged under the National Environmental Policy Act with providing policy advice to the President on all human activities that endanger the environment. By the time of the Rogers hearings, CEQ had already been asked by the Food and Drug Administration (FDA), and had agreed, to assemble governmental representatives to determine a lead agency to prepare the Environmental Impact Statement (EIS) for whatever Federal actions might be required.

Stever and Peterson then joined forces, and in January 1975 announced the formation of an *ad hoc* Interagency Task Force on Inadvertent Modification of the Stratosphere (IMOS) to which agency heads were invited to appoint official representatives. The purpose of this group, later set forth in an official charter, was "to provide a vehicle for a coordinated Federal assessment of possible effects upon the stratosphere and related adverse effects resulting from the release to the atmosphere of certain man-made chemicals or from other human activities, and for development and initiation of a plan of coordinated action and research by the Executive Department agencies of the Federal government in response to such issues". Muir was appointed Co-Chairman representing CEQ and Ms. Carroll Leslie Bastian* of National Science Foundation was appointed Co-Chairman representing the Federal Council for Science and Technology.

In addition to CEQ and FCST the IMOS membership included all Federal agencies with a substantial research, policy, or regulatory interest in fluorocarbons or stratospheric protection: the Departments of Agriculture; Commerce; Defense; Health, Education, and Welfare; Justice; State; and Transportation; Consumer Product Safety Commission; Energy Research and Development Administration; Environmental Protection Agency; National Aeronautics and Space Administration; National Science Foundation; and the Interdepartmental Committee for Atmospheric Sciences. This wide representation aided the group's effectiveness despite the somewhat unwieldy size. Policy questions were thrashed out internally during the frequent meetings rather than through complex coordination with nonmember agencies.

The committee had the authority to act because all affected government groups were included. The only pertinent agency that did not fully participate in the initial stages of IMOS was the Office of Management and Budget (OMB), which was an "official observer" agency but seldom sent representatives. This was regrettable. Because of the internal organization of OMB and the complexity of the stratospheric protection issue, the subject cut across the responsibilities of several OMB offices so that six budget examiners were somewhat interested, but no single one was responsible. This later resulted in the need for briefings of OMB officials when the policy implications of the IMOS work became apparent, and led to some difficulties when research budgets were being developed. The other relevant part of the Executive Office of the President, besides CEQ and OMB, was the White House staff (the "Domestic Council" in those days), which was kept informed of IMOS activities by Co-Chairman Muir.

The first task for IMOS was to prepare an assessment of the fluorocarbon problem and a plan of action, which Stever and Peterson directed the task force to produce within about 4 months. The committee first met on January 30, 1975, and adopted a

* For convenience, in this article the author will use her present name, Carroll Bastian, throughout, although during much of the time covered by this report (and therefore in references from that time) she was Carroll Pegler.

tight timetable leading to a June report, an outline of the report, and agency assignments for the initial drafting of chapters. Fortunately, member agencies were willing to commit large amounts of staff time to the effort, which is frequently not the case for interagency committees. Some sections of the report could be prepared simply by compiling information on work already completed; for other portions, on-going research presented a moving target. Informal exchanges of information occurred with industrial groups, public interest groups, Congressional staff, and individual scientists.

During this same period the National Academy of Sciences was organizing its more detailed study, which was funded initially by four Federal agencies: NSF, NASA, NOAA, and EPA. The NAS study group was publicly announced in March, with the appointment of Professor H. S. Gutowsky, head of the Department of Chemistry at the University of Illinois, as Chairman of the Panel on Atmospheric Chemistry that would perform the detailed atmospheric part of the study.[25] The overall assessment was made by the Committee on Impacts of Stratospheric Change (formerly the Climatic Impact Committee, which prepared the SST study) of which Dr. John W. Tukey of Princeton University and Bell Laboratories became Chairman.

IMOS and the NAS study groups maintained a somewhat aloof and awkward liaison with each other that involved a few formal briefings. The NAS, a nongovernmental organization, normally does not reveal much about its deliberations to the sponsoring government agencies it advises prior to completion of a study. IMOS, for its part, could not legally include NAS in its meetings without opening them to other nongovernmental groups as well. Informal communication was maintained through a few personal contacts between the two groups, however. With hindsight, it can be realized that the aloofness was beneficial in preserving the independence of the NAS assessment as a check on the government analysis. At the time, the situation presented a nuisance for the various investigators who had to report separately to both bodies, and at times it was annoying to the government agencies paying for a study for which they had little knowledge or control.

In February 1975, IMOS held a formal public meeting to hear the testimony of several academic scientists associated with the fluorocarbon/ozone depletion theory (Rowland of the University of California at Irvine, Cicerone of the University of Michigan, and McElroy of Harvard); industry representatives (Dr. Frank Bower and Dr. Raymond L. McCarthy of E. I. du Pont de Nemours & Company, Mr. Ralph Engel of the Chemical Specialties Manufacturers Association, Mr. John W. Dickinson, Jr., of the Gillette Company [representing the newly-formed Council on Atmospheric Sciences (COAS), a counterpart to IMOS established by the aerosol and related industries], and Mr. Herbert Gilkey of the Air Conditioning and Refrigeration Institute); and the Natural Resources Defense Council (Dr. Karim Ahmed, Mr. David Michaelman, and Mr. Tom Stoel).[1] The meeting brought all IMOS members up to date on various aspects of the scientific facts and identified areas of agreement and disagreement in the controversy. It also marked the beginning of wide public interest in the government's deliberations. Announcement of the meeting in a *New York Times* article[26] resulted in over 100 observers attending, representing a variety of affected industries and trade associations, Congressional staff, the Canadian and British governments, press wire services, newspapers, and magazines. At this time the Canadian government was beginning to consider some type of action to reduce fluorocarbon emissions. However, the British from the beginning were more skeptical. Lovelock was quoted in December 1974 as expressing this view:

"Some scientists are worried about certain gases used in aerosols....This is one of the more plausible of the doomswatch theories but it needs to be proved. The Americans tend to get into a wonderful state of panic over things like this....I think we need a bit of British caution on this."[27]

At the IMOS public meeting, industry representatives began to indicate the complexities of switching to alternatives for fluorocarbons or of assessing economic impacts of any restrictions. They also repeated their request from the December Congressional hearings for more time to perform research. The NRDC witnesses stressed the importance of sorting out the Federal jurisdictional confusion so that regulations could be developed expeditiously. The academic scientists presented a variety of interesting testimony on what was known and unknown about the theory, but the comments picked up by the press were McElroy's controversial comments about the possibility of bromine "weapons" used to reduce ozone deliberately, for which he was criticized by colleagues.[24,28,29] Wrap-up remarks by Rowland and Dr. Edward S. Epstein, NOAA's Associate Administrator for Environmental Monitoring and Prediction, summarized the easy and difficult aspects of the IMOS task. Epstein stated that the task of the government was easier because the industry admitted that the threat was worthy of evaluation and they intended to act responsibly. Rowland observed that the difficulty was one of time scale.[1]

As preparation of the IMOS report progressed, many problems had to be faced. The Stratospheric Effects portion was prepared by ICAS and was essentially a summary of the ICAS report that was already underway and published later in the summer of 1975.[30,31] This chapter was especially difficult to prepare because of the rapidity with which new research, monitoring, and measurement results were being reported. In January 1975, it was impossible to predict what might be the conclusions in June, unlike the human health and biological effects sections where little new information was expected. Not only did the new information have to be reported, but it had to be evaluated in terms of its effects on the original Rowland hypothesis and on the modelers' predictions of eventual ozone reduction at steady state if fluorocarbon production were to continue.

Climate effects took only a few paragraphs in the first IMOS report which could say little more than that effects of some nature were possible but ill understood, and might be serious if they occurred, at least in regard to effects on certain regions and agricultural crops. IMOS only concerned itself with climatic changes that might result from ozone reduction itself. The theory of a warming trend comparable to the "greenhouse effect" from CO_2, resulting from the accumulation and infrared absorption of fluorocarbons in the lower atmosphere, was not put forth until the following fall.[32] That concern coupled with new awareness of major perturbations in the temperature-altitude profile of ozone became much more significant in later assessments, such as the 1976 NAS study, but has subsequently decreased in significance.

Human health effects were difficult to address because little work had been done on the effects of UV radiation on skin cancer since the CIAP and NAS work related to the SST, which had left many unanswered medical and statistical questions. Much debate was beginning to occur within NAS, especially over the effect on melanoma, the most often fatal type of skin cancer. It became obvious that this debate was not going to be settled in time for the IMOS report, or for the April meeting in Philadelphia of the American Chemical Society, where a special symposium was held on Chlorine Reactions and Stratospheric Ozone. The symposium was well-attended by IMOS and NAS committee members, and no conclusion about melanoma was reached.

There was no real expertise on these matters within IMOS. A unified assessment could not be obtained from government agencies or the scientific community, either on the melanoma question or on the most accurate magnification factor to be used for the relationship between ozone reduction and increased UV-B* at the ground, or

* Biologically active solar ultraviolet radiation in the wavelength range 280 to 320 nm. UV-B is alternatively defined as 290 to 320 nm.

the biological amplification factor to indicate the relationship between UV-B exposure and increased cancer. Therefore, IMOS chose to be conservative on all three matters (i.e., downplaying the likelihood of increased melanoma and using low multipliers for increased cancer as related to ozone reduction).* Members knew that public concern and justification for any regulation would probably rest mainly on the skin cancer threat, no matter how small; however, most members were more concerned about the possibility of other less well understood but potentially more serious biological, ecological, and climatic effects.

The Biological Effects chapter also relied upon research results from CIAP that were very limited and not at all definitive. Although there was sufficient qualitative reason for concern, little hard data were available that accurately represented natural conditions. The report suggested areas of possible problems: effects on nucleic acids, proteins, chromosomes, and other cellular constituents; phytoplankton and other marine organisms; livestock eye diseases and insect behavior; the effectiveness of agricultural chemicals; and the indirect effects resulting from any climatic change that might occur.

IMOS also wished to characterize the industries related to fluorocarbon production and use, describe what substitute products and other adjustments would be likely to result from possible restrictions, and characterize the costs of such restrictions to the individual industries, labor, the consumer, and the economy as a whole. Much of the information was simply not available within industry or otherwise. Most of the information used was supplied to the Commerce Department by industry; however, it was incomplete and not conducive to detailed analysis. Because this information was obtained at the last moment, there was no time for investigation of any apparent anomalies. Most available aerosol data were not specific for aerosol products using fluorocarbons, and the trade associations did not have desired details. The Manufacturing Chemists Association represented all of the fluorocarbon producers concerning atmospheric research, but not concerning economic matters. Economic data were provided by the largest producer, Du Pont, but this data did not always hold true for the other five U.S. producers. EPA had begun its own economic fact finding through a study conducted by Arthur D. Little, Inc., from which some preliminary results were made available to IMOS.[33] IMOS mainly reported a collection of assorted facts rather than analyzing the elusive questions of substitutes, leakage control, and the timing and scope of dislocations and adjustments. By June 1975 some switching away from fluorocarbon aerosol products was occurring.[34,35] However, it was difficult to predict whether the trend toward nonaerosol products would continue or whether an acceptable substitute propellant, perhaps even an environmentally acceptable fluorocarbon, might be developed.

IMOS coverage of the jurisdictions of the three potential regulatory agencies seems straightforward enough now, but at the time was a major puzzle. In early 1975 no one was certain what agencies could regulate fluorocarbons under which legislation; the authorities appeared to be both overlapping and incomplete. The Clean Air Act had not yet been amended to clarify the authority of EPA. The 1970 version would have been a legal nightmare to utilize for controlling this type of pollutant because the law focused upon direct threats to public health rather than the general environment and it made the courts the initial triers of fact. The Toxic Substances Control Act had not yet been passed. A half dozen fluorocarbon-specific bills had emerged in Congress, and the Administration needed to develop a position on them. Finally the Justice Department, working closely with lawyers in each of the regulatory agencies, prepared

* For discussion of the amplification problem, see testimony and material for the record of Urbach and Green.[23,62] Later reports by the National Academy of Sciences,[66,77,106] treated these matters in greater depth. NAS[106] later used an overall amplification factor twice as large as that used by IMOS.

an opinion on the present authorities. This opinion was included in the IMOS report: EPA could regulate pesticide aerosol products; the Food and Drug Administration could regulate aerosolized foods, drugs and cosmetics; and the Consumer Product Safety Commission could regulate all other consumer aerosol products plus home and school refrigeration and air conditioning. However, no agency could regulate the many industrial and commercial uses of fluorocarbons or automobile air conditioning. IMOS decided not to endorse any fluorocarbon-specific bill, believing that regulatory authority on a chemical-by-chemical basis would set a cumbersome precedent. Instead, IMOS endorsed passage of comprehensive toxic substances legislation, skirting the issues of the specific controversial proposals then under consideration within Congress and the Administration. Since the IMOS report became "the Administration position", this endorsement was quite important in promoting passage of the Toxic Substances Control Act.*

The portion of the Conclusions and Federal Actions Chapter that was easiest for IMOS members to agree upon was the recommendation to seek international cooperation on any actions that proved necessary. The U.S. produced slightly less than half of the fluorocarbons worldwide. At the time of the IMOS report only the beginnings of international consultation were occurring within the Organization for Economic Cooperation and Development (OECD), the United Nations Environment Program (UNEP), and the World Meteorological Organization (WMO). The IMOS recommendation was used as a prod to get the Federal agencies and the State Department working together on more aggressive actions.

To nongovernmental readers, the rather bland IMOS recommendations for critical areas of research needs and its description of present programs probably also appear simple to prepare, but this is what nearly delayed publication of the report. A proposed plan for research was clearly vital to the success of the IMOS recommendation for regulatory action if necessary, following a period of further study. However, Federal agencies are notoriously reluctant to cite any specifics concerning present (or "baseline") research programs on a particular topic, especially when the program is not a distinct item separately identified in the budget but a series of projects dispersed among many different programs of varying relevancy to an issue. The reason for this reluctance is a practical one: if the amount of relevant research cited is too low, concerned citizens and Congressmen will berate the agency for neglecting its duty or lacking foresight; if the amount is too high, justification of the need for additional research funds will be difficult. Finally the IMOS agencies were persuaded to develop some general information, aided by pressure from Congressman Rogers' committee, which desired the same information. Office of Management and Budget staff acquiesced in the plans, but it was unclear how much attention they gave the matter, or whether they would later support proposals in accordance with the recommendations. This caused some difficulties later, but there was no way to get a firm budget commitment from OMB during the time available.

The principal recommendation of the IMOS report concerned whether or not, and when, regulation might be necessary. There was easy unanimous agreement that the matter was potentially serious and that its seriousness must somehow be clearly conveyed. Midway in the IMOS study, the NAS and CIAP reports on the SST were issued.[6,7] The CIAP report had found that the amount of supersonic transport operations presently scheduled would cause climatic effects smaller than minimally detectable, and that the adverse effects from future SST operations could be avoided if certain technological changes were made. The report also recommended a continu-

* An Administration position is developed on all proposed legislation through consultation among relevant agencies and coordination by the Office of Management and Budget.

ous atmospheric monitoring and research program. Although the body of the report took the ozone reduction threat quite seriously, the Executive Summary released earlier was less cautionary and totally obscured the question of the ozone-skin cancer connection. The press conference that accompanied release of the CIAP report was handled in such a way that all of the reassurances but few of the warnings were reported. Many resulting press accounts reported that CIAP had cleared the SST on the ozone question. This caused much consternation among scientists involved with CIAP. The NAS was more careful and, for the most part, succeeded in avoiding such distortions of its message, so the public perceived more difference between the two reports than actually existed.*

Although Dr. Alan Grobecker, head of the CIAP program, was the IMOS member from the Department of Transportation, and several other government scientists were deeply involved in both the SST-related and fluorocarbon-related assessments, the SST work had relatively little direct effect on the outcome of the IMOS report. IMOS, by its own choice, steered away from the controversy surrounding the SST issue and did not attempt to achieve any consensus on that matter, which was considered much more an economic, technological, and political issue than a scientific one.

The April 1975 symposium of the American Chemical Society served to crystallize the seriousness of ozone depletion by summarizing the mounting atmospheric measurements and laboratory evidence, albeit circumstantial, supporting the Rowland hypothesis. Some, but not all, of the recent information tended to lower the original estimates of ozone depletion. IMOS and ICAS decided to accept 7% as a conservative best estimate of eventual ozone reduction,** which exceeded the 5% that industrial representatives on several occasions indicated might be considered an unacceptable amount.

It was one thing to agree that the matter was "a legitimate cause for concern", as asserted in the IMOS report, and another to develop a course of action. During the first discussions, there were differing opinions. Interestingly enough, especially in the early meetings, the atmospheric scientists expressed the most alarm. The policy and regulatory members, on the other hand, were more reluctant to recommend drastic action. Perhaps this was because they were acutely aware of the legal difficulties of supporting regulatory positions.

Some IMOS members believed that more research would only further confirm the theory and that regulation should occur soon since we could not wait for the final proof of actual significant ozone reduction. Others believed that a time for study should be provided, in hopes that the research would prove the theory false, or that in the meantime industry would develop cost-effective substitutes. A few favored the position of further study more out of timidity; it seemed incredulous to them that an interagency committee could come up with anything more dramatic than a plan for needed research.

There also was a variety of opinion on the significance of any particular amount of ozone decrease that might occur. At that time, no IMOS members considered that the change would be so great as 15%. This amount was cited several times as one that

* For an entertaining account of the difficulties surrounding the release of the CIAP report, see Reference 24, Chapter 4, "The Great Executive-Summary Snafu" (situation normal, all [fouled] up). See also Chapter 5.

** The 7% selected by IMOS as a current best estimate of eventual steady state ozone reduction if 1972 production levels were to remain constant was considered by IMOS to be somewhat conservative, since it assumed a fast reaction constant for the $OH + HO_2$ reaction, which would underestimate ozone reduction. This turned out to be the case after OH stratospheric measurements were taken and analyzed. On the other hand, other unknowns also affected the calculations in the opposite manner.

would clearly pose too great a risk, whatever the particular consequences, on the assumption that "you can't fool Mother Nature that much." (Note that later predictions did exceed that amount.*) But what did 7% mean, or 5%, or 2%, as some industry scientists were then predicting?

Just at the point when agreement on any particular wording appeared unlikely, a very helpful special meeting was held to discuss a paper developed by Mr. Tom Waltz of NOAA.[36] The paper used decision analysis theory to chart different levels of possible environmental costs against levels of confidence that these costs would occur, in order to determine if the "correct" decision were a total ban, partial ban, or no ban. Although the paper did not recommend a particular solution, applying its principles to IMOS estimates led to the conclusion that partial regulation (i.e., just of aerosol products) was the logical position both for those members who had great confidence in the theory that the ozone would be depleted, but little confidence in the knowledge about the consequences, as well as those members who had less confidence in the theory but more confidence about the seriousness of the risks, if ozone depletion were to occur. There was no acceptance or rejection of the validity of the paper's analysis by IMOS and no formal vote, but an around-the-table unveiling of each member's views considerably helped to crystallize issues and identify areas of agreement.

The Co-Chairmen then prepared a draft which reflected the apparent consensus of the IMOS members. This was done by being specific in areas of agreement and hedging on other aspects. The areas of agreement were the seriousness of the concern, the desirability of postponing regulatory action until completion of the NAS study, and the belief that some type of restrictions at some time were highly probable. Diversity of opinion was handled in one instance by indicating the type of restrictions that were likely to occur at some time while not specifically indicating the content of any restrictions that might be implemented as early as January 1978. Three reasons for the 1978 proposed date for regulation were given: (1) opportunity for further research, (2) time for industry preparation and adjustments, and (3) the time it would take agencies to complete the rulemaking process.

The origin of the January 1978 date for regulation was a guess by the two Co-Chairmen as to what might be acceptable to the greatest number of members. Actually, each co-chairman independently prepared a first draft of the conclusions and recommendations, and they differed from each other by one day — December 31 and January 1. The date happened to coincide with what industry had requested in December 1974. After further modification and fine tuning by IMOS members, the final wording of the recommended timetable for a regulatory decision read as follows:

> The task force has concluded that fluorocarbon releases to the environment are a legitimate cause for concern. Moreover, unless new scientific evidence is found to remove the cause for concern, it would seem necessary to restrict uses of F-11 and F-12 to replacement of fluids in existing refrigeration and air-conditioning equipment and to closed recycled systems or other uses not involving release to the atmosphere.
>
> The National Academy of Sciences is currently conducting an in-depth scientific study of man-made impacts on the stratosphere and will report its findings in less than a year. If the National Academy of Sciences

* The December 1979 National Academy of Sciences report gave a most probable value of 16.5% for the eventual ozone depletion due to continued release of chlorofluoromethanes at the 1977 level.[105] The central value accepted by the UNEP Coordinating Committee on the Ozone Layer at its 1979 meeting was 15%.[108] More recent calculations, based on revised reaction rates derived from laboratory experiments, have again lowered the numbers. The November 1980 meeting of UNEP/CCOL reported 10%, and the central value to emerge from a series of spring 1981 scientific meetings is likely to be closer to the 7% of the 1975 IMOS report, although uncertainties for certain reactions may result in later revisions upwards again. These lower values reflect a closer agreement between certain atmospheric measurements and modeling calculations than in 1979.

study confirms the current task force assessment, it is recommended that the Federal regulatory agencies initiate rulemaking procedures for implementing regulations to restrict fluorocarbon uses. Such restrictions could reasonably be effective by January 1978 — a date that, given the concern expressed now, should allow time for consideration of further research results and for the affected industries and consumers to initiate adjustments.[31]

IMOS also recommended immediate action to label products containing fluorocarbons as a guide for consumer choice, but did not go so far as to suggest a voluntary boycott.

Contrary to a statement overseas[37] that the IMOS vote had been "by a narrow margin of 8 to 6," this conclusion and the whole report received unanimous approval. The only split vote of individual members occurred at an earlier stage in which the losing side favored *more* immediate regulatory action, not a weaker position. Once the above wording was agreed to in principle by the individual members, the report went to the agencies and the IMOS members for formal concurrence, generally at the Assistant Secretary level or above.

The only near break in unanimity by IMOS member agencies came at the last moment. While the rest of the report was being printed, the Commerce Department completed a further economic analysis of its own (but based on industry data). The Commerce Department insisted on inclusion of the economic analysis as a condition for its concurrence in the report. Although there was no time for other IMOS member agencies to review and comment in detail upon the Commerce report, a brief perusal by IMOS members indicated that at least several of them would not endorse its content without substantial revision. Rather than lose the consensus on the IMOS report, a compromise was reached whereby the Commerce material was included with copies of the IMOS report as an insert[38] prepared by the Commerce Department, with a cover letter indicating it had not been reviewed or endorsed by IMOS. Some of the factual information in the Commerce material was helpful despite some over-simplifications in its analysis. Later this same material was used in a more in-depth economic analysis by Commerce.[39] It should be noted that throughout the IMOS deliberations, the Commerce Department had a special difficulty of internally reconciling the various viewpoints of its Bureau of Domestic Commerce (protective of industry, concerned about economic impact of regulations), NOAA (a leading spokesman on IMOS and ICAS for the significance of the atmospheric threat), and the National Bureau of Standards (not involved in the policy debate, but performers of some of the relevant research and generally supportive of time for further study). By the end of May, pressure mounted for release of the IMOS report before the end of June, when finished copies would be back from the printers. Every day more radio, TV, and press accounts discussed the issue, many in an exaggerated and inaccurate fashion, and the affected industries began showing signs of nervousness. When it was accidentally discovered that one of the fluorocarbon-producing companies had obtained or been given a copy of the IMOS galley proofs, IMOS moved up the day of its press conference by 2 weeks and decided to release a preprint of the conclusions before the full published report. A last-minute attempt by industry to stall or modify the report led to a hurried White House consultation, at which the IMOS Co-Chairmen were given the official blessing to release the report by Mr. Douglas Costle, later to become Administrator of EPA, but then on the Domestic Council staff.

Reaction to the IMOS report, *Fluorocarbons and the Environment*,[31] was generally favorable, although it was criticized by industry as being too precipitate and prejudging the research results[40] ("another damper on the rational mood required to conduct studies"[41]), and by environmentalists as being too cautious. The headline in *The Christian Science Monitor* was "Aerosol spray debate spawns compromise." The article called it a "delicate balance between industrial interests and the possibility of serious damage to the earth's ozone layer."[42] Many of the press accounts assumed an inevitable ban in 1978, a distortion of the IMOS recommendations which portrayed them as more radical than they were.* The correct interpretation was given by Co-Chairman Bastian in Congressional hearings in July 1975: rulemaking was to begin after the NAS report, although the nature and timing of the restrictions to be adopted were not yet clear and might change as the results from further research were evaluated.[43] Editorial opinion varied. *The Washington Star* said the report was "truly sobering" and supported prompt curbs on fluorocarbon aerosol products if the NAS report did not contradict IMOS.[44] A feature article in *Fortune,* on the other hand, praised the report but was much more relaxed about the timetable:

> For the moment, it should be enough to have served notice upon the industries involved that there is good reason for concern. If we avoid hasty legislation, nature or the marketplace may continue to solve the ozone problem without the usual havoc from clumsy government intervention. After 3 to 5 years, the whole matter can be reexamined once more. Who knows, it might even turn out by then that there's too *much* ozone up there and that everybody will have to get busy with his aerosol cans again.[45]

Industry reaction to the IMOS report and to the flurry of public interest in the issue during the summer of 1975 varied among different companies and trade associations; some were relieved and some disturbed. Most looked toward the new scientific findings being uncovered at a rapid pace and hoped that the problem would turn out to be less severe than predicted. Some industries prepared informative presentations of the controversy, such as the booklet *At Issue: Fluorocarbons* published by Kaiser Aluminum and Chemical Corporation.[46] Du Pont ran a series of double-page newspaper advertisements beginning on June 30, 1975, addressing the major elements of the ozone reduction theory and the uncertainties as they viewed them. In January 1976, Du Pont began publishing a series of "Science Update" flyers for wide distribution.[47] Less successful was the sponsorship by the industry group COAS of a U.S. speaking tour by Dr. Richard S. Scorer, professor of theoretical mechanics at London University's Imperial College of Science and Technology. Scorer discussed the fluorocarbon issue on radio and T.V. throughout the country without mentioning that COAS was paying his expenses, but then the *Los Angeles Times* wrote a devastating and widely quoted editorial about the "scientific hired gun".[48] Scorer's extreme position that the whole issue was "utter nonsense" and his abrasive manner did little to advance industry's position. In fact, he was quietly disowned by many of the Du Pont and other fluorocarbon industry scientists.[49]

Some companies, such as Gillette and Bristol-Myers (Clairol products), began advertising their environmental soundness by labeling their aerosol products as not containing fluorocarbons or by bringing out a line of nonaerosol alternatives such as pump sprays or roll-ons.[50-52] The aerosol business was said to be down about 25% with perhaps one-third of that attributed to the ozone scare and the rest to recession.[50] The Johnson Wax Company gave a psychological blow to the industry when in June 1975 it announced a total withdrawal from the fluorocarbon market, even though less than 5% of its products used fluorocarbon propellants at the time.[53]

IV. RESEARCH AND MEDITATION, JULY 1975 TO OCTOBER 1976

After a flurry of press activity and public interest in the fluorocarbon question during the summer of 1975, the public aspect quieted down somewhat, and government

* Two reasonably thorough and accurate accounts of the IMOS report and the industry and environmental reactions to it are in "Ban on fluorocarbon gas in aerosol cans by January 1978 is broached by panel", *The Wall Street Journal,* June 13, 1975, and "End of aerosol age? Federal report says probably", *Science News,* 107, 396, June 21, 1975.

and industry settled into a period of research and reassessment until completion of the NAS report in September 1976.

There was some regulatory-type activity during the period, however. In July 1975 FDA issued a "Call for Information" asking for submission of detailed information from aerosol manufacturers and distributors and advising of FDA's intent to regulate in the future if scientific reports confirmed that ozone depletion might occur.[54] In December 1975 EPA issued a warning to pesticide manufacturers urging them to seek substitute propellants if they expected their registrations to be approved or renewed.[55]

In August 1975 the Consumer Product Safety Commission rejected the NRDC petition for a ban on fluorocarbon aerosol products by a vote of three to two, with two strongly-worded dissents, on the grounds that the additional information to be provided by the NAS study was desired. A similar petition by NRDC, New York, Michigan, and Oregon was filed with FDA in July 1975 and rejected in September 1975.[62]

IMOS worked with the three regulatory agencies on the recommendation for immediate labeling, as well as with the Federal Trade Commission (FTC), which seemed to have some jurisdiction. The process was slow and difficult. FTC never acted one way or the other, apparently giving the matter a low priority within its workload. On October 18, 1976, EPA promulgated a requirement for identification labeling of pesticides using fluorocarbons, to be effective on April 15, 1977.[56] The CPSC labeling requirement was first proposed in April 1977, made final in August 1977, and took effect on February 20, 1978.[57] FDA required a warning rather than mere identification: "WARNING — Contains a chlorofluorocarbon that may harm the public health and environment by reducing ozone in the upper atmosphere". This was first proposed on November 26, 1976, made final the following April, and took effect on October 31, 1977.[58]

Many states were considering their own regulatory action, a prospect which dismayed industry as well as many Federal officials. Ten states joined in the second NRDC petition to CPSC in December 1975, and 11 states in the July 1976 petition. Oregon and New York passed legislation in June 1975 providing for future bans on aerosol products using fluorocarbons in March 1977 and October 1978, respectively, under certain conditions based on further assessment of scientific results. By June 1975 at least 13 other states had similar bills pending,[50] but no further state actions preceded the NAS report.

The $22 million government atmospheric research program for fiscal year (FY) 1976 proceeded reasonably well. The lead agency role in development of instrumentation and measuring systems was assigned to NASA by Stever as Chairman of FCST, upon recommendation of ICAS. Dr. James King* of NASA chaired an interagency subcommittee of ICAS set up to assist NASA in its coordinating role. During early 1975 NASA re-focused its program to emphasize stratospheric research, establishing a $7 million program. NASA also set up a Stratospheric Research Advisory Committee from the academic community and sponsored a major workshop on stratospheric research in May 1975.[59] Congress expanded the role of NASA in its FY 1976 Authorization Act by amending the National Aeronautics and Space Act of 1958 to establish a new Upper Atmospheric Research Office (in the Office of Space Science) with an expanded stratospheric research program.[60] Interagency coordination of monitoring programs was accomplished through the Federal Coordinator for Meteorological Services and Supporting Research, located in NOAA, as established by OMB.

A number of interagency and advisory committees and subcommittees under NASA, NOAA, ICAS, or other leadership had somewhat overlapping roles in developing and

* King was Special Assistant for Stratospheric Research and later Director of the Upper Atmospheric Research Office. He was also a NASA member of IMOS.

coordinating various pieces of the research program. This led to some confusion and delays in the issuance of written reports describing research programs, which occasionally frustrated the government policymakers and regulators as well as nonfederal researchers. On the other hand, ICAS did provide excellent opportunity for regular communication among all the principal relevant offices involved, and the research programs that resulted were better in reality than articulated during this rather frantic period. The dialogue between government researchers and regulators improved greatly with time. NASA and EPA eventually consulted closely on policy-relevant research needs, which was both unusual and extremely beneficial.

Additional research, including laboratory experiments, atmospheric measurements, and theoretical work, was carried out by the Fluorocarbon Technical Panel of MCA — about $5 million for 1975 to 1977,[47] and $11 million by mid 1981.[12] A July 1975 interim report from the NAS study concerning research needs was quite consistent with the research underway or planned in the Federal government and MCA programs.[47,61]

IMOS became very worried about the need for research on the biological and climatic effects of ozone perturbations and increased UV-B radiation, however. Little advance in knowledge of these effects had occurred since the small program undertaken as part of CIAP. In March 1976, Senator Dale Bumpers, Chairman of the Senate Subcommittee on the Upper Atmosphere, complained, "I am appalled to find there has been so little research done in this field."[62] Improved information was considered necessary to support any rule-making process in the next several years relating to aerosol products, especially in regard to skin cancer effects and agricultural crops. In addition, more basic knowledge about the mechanisms of UV radiation, UV effects on whole ecosystems, and characterization of any climatic perturbations, was deemed vital not only for dealing with the more essential uses of the fluorocarbons, but also for considering what to do about other possible future modifiers of the stratosphere.

During 1975, ICAS prepared for IMOS a brief overview of "Potential Modifiers of the Stratosphere Other Than Fluorocarbons", released in March 1976.[63] The report steered away from evaluating the NO_x effects from high altitude aircraft, which were being studied through the Federal Aviation Agency (FAA) High Altitude Pollution Program (HAPP) and later as part of the Tripartite Agreement on Stratospheric Monitoring signed in May 1976 by the U.S., France, and Great Britain (see Chapter 1). Likewise IMOS and ICAS left the assessment of chlorine effects from the Space Shuttle to NASA. The IMOS/ICAS report concentrated on stratospheric effects from nitrogen fertilizers, brominated compounds, other chlorinated compounds, particles, and carbon monoxide. The report concluded that "in each case there appears to be no immediate serious problem either because the concern is highly speculative or because it is based upon compounds that are not as yet released to the upper atmosphere in quantities believed sufficient to produce a hazardous effect." But the report warned that the additive effects of several stratospheric modifiers might become significant in the future and there may be other potential modifiers that have not yet been identified or produced. More research was needed as well as improved information on industrial production, uses and emissions, and predictions of future trends worldwide.

The Federal-wide Upper Atmospheric Research Program under development by ICAS was to cover these research needs, but projects were slow to develop (Chapter 5). Fortunately, some of the threats that seemed most difficult, such as effects from nitrogen fertilizers, are now expected to be less significant than originally feared, and to occur sufficiently far in the future to allow for more study and appropriate action if needed.

Nevertheless, it was embarrassing, to say the least, for scientists to be calling for restrictions on products based on their suspected health and environmental effects but not be engaging in any meaningful research program to characterize those effects or discover whether these effects might be mitigated or prevented. The embarrassment became even more acute as the U.S. became more aggressive on the international scene in urging the importance of such research without itself having a viable program. The statements in support of effects research made by U.S. representatives at the March 1977 UNEP meeting (see Section V) are particularly illustrative of this dichotomy between what was practiced and preached by the U.S. The reason was, of course, that representatives to international meetings on the stratosphere tended to be scientists rather than budget analysts. To the outside world, however, the "official U.S. government position" on this aspect of the subject must have seemed unclear.

In November 1975 IMOS formed a chartered Subcommittee on Biological and Climatic Effects Research to develop a program that became known as BACER. Working with the Federal Council for Science and Technology, the IMOS subcommittee developed a working paper that was approved in February 1976 by FCST as the basis for further development of a detailed program, discussions with the scientific community, and budget proposals.[64] Agency representatives had little difficulty agreeing on the basic research needs to be included, which were similar to ones cited in earlier NAS and government reports and Congressional hearings. Varying opinions on funding levels were resolved by using the lowest estimates that anyone suggested; further reductions were made by the IMOS Co-Chairmen and BACER subcommittee chairman (Dr. Donald Krizek of the Department of Agriculture) in order to make the program more palatable to OMB in a tight budget year.

The program consisted of a short-term program to acquire some policy-relevant data within 2 years as the first stage of a larger program "that will address in a longer time frame many basic questions concerning the effects on earth of ozone radiation and accompanying increased UV radiation, both from fluorocarbons and from the many other postulated causes of such ozone reduction." The short-term program emphasized instrumentation needs for measuring and simulating UV, human health effects, effects at the molecular and cellular level, and effects on plant organisms. The longer term program added research on effects of UV radiation at the ecosystem level and research on climatic changes resulting from fluorocarbon use.

Considerable discussion took place first within the BACER subcommittee and IMOS, and later within FCST concerning the best management scheme for the program. Alternatives that were considered included: (1) a CIAP-type lead agency with the pros and cons of several possible lead agency candidates debated, (2) establishment of a new program within a single federal agency, (3) a loosely coordinated program spread among many federal agencies, and (4) a triumvirate-type scheme with three agencies managing three major parts of the program. The second scheme was considered too time-consuming, the third to be an insufficient improvement over the present situation, and the fourth to be difficult to manage and too resource-intensive. Therefore, the first was chosen. On February 23, 1976, the FCST established BACER with EPA assigned as lead agency, supposedly in an arrangement similar to DOT's role in CIAP, where new funds would be provided the lead agency for pass through to other Federal agencies. EPA provided the management staff, while research projects were developed and reviewed in detail by a series of interagency committees working closely with nongovernment scientists. Only minimal funding was forthcoming, however, and most had to be reprogrammed by EPA away from its other research. The original BACER plan had called for $12.6 million for FY 1977 above existing programs; EPA requested $8 million and received nothing. It had to reprogram $4 million to continue a minimum program. In December 1976 Congressional hearings, agencies were asked to comment on rumors that OMB had refused funding for the program because of a sense of lack of urgency once a regulatory course of action had been set.[65] None of the witnesses admitted to any knowledge of this, but oral remarks of OMB staff from time to time tended to support this view. In the author's opinion, OMB short-sighted in its lack of sympathy for the research needs in this area in contradiction to the pleas from many government and nongovernment scientists, NAS, and Congressional members and staff.

The short-term program, despite its insufficient resources, was welcomed by NAS as a good beginning that provided useful insights that should be pursued in a longer term sustained program.[66] The longer term program did not get off the ground, however, and some of the short-term projects were later discontinued. Although EPA was assigned the lead role for the long-term program in July 1976, it later backed away from the responsibility because of the management and budget burden it had been forced to bear (although it subsequently re-volunteered). By the winter of 1978, the future of BACER was still uncertain, although attempts at resurrection were in progress under the jurisdiction of two separate committees with a partially overlapping membership: the Interagency Committee for Stratospheric Ozone Protection (ICSOP), chaired by EPA and charged with coordinating such research by Section 153(d)(1) of the Clean Air Act Amendments of 1977 (P. L. 95-95), and the Committee on Atmosphere and Oceans, chaired by NOAA, of the Federal Coordinating Council for Science, Engineering, and Technology (FCCSET) that replaced FCST.*

There were several reasons for these difficulties:

1. The need for such research was more relevant to the good of scientific advancement, the government and public welfare as a whole rather than to the "mission" or self-interest of any particular agency, and its advocates within Congress.
2. Expertise to conduct or sponsor the research is spread throughout a large number of Federal agencies so that the dollar share of any one agency is small and the research plans get lost in the "noise" of the agency's budget (which is why the pass-through budgeting mechanism had been advocated by IMOS).
3. Once a regulatory course of action is determined, even if only tentatively, some government officials apparently see no need for further study of the question.
4. Some persons believed that a regulatory agency such as EPA should not operate a program of basic research as opposed to policy research, although EPA evidenced certain management capabilities to do so that were lacking in the other agencies (some of the other agencies were deemed incapable or unwilling to coordinate an interagency research program).
5. The small base program of on-going research relevant to the effects of stratospheric modification, to which BACER was intended as an increment, was never clearly defined in a manner agreed upon by the sponsoring agencies, the Office of Management and Budget, and the Office of Science and Technology Policy.** This led to situations in which programs were sometimes described as meeting the needs of BACER which they did not do in the minds of IMOS, the new BACER managers, or the sponsoring agencies.

During this same 1975 to 1976 period more detailed economic studies were underway, principally by EPA, the Department of Commerce, and OECD. These were all

* By 1981 the program, now re-named Stratospheric Impact Research and Assessment (SIRA), was barely surviving at a level of $1.4 million per year funded entirely by EPA and was in jeopardy of further budget cuts.

** The Office of Science and Technology Policy (OSTP), the office of the Science Adviser to the President created in 1976, is not to be confused with the earlier Science and Technology Policy Office (STPO), the office of the Science Adviser when that position was in the National Science Foundation. Both were also staff offices for the FCST, later to become the FCCSET.

imperfect because of the many uncertainties involved in foreseeing replacements for fluorocarbon products, the limitations of the state of the art of cost-benefit analysis, and the difficulty of comparing short-term economic disruptions to environmental benefits that would not occur until decades in the future.

More shifts to substitutes were occurring, both to nonaerosol products and alternative propellants such as hydrocarbons and carbon dioxide. The fluorocarbon industry continued to work on fluorocarbon substitutes, suffering several setbacks when some were found to have potentially toxic properties or to be otherwise inappropriate. Thus, much of the answer to the aerosol economic question was taking place in the marketplace. By the fall of 1978 the Chemical Specialties Manufacturers Association estimated that 87% of all aerosol products used nonfluorocarbon propellants as opposed to about 50% in 1974.[67] The more that IMOS and others studied the economic, social, and technological aspects of the refrigeration and air conditioning problem, however, the more difficult it appeared, since no completely satisfactory substitutes were in sight and no practical economic plan was evident for preventing eventual leakage of fluorocarbons from these uses into the atmosphere.

Congress also focused more of its attention on research than on regulation during this 1975 to 1976 period. Hearings were held in May and July 1975 by the House Subcommittee on Environment and Atmosphere (of the Committee on Science and Technology), chaired by Congressman George Brown.[43] Hearings were also held in September 1975 and February 1976 by the Senate Subcommittee on the Upper Atmosphere, a new subcommittee of the Committee on Aeronautical and Space Sciences, chaired by Senator Bumpers.[58,68] The 1975 Bumpers hearings lasted 7 days and were the most extensive ever held, making that record the most up-to-date account of the state of science between the IMOS and ICAS reports and the 1976 NAS report. The hearings were also extremely helpful in establishing communication among the scientists, regulators, and legislators. Senator Bumpers called the hearings because of the oversight role of the subcommittee for NASA and its concerns about the adequacy and appropriateness of the NASA atmospheric research program. Senator Bumpers made it clear that he personally favored prompt regulation of fluorocarbons in aerosol products, although his subcommittee did not have legislative authority over the regulatory agencies. He was also very worried about the lack of effort on biological effects research. Senator Pete V. Domenici, ranking minority member on the subcommittee, shared a similar sense of urgency. During the September 1975 hearings, an immediate aerosol ban was supported by Senator Robert Packwood of Oregon and researchers Rowland, Cicerone, and Dr. James G. Anderson, Cicerone's colleague at the University of Michigan Space Physics Research Lab. Among those expressing a desire to wait for more information were FCST Chairman Stever and Dr. James Fletcher, Administrator of NASA.

International interest was also increasing. The U.S. was quick to recognize that it could not protect ozone on its own so long as foreign fluorocarbon production continued to increase. It made little sense for the U.S. to consider restrictions on refrigeration and air conditioning uses without first trying to persuade other nations to reduce use of fluorocarbon aerosol propellants or otherwise to control fluorocarbon emissions. Furthermore, some research and monitoring could be carried out more effectively on a multinational basis. Finally, improved calculations of what amount of ozone reduction might occur were dependent on better data on worldwide production, uses, and expected future trends.

The other most interested nation was Canada, where regulatory action had been proposed as early as the winter of 1974. In May 1975 Canada raised the fluorocarbon problem before the OECD Sector Group on Unintended Occurrence of Chemicals in the Environment (which was later changed simply to the Chemicals Group). OECD is an organization of major industrialized nations. Since its members accounted for about 84% of world fluorocarbon production,[69] OECD seemed to be a logical organization to take an interest. Originally planned as an information item for the June 13 meeting of the Chemical Group's parent Environment Committee, the fluorocarbon matter was raised on June 12 by Canadian and U.S. representatives, who summarized the conclusions of the IMOS report released that morning at a press conference. Canada and the U.S. volunteered, and were formally asked, to take the lead in gathering more information and to prepare a paper for the November meeting of the Environment Committee. A paper prepared by a U.S.-Canadian working group summarized the scientific findings and new data on worldwide production, gathered through an OECD inquiry to member nations.[70] Some very incomplete information was also gathered for the report from the non-OECD nations by UNEP. By the spring of 1976, OECD adopted a work program to study the fluorocarbon question, especially regarding production, uses, emissions, possible substitutes, laws under which countries might take restrictive actions, and the likely economic impacts of any restrictions.[71] This work proceeded slowly and was inconclusive; economic analyses were made[72] but to date no OECD-wide policy position has been developed. However, the various OECD activities and meetings provided an opportunity to educate all the member nations about the ozone reduction problem and establish a groundwork of mutual understanding which later broadened into discussions of cooperation on other global chemical problems and "harmonization" of approaches to toxic substances control in general.

Other organizations also became involved. UNEP officials were discussing an appropriate role for UNEP as early as the spring of 1975, but at first were overly-ambitious compared to their available management, staff, and financial resources. U.S. officials urged UNEP to concentrate on the research aspects, since UNEP was the logical organization to coordinate the research activities of the several appropriate U.N. organizations such as the World Meteorological Organization (WMO), World Health Organization (WHO), Food and Agricultural Organization (FAO), and United Nations Educational, Social and Cultural Organization (UNESCO). Also, the many underdeveloped countries which were members of UNEP but not of OECD presumably would have an interest in the biological and climatic effects of ozone reduction that would occur in nonproducing nations as well as producing ones. By the summer of 1976, UNEP began planning for a UNEP-sponsored world conference of experts to develop a plan of action.

Other international activities during this period included those of WMO. For many years, WMO had been conducting scientific conferences, collecting data, and reviewing research findings relevant to ozone variations and trends, atmospheric chemistry and circulation, and atmospheric pollution. WMO began to develop detailed plans for coordination of international stratospheric monitoring and research called the "Global Ozone Research and Monitoring Project". Stratospheric ozone protection was also the subject of discussions held under the U.S./U.S.S.R. Bilateral Agreement on Cooperation in the Field of Environmental Protection. The meteorological and climatic aspects were of particular interest in the Working Group on Influence of Environmental Changes on Climate chaired by Epstein of NOAA.

The policy aspects of ozone reduction were the subject of annual "roundtable" discussions at CCMS (Committee on the Challenges of Modern Society), a North Atlantic Treaty Organization (NATO) group which seeks to achieve cooperation among NATO nations on environmental and other matters affecting the quality of life. The U.S. representative to CCMS, EPA Administrator Russell Train, took a strong position in October 1975 calling for an international mechanism to control fluorocarbon emissions if this should prove necessary.[73]

Although there was much international discussion, few other countries were as concerned as the U.S. or as interested in consideration of government restrictions. Some were willing to encourage voluntary action by industry not to expand fluorocarbon production and to search for alternatives. Most countries desired a period of further study before deciding what action, if any, might be appropriate.

The greatest amount of relevant research was being performed in the U.S., but other countries such as Canada, the U.K., France, Federal Republic of Germany, and Belgium were making significant contributions. The U.K. made an overall assessment, "Chlorofluorocarbons and Their Effect on Stratospheric Ozone," which predicted an eventual ozone depletion of about 8% assuming a continued 1973 rate of usage, but concluded that further research should be conducted to reduce uncertainties before any regulatory action was taken.* In the meantime, industry was advised to search for alternatives to aerosol and foam uses of fluorocarbons and to reduce leakage in air conditioning and refrigeration.[74]

IMOS asked the State Department to take the lead in coordinating the various U.S. activities relating to the international aspects of stratospheric modification and the development of governmental positions for presentation at international meetings. As the number of international organizations interested in contributing to the solution of the fluorocarbon and high altitude aircraft problems became more numerous, such a mechanism became a necessity. Therefore, in the late fall of 1975, the State Department convened and chaired a Subcommittee on Stratospheric Pollution (of the Committee on International Environmental Affairs — CIEA), a permanent interagency body for international environmental coordination. This CIEA Subcommittee became an extremely active and effective group for developing a strong U.S. leadership role in stratospheric protection. Besides supporting the U.S. position abroad, the subcommittee became an unexpectedly valuable tool for coordination within the U.S. Many different individuals and agencies served as U.S. representatives or lead agencies in the many international programs. Their positions and activities were frequently duplicative or even conflicting. Through this CIEA subcommittee, staff resources were allocated more effectively, and international policy and domestic policy were made more mutually supportive.

As the time for the scheduled release of the NAS report in the spring of 1976 neared, the research and monitoring programs were generating a rapid accumulation of new data, in some cases serving to reduce uncertainties and in others to raise new questions. In a letter to the President dated March 2, 1976, Peterson asserted that the research results tended to "reinforce rather than diminish the concern."[75] Stratospheric measurements by balloon, aircraft, and satellite were being made of HCl, OH, F-11, F-12, ClO, and other atmospheric constituents critical to the theory. Atmospheric measurements of F-11 and F-12 showed an increased accumulation with time that was fairly consistent with production trends. Their concentrations in the stratosphere at different altitudes and compared to each other were consistent with the calculated differences in their photodissociation rates. Numerous laboratory experiments resulted in new and improved calculations for many of the 100-plus important chemical reaction rates. The breakdown of fluorocarbons by UV radiation and the resulting catalytic reaction of chlorine atoms with ozone had been demonstrated in the laboratory. No major tropospheric sinks had been discovered for fluorocarbons nor any large natural sources of chlorine sufficient to upset the calculated chlorine budget and overwhelm the manmade effects. The major unanswered question at this point was whether a sink could be discovered for chlorine atoms in the stratosphere.[62,68,76]

* A second report, Pollution Paper No. 15, published in 1979, updated the estimate to 13% at a continuing 1975 rate of release, but also asked for more research while voluntary reductions were made by industry.[107]

The modelers' calculations of expected ozone reduction went up and down many times. The quantitative picture of the extent and timing changed monthly, but qualitatively most of the information tended to substantiate Rowland and Molina's "hypothesis". That "hypothesis," now considered by most scientists to be at least a full-fledged theory, had not been proved and was not provable at that time but was supported by an increasing weight of convincing circumstantial evidence. A gradual shift was beginning to take place that has continued to the present. Fewer persons criticized the plausibility of the basic theory or doubted the corroborative evidence. They instead hoped to discover that present understanding of stratospheric processes was somehow wholly mistaken. In the words of the 1976 NAS report, "Further allowance for uncertainty would be required if some so far unallowed for contingency were to materialize, thus changing the bases for calculations."[77]

As the April date for the NAS report neared, more attention focused on it, both domestically and internationally. Rumors were rampant during the closely-held circulation of early drafts, which were somewhat behind schedule. Then great consternation struck: Rowland and Molina, the originators of the concern, brought chlorine nitrate ($ClONO_2$) to the attention of the scientific community as a possible connection between the NO_x and ClO cycles that would serve as a partial sink for stratospheric chlorine before it could make substantial inroads on ozone:

$$ClO + NO_2 + M \rightarrow ClONO_2 + M; \text{ and } ClONO_2 + h\nu \rightarrow ClO + NO_2$$

In raising this point in January, they were quick to point out that further investigation was needed to determine the effect of other chemical reactions that occur simultaneously and to explain the apparent incompatibility of a major role for chlorine nitrate with the actual stratospheric measurements of HCl. The news was out, however, and first reports were that the chlorine nitrate role would minimize the effect of fluorocarbons to a trivial level and that the NAS report would be revised completely.* At first, chlorine nitrate appeared to be a possible explanation for the unexpectedly high measurements of HCl reported by Dr. Allan L. Lazrus, a research chemist at the National Center for Atmospheric Research, until NCAR corrected the measurements, reporting a laboratory error in analyzing the stratospheric samples.[78]

As the old modeling calculations were reexamined in the light of the new chlorine nitrate theory, the modelers predicted a decrease in the probable ozone reduction by one third or one half, accompanied by a greater effect on climate than was previously anticipated.[79,80] By May 1976, Rowland and Molina were attempting to persuade everyone that chlorine nitrate was not of major importance after all, and were insisting that ozone depletion would be at least as severe as their original 7 to 13% estimate.[81] But the confusion and psychological shock to the scientists' self-confidence and their public credibility were already established. Whatever the effect of chlorine nitrate upon ozone reduction, it was the first major challenge to the basic operative assumptions and served to increase the uncertainties rather than to narrow them.

Meanwhile, progress on the NAS report had screeched to a halt. The committee and the panel decided that all of the calculations would have to be reexamined and redone where necessary. A leak of the conclusions of the draft report (before chlorine nitrate) was reported in *The Philadelphia Inquirer,* which also reported a delay in the study.[82] Without consulting with the government agencies providing funding for the study, NAS issued a press release announcing the delay of the report until new information (unspecified, but known to concern chlorine nitrate and certain climatological information) was considered.[83] The Federal agencies were aghast, fearing that there would

* For press coverage of the $ClONO_2$ debacle, see Reference 24, Chapter 10, " Ten Plus Ten Equal Twelve."

always be new scientific information to delay the report once the precedent for delay was set. NAS had given no clear promise of a completion date upon which regulatory decisions and international discussions could be targeted. The only certainty was that an extension of the study would mean requests for more Federal funds for completion.* IMOS and several individual agencies wrote letters to NAS expressing their disappointment at this turn of events.

On May 3, a special meeting of IMOS was called to hear the testimony of Chairman Tukey of the NAS committee and atmospheric scientists Crutzen and Cicerone. The conclusions of the meeting were that fluorocarbons were still a serious concern, even if chlorine nitrate somewhat lessened the amount of eventual ozone reduction and provided a longer time for decisionmaking. According to the principles of the modified theory, although total ozone reduction might be less, dramatic perturbations in the distribution of ozone and temperature at various altitudes would result in an increased potential for greater climatic disturbances than was originally believed. In Crutzen's dramatic words, "it looked like a different planet." By this time it was also clear that the NAS study would support the Ramanathan theory of a major problem from direct infrared absorption by fluorocarbons accumulating in the troposphere, aside from their effects on ozone.

Although the consternation of the Federal agencies over the delay in publication of the NAS study was perfectly understandable, in the end everything turned out fine. The delay of 5 months resulted in a more credible and conclusive report than if it had been issued when originally planned. It is interesting to note, however, that new calculations and atmospheric measurements had already made part of the report out of date by the time of its publication.[84-86] Since NAS had included a thorough discussion of the broad range of uncertainty, the report's conclusions remained valid despite the subsequent changes in knowledge, however.

The summer of 1976 was one of tension and waiting. The CPSC stalled in replying to the latest NRDC petition, hoping for issuance of the NAS report. An EPA-sponsored international meeting of environmental regulatory officials for major fluorocarbon-producing nations was postponed from September 1976 to the following spring because of the delay in issuing the NAS report.

In the meantime another international scientific meeting was assuming an increased importance because of its timing — one to be held in Logan, Utah, on September 15 to 17, 1976. This "International Conference on the Stratosphere and Related Problems" was sponsored by the International Association of Meteorology and Atmospheric Physics, the International Association of Geomagnetism and Aeronomy, the American Geophysical Union, the American Meteorological Society, NASA, the Jet Propulsion Laboratory, and Utah State University. Including leading scientists and policymakers from both the public and private sectors, the conference was convened to review the latest experimental results and atmospheric measurements, and to discuss the relationship of these findings to the decisionmaking process and the weighing of social, economic, legal, environmental, and international policy considerations. NAS was persuaded to promise that its report would be available before the conference so that this gathering of eminent leaders from the U.S. and about ten other nations could address the significance of the findings. By undertaking extraordinary measures to shortcut the usually lengthy NAS review and clearance procedures and by printing a limited number of copies prior to availability of the published documents, NAS was able to hold its press conference in Washington on Monday, September 13, just in time for Tuesday's meeting of the MCA Fluorocarbon Technical Panel in Salt Lake City and the opening of the Logan meeting on Wednesday.

* In fact the final cost of the study was about double the original cost.

Few copies of the NAS report were available at the Logan meeting and the report was difficult to digest quickly, but some extra copies were duplicated and some of the study participants were on hand at Logan to offer their interpretations. The tone of the report was clearly sobering, particularly when addressing the effects of increased UV radiation on melanoma skin cancer ("...a serious health hazard") and several types of possible climatic disturbances: warming of the climate of the earth resulting from tropospheric accumulation of fluorocarbons, climatic effects of unknown direction or magnitude resulting from ozone reduction in the stratosphere and ozone redistribution at various altitudes resulting in changes in radiation balance, and complex feedback effects involving temperature, precipitation, wind, and other climatic mechanisms. The best estimate for expected ozone reduction again was cited as "about 7%", but NAS included a range of uncertainty of 2 to 20%, which provided hopes and fears for everyone.

The detailed report of the NAS Panel on Atmospheric Chemistry, *Halocarbons: Effects on Stratospheric Ozone,*[87] was so exhaustive in its evaluation of all scientific findings to date that it received high praise from both the press and participants at Logan. The report of the NAS Committee on Impacts of Stratospheric Change, on the other hand, received somewhat mixed reactions. That report, *Environmental Effects of Chlorofluoromethane Release,* included the expected discussion of biological and climatic effects of stratospheric change and future research needs. The Committee also commented on some economic aspects of the issue (despite their lack of economic expertise or any in-depth study in this area), and offered judgments about whether and when to regulate. For this, the Committee attracted criticism both from those who disagreed with its conclusions and those who believed that the committee had gone beyond its mandate by offering such advice. The report advised that the magnitude of the scientific uncertainties and the expected economic impact of restrictions were sufficiently great that restrictions should be selective, according to use. The report recommended that labeling requirements be imposed immediately and legislation be passed to allow regulation to occur later, but implementation of regulations should passed to allow regulation to occur later, but implementation of regulations should await a period of further study. The ambiguity of certain NAS report statements concerning the decision to regulate and the timing of regulation, and slight differences of wording among these statements, provided room for different interpretations as to what, in fact, was recommended. For example,

> "...a one or 2-year delay in actual implementation of a ban or regulation would not be unreasonable."
> "Selective regulation of CFM uses and releases is almost certain to be necessary at some time and to some degree of completeness. Neither the needed timing nor the needed severity can be reasonably specified today. Costs of delay in decision are small, not more than a fraction of a percent change in ozone depletion for a couple of years' delay."
> "The impact on the world of waiting a couple of years before deciding whether or not to regulate...is small...."
> "As soon as the inadequacies in the bases of present calculations are significantly reduced, for which no more than 2 years need be allowed, and provided that ultimate ozone reductions of more than a few percent then remain a major possibility, we recommend undertaking selective regulation of the uses and releases of CFMs on the basis of ozone reduction."

The different possible interpretations of the NAS report can be illustrated by two articles in the September 13 *Washington Star* and the September 14 *Washington Post.* The two papers carried similar accounts of the content of the NAS report, yet the *Post's* headline read, "Aerosol Ban Opposed by Science Unit", whereas the *Star's* front page declared, "Spray Can Use Threat to Ozone, Science Panel Says". In December 1976 when Bumpers criticized NAS for recommending a 2-year delay, Tukey clarified the matter by explaining that the report recommended further study for up to 2 years. Tukey further stated that he believed the NAS group did not think the study would require the full 2-year period.[65]

The representatives of Federal agencies and industry at Logan did not agree among themselves on the wisdom of the NAS recommendations. The greatest area of agreement was on the science; most of the debate centered on cost-benefit questions and the timing for initiation and completion of regulatory action.

The policy seminar at Logan included an Assistant Administrator of EPA, a Commissioner of CPSC, the Chairman of CEQ, the Director General of the Ministry of Environment of Norway, the Science Editor of the *New York Times,* a Congressional staffer, and other prominent persons representing industry, NRDC, the Canadian government, and the academic community. No agreement was reached but the discussion helped to articulate reactions to the NAS report and plans for future action. Peterson's speech was the most aggressive, calling for immediate initiation of rulemaking, voluntary cutbacks by industry, and a consumer boycott.[88]

After the Logan meeting, the pace of events moved quickly toward regulatory action. Although the NAS report seemed to recommend waiting up to 2 years before implementing regulations, it was completely silent on the question of when rulemaking should be initiated, a critical ingredient of the June 1975 IMOS position. On September 21, 1976, IMOS met and prepared a policy statement.[89] IMOS stated that the NAS findings substantiated those of the IMOS report, and that the most "prudent" course of action was for the regulatory agencies to commence proposed rulemaking. "This recommendation is not inconsistent with the conclusion of the Academy committee, which urged time for further scientific work prior to a final decision to regulate." Measures were taken almost immediately to begin rulemaking. The U.S. decision to prepare regulations on aerosol uses of fluorocarbons was announced by EPA Administrator Train at the October 12, 1976, meeting of CCMS in Brussels, where he asked other nations to take similar regulatory action.[90] Also, on October 12, FDA announced the initiation of rulemaking by the three U.S. regulatory agencies, which would include labeling plus a phaseout of nonessential aerosol uses. FDA justification for its initiation of rulemaking included not only the lead time required before the regulation would take effect, but their belief that even if eventual ozone reduction were no greater than 2% (the bottom of the NAS range of uncertainty), the effects would be intolerable when compared to the nonessential nature of most aerosol products.[65] Industry representatives were quoted as expressing surprise at the FDA announcement, which they believed to be contrary to the NAS recommendation.[91] However, there was surprising unanimity among key government officials that this was not the case. Certainly, the choice of effective dates for the regulations was made with the NAS recommendations in mind (see Chapter 10). On October 18, 1976, EPA issued a notice to producers, formulators, and registrants of pesticides urging voluntary use of propellants other than F-11 and F-12, and warning that failure to do so might result in the cancellation of product registration.[92] On the same day as the FDA Notice of Proposed Rulemaking, November 22, 1976, CPSC granted the NRDC petitions.[65]

V. REGULATORY DEVELOPMENT AND INTERNATIONAL POLITICS, OCTOBER 1976 TO 1978

The decision to proceed with regulation coincided with the October 1976 passage of the Toxic Substances Control Act (TSCA). Until the last moment, it was not certain whether TSCA or the fluorocarbon amendment to the Clear Air Act would pass first. As it turned out, the regulations for nonessential aerosol uses were promulgated under TSCA authority. Any future restrictions on other fluorocarbon uses will probably occur under the authority of the Clear Air Act Amendments of 1977, signed into law on

August 7, 1977. This law *requires* the Administrator of EPA to promulgate regulations, if in his judgment, any human activity that may affect stratospheric ozone may reasonably be anticipated to endanger the public health or welfare.

Following the decision to initiate rulemaking, FDA promptly requested CEQ to sort out remaining questions of overlapping jurisdiction and requirements for an Environmental Impact Assessment.[93] Several meetings chaired by CEQ led to agreement on cooperative rulemaking and to the formation of an Interagency Chlorofluorocarbon Work Group, whose activities are described in detail in Chapter 10.

From then on, the center of Federal activity shifted away from the general policymaking efforts of IMOS to the activities of the regulatory agencies. Those agencies became the centers of government-industry discussions, both through informal and public meetings. IMOS concentrated on Federal coordination, placing much of the responsibility upon other related organizations: the CIEA Subcommittee on Stratospheric Pollution, the various BACER committees, and ICAS. NASA worked on its Congressionally-mandated assessment, due September 1977.[94,95] By that date, new reaction rate constants had resulted in higher estimates of ozone reduction — 10.8 to 16.5% at continued 1975 release rates, with an uncertainty range of 4 to 30%.

EPA had contracted with NAS for further assessments of new scientific results. The December 1977 NAS report[66] concurred with the NASA findings and reported that the increased calculated effectiveness of Cl/ClO led roughly to a doubling of estimated seriousness of ozone reduction cited in the 1976 NAS report. At the same time, the estimated effectiveness of NO_x in reducing ozone had resulted in the elimination of any serious concern about high-altitude aircraft. Significant changes, although perhaps not "major upheavals," have continued to occur, however. More significant than the particular quantitative estimates reported were the NAS remarks about the considerable progress in understanding that had occurred in the past year, especially the greatly improved and more numerous laboratory measurements of reaction rates and direct stratospheric measurements that could be used to check the atmospheric models. NAS considered less and less likely any further "major upheavals" in understanding such as chlorine nitrate chemistry or the rate constant for HO_2 + NO that had resulted in the latest revisions of the effectiveness of ClO and NO_x. By late 1980 the situation had reversed again, with improved understanding of the chemistry of the lower stratosphere: SSTs are considered as much of a potential problem (theoretically) in regard to ozone depletion as in the days of CIAP. High altitude subsonic aircraft have the effect of increasing ozone in the lower stratosphere, however. Other findings of the report included a discussion of the complexities of possible climatic effects, concern about the failure of government to institute a long-term biological effects research program, and warnings about the possible effects of future growth in atmospheric release of certain other halocarbons, especially methyl chloroform.

The regulations on nonessential aerosol uses were announced in May 1977 by the three regulatory agencies ("an unprecedented joint announcement", according to *The Washington Post*[96]), which promulgated final regulations in March 1978 with effective dates for controlling various activities being October 15 and December 15, 1978, and April 15, 1979 (see Chapter 10 for details).[97,98] The original timetable of spring 1978 for proposed restrictions on other uses, including air conditioning and refrigeration, was abandoned in view of various factors, including the economic, social, and public health significance of these fluorocarbon uses, the complexities of designing an appropriate regulatory mechanism to control fluorocarbon releases from these uses, and the obvious advantage in working for greater worldwide restrictions prior to further U.S. actions.

Until recently, the public has never seemed particularly upset about the air conditioning, refrigeration, and other uses of fluorocarbons. The media had so stressed the aerosol problem that many people are surprised when told that other fluorocarbon uses are even involved. Now that EPA is considering regulation of these uses (see Epilogue) the situation is changing and the matter may again be debated at the White House and Cabinet level. IMOS and ICAS in the meantime have faded away. For some time they were caught up within the intricacies of Federal reorganization. The Federal Council for Science and Technology (FCST) had been abolished by legislation in May 1976, and replaced by the Federal Coordinating Council for Science, Engineering, and Technology (FCCSET). The structure of FCCSET was slow in getting organized because of the change in administrations. A year after its formation, discussions were proceeding on the role of the new Committee on Atmosphere and Oceans (CAO), which was to take over several previous FCST committees including ICAS. Because IMOS was not solely subordinate to FCST but was a joint FCST-CEQ creation, it became an *ad hoc* FCCSET-CEQ committee that would continue in existence until certain ongoing activities were completed. ICAS kept scheduling its "last meeting," but was not officially abolished until June 30, 1978.

IMOS never had a definite ending point. Because of numerous agency reorganizations and staff changes, IMOS just faded away by no longer meeting. A final report was planned to summarize IMOS history and make recommendations for future actions needed for stratospheric protection. A partial first draft was prepared by the Co-Chairmen, but never circulated to members for comment since so many IMOS members (including the Co-Chairmen) had left the agencies they had represented on IMOS by the fall of 1977. The partial draft concluded that more work was needed:

> We are interested not only in the Federal government's effective pursuit of the fluorocarbon problem but *also* in its development of more effective means of predicting and reacting to future global environmental issues of a similar nature. A continuing oversight of man's activities in relation to the upper atmosphere is clearly required, whether by a single institution or several.

That conclusion holds true today. Whether or not effective oversight of the upper atmosphere will be maintained, either by Congress or the Executive agencies, is uncertain at this time.

During this same period 1977 to 1978, international activities received increased attention, with the CIEA Subcommittee on Stratospheric Pollution playing a significant role in the formulation and pursuit of U.S. international policy on stratospheric protection.*

The fall and winter of 1976 to 1977 was occupied with preparations for the spring 1977 UNEP meeting to be held in Washington, co-hosted by the State Department and NOAA. A special interagency task group headed by those two agencies was appointed by the CIEA Subcommittee to assist UNEP with arrangements and develop the U.S. position for the meeting. The frustrations of working with a cumbersome international organization that has extremely limited personnel resources soon became apparent. The task group had to do much of the work itself, as well as send representatives to Nairobi in the fall and spring to augment the UNEP staff preparations.

The UNEP meeting was attended by scientific experts designated by 32 UNEP countries and representatives of WHO, WMO, FAO, the International Civil Aviation Organization (ICAO), the U.N. Department of Economic and Social Affairs (UNESA), UNESCO, the European Economic Community (EEC), the International Council of Scientific Unions/Scientific Committee on Problems of the Environment (ICSU/SCOPE), and the International Chamber of Commerce (ICC). Papers were presented on the risks to ozone and possible effects of ozone reduction, and reports were made

* More recently, the CIEA Subcommittee became inactive.

of ongoing and planned research activities of governmental groups and other organizations.

The recommendations from the meeting took the form of a "World Plan of Action on the Ozone Layer" for coordination of international research.[99] The plan was subsequently adopted by the UNEP Governing Council. A Coordinating Committee on the Ozone Layer was formed, composed of representatives of international agencies, nongovernmental organizations, and countries participating in the Action Plan. Annual reports are prepared of relevant research underway.[100,108]

This UNEP meeting preceded the April "International Meeting on Regulation of Chlorofluorocarbons" postponed from the previous September. Some public misunderstanding occurred about the purposes and accomplishments of the two meetings. Even Walter Sullivan, Science Editor of *The New York Times* and one of the best informed and persistent members of the press in covering ozone news, wrote a misleading article describing the UNEP meeting as a failure for not endorsing regulatory action.[101] He also mistakenly reported that the EPA-sponsored meeting might be cancelled as a result. The UNEP meeting was mainly educational and was never intended to be a policy forum or to result in any agreement on the need for regulation. Participants were scientific specialists for the most part, not those who would make regulatory decisions for their countries. UNEP management seemed somewhat confused about what role UNEP could or should play, in view of its limited financial resources and staff expertise. The effectiveness of the very general World Plan of Action beyond that of encouragement and communication was considered minimal initially, although the UNEP role has recently taken on a greater importance (see Epilogue). In any case, the UNEP meeting was a prelude to the April regulatory meeting only in the sense of providing a scientific summary.

The April 26 to 28, 1977, meeting was chaired by EPA, with CPSC and FDA serving as co-hosts. It was an informal, closed meeting among government policymaking and regulatory officials from 13 major fluorocarbon producing and consuming nations, plus representatives from EEC, OECD, UNEP, WMO, and WHO.[102] The objectives were the following:

1. To establish a preliminary dialogue on whether/how/when fluorocarbon emissions might be controlled or otherwise reduced
2. To discuss a continuing mechanism for information exchange on what countries are planning to do and for assessment of the socio-economic effects of actions
3. To present the detailed plans for proposed U.S. and Canadian regulatory actions and obtain viewpoints of other countries that might be useful to the U.S. and Canada while finalizing those regulations
4. To lay the initial groundwork for continuing discussions on international cooperation on stratospheric protection and possibly on similar global environmental matters

The first three objectives were achieved. The fourth was only touched upon lightly at this meeting but was seen as a topic for a second meeting, later scheduled for December 4 to 6, 1978, in Germany.

At the April 1977 meeting the three U.S. regulatory agencies announced the specifics of their proposed regulations, which were not publicly released for another 2 weeks, and also presented the findings of the EPA-sponsored study on the expected economic impacts of regulation.[103] Canada also presented its proposed actions, which called for an agreement with the aerosol industry to cut back 50% of its fluorocarbon usage by the end of 1977 and regulations to end nonessential propellant uses by the end of

1978.* Other countries more briefly explained what they were planning to do at present or in the future, or why they were not yet prepared to choose a course of action. Although there was strong agreement on the basic scientific information, there were considerable differences among countries in regard to environmental pollution control philosophies, legislative and administrative mechanisms, characteristics of the affected industries and of the economic systems of the countries, uses of fluorocarbons and views on the essentiality of various uses, composition of products, and perceived feasibility of potential substitutes. Clearly, no single analysis of possible control measures and their effects could be applicable to all countries. All countries did agree that although individual nations would be making their own decisions about how, whether, and when to act to reduce fluorocarbon emissions, international cooperation and coordination is essential in regard to ozone, a global resource.

The December 1978 meeting hosted by the Federal Republic of Germany in Munich was a disappointment to U.S. participants. Only a few countries had taken measures to reduce fluorocarbon emissions in addition to the U.S. and Canada. Sweden had enacted legislation to prohibit most fluorocarbon propellant uses after June 30, 1979. Norway had announced its intention to regulate but had not yet done so.** The Netherlands had a labeling law effective February 1979. The Federal Republic of Germany was working with industry to achieve voluntary cutbacks and supported agreements among nations to achieve a set percentage reduction in emissions. An August 1977 draft position of the Commission of European Communities had endorsed cooperative studies, encouraged industry to develop alternatives and eliminate unnecessary fluorocarbon discharges, and instructed that measures be taken not to increase fluorocarbon production capacity. That policy had not been formally adopted by the European Communities, however. The Munich conference recommended that a significant reduction in the global release of chlorofluorocarbons be achieved over the next few years, as a precautionary measure; however, attempts to reach agreement at this meeting on some uniform level of cutback were unsuccessful.*** In summary, therefore, by the end of 1978 the international problem still remained because the major European producers (and Japan and the U.S.S.R.) were not yet prepared to take regulatory action or otherwise implement effective controls. Without such controls, increased fluorocarbon use abroad could more than offset the reductions in emissions achieved through U.S. and Canadian action.

VI. EPILOGUE

Since the end of 1978, when the bulk of this account was prepared, the major developments have been the appearance of numerous voluntary or regulatory measures taken in other nations to begin controlling fluorocarbon emissions, some progress toward international cooperation, and the announcement by the U.S. government of intent to propose further regulations to limit growth in nonaerosol uses.

In addition to regulatory actions by the U.S., Canada, Sweden, and Norway to reduce or eliminate aerosol propellant uses, many other nations have achieved reductions in such uses through voluntary agreement with industry or by a decline that resulted from a combination of economic conditions, changed consumer preference (e.g., hair styles not requiring hair spray), and concern about the ozone layer.

* The Canadian regulations, which took effect on May 1, 1980, banned the use of any chlorofluorocarbon in aerosol hair sprays, deodorants, and antiperspirants.

** The Norwegian regulation which prohibits most fluorocarbon propellant use takes effect July 1981.

***Limited cutbacks related to aerosol use were agreed to in 1979.

Some actions have now been taken to control all uses of CFC's, not just aerosol propellants. A Decision by the EEC Council, adopted in March, 1980, required each of its nine (now ten) member states to ensure that its industries do not increase production capacity for CFC-11 and CFC-12 and to achieve on or before December 31, 1981, a reduction of at least 30% in the use of these CFC's in aerosols compared with 1976 levels. Several of the member nations will achieve reductions of 40 to 50% or possibly more, although much of these decreases will be offset by increases in nonaerosol uses. A reexamination of scientific and economic data was also undertaken. The March 1980 decision called for the Council to adopt, by June 1981, such further measures as may be necessary in the light of that reexamination. At the December 1980 meeting of the OECD Environment Committee, Japan announced its intent to prevent any increase in production capacity for CFC-11 and CFC-12. Canada and several other nations are reported to be actively considering further voluntary or regulatory measures to control all or some fluorocarbon uses.

Another conference on international regulation of fluorocarbon emissions was held in Oslo, April 14-16, 1980.* Participants agreed that a "wait and see" policy was unacceptable and that all major producer and user nations should take adequate steps to reduce emissions from both aerosol and nonaerosol uses of chlorofluorocarbons. At the meeting, the U.S. government announced its intention to propose limiting production to present levels. This announcement, and one from Canada to investigate ways of achieving the same objective, were welcomed by other nations, who agreed to consider similar actions. The possible desirability of an international convention for the protection of the ozone layer was also considered.

The UNEP Coordinating Committee on the Ozone Layer (CCOL) is performing a useful function in providing a truly international forum for reaching agreement on the results of research undertaken by governments, industry, and international organizations. This role has been formally recognized by OECD, EEC, and participants at the various international conferences. In April 1980 a decision of the UNEP Governing Council recommended that member governments achieve significant reductions in fluorocarbon use, not increase production capacity, pursue further research, and reexamine the adequacy of control measures already taken in the light of available data. It also requested recommendations for means of international cooperation and a special assessment report. The report, which is underway but stalled at the present time, will be the first global environmental assessment to be prepared under UNEP's EARTHWATCH program.

In February 1981 Sweden surfaced a proposal to begin developing an international convention to protect stratospheric ozone, under the auspices of UNEP. A resolution supporting the general idea was adopted at the May 1981 meeting of the UNEP Governing Council, despite some skepticism about the capabilities of UNEP and a variety of views on the proper content, timing, and administrative mechanisms for such an activity. Even the reasons for supporting a convention are varied, with some believing that it is the most effective way to obtain further control actions at this time, and some seeing it as an excuse to delay further actions by the European Economic Community and other producing nations. Others see it as the inevitiable long-term solution to the problem of "runaway plants" relocating from nations with regulations in place (mostly industrialized nations) to nations without such regulations (mostly developing nations).

In 1980 OECD undertook its own comprehensive report on the scientific, economic, technological, and legal aspects of the fluorocarbon problem. The scientific assessment

* Attendees were representatives of Canada, Denmark, the Federal Republic of Germany, the Netherlands, Norway, Sweden, U.S., and the Commission of the European Communities.

draws heavily on the UNEP/CCOL meetings and reports of 1979 and 1980, but OECD is contributing original work in the area of economic analysis and development of international scenarios of what might happen to the ozone layer under different conditions of emissions and controls. A report is expected in the fall of 1981.

A major concern of the U.S. government, as expressed in its statements at international meetings and in its announced domestic regulatory plans, is the need to limit growth in fluorocarbon use while continuing to evaluate scientific and technological information. On October 7, 1980, EPA issued an Advance Notice of Proposed Rulemaking,....which is an early announcement of possible regulatory options for limiting nonaerosol uses of chlorofluorocarbons, and which solicits comments from interested parties.[109] The Advance Notice outlined options ranging from no action at this time to reductions in use and recommended an intermediate "no-growth" strategy. The preferred option for accomplishing this was a marketable permits system that would restrict production of all CFC's such that the total potential for depleting ozone would not increase over present levels. The reasons given for adoption of a "no growth strategy" include (1) continued concern about the probability of significant ozone depletion and its serious effects, (2) the expectation that actual ozone depletion will not be able to be detected and attributed to fluorocarbon emissions for a decade or more, and (3) the recognition that continued growth in nonaerosol uses in the meantime would result in greater ozone depletion that would continue even after emissions were eventually drastically reduced, and in greater social and economic disruption as a result of the later actions to reduce emissions. Data prepared for the OECD show that nonaerosol uses of CFC's have been growing at a rate of about 7% per year worldwide for CFC-11 and -12, with an even higher growth rate for some uses, for some countries, and for some other halocarbons.[110]

Over 2,000 public comments were received in response to the Advance Notice of Proposed Rulemaking. Several factors led EPA to relax its original timetable of issuing a proposed rule in the spring of 1981: (1) complex economic and administrative considerations associated with a production cap, which were raised in the comments and by EPA's own internal studied; (2) scientific developments, including the new reaction rates that resulted in calculations of potential ozone depletion that were considerably lower than those upon which the Advance Notice was based, and promise of continued improvements in statistical analysis of actual ozone trends; and (3) need for complete review of the policy issues by the new Administration.

Although the timing and content of any proposed rule has not been determined, it is clear that domestic and international considerations are now interrelated. The new ingredient this time is the improved climate of international scientific and political cooperation and understanding, which will move some of the controversy into the broader arena of international policy. Whether or not the "success" of early Federal policymaking on fluorocarbons will be repeated for nonaerosol uses is unknown at this time, but it is likely that the stratosphere will be a permanent concern in future public policy. It was an extremely interesting experience to have participated from the early days in the consideration of this issue for which a new type of decision process had to be developed in a short period of time. Some observers have praised the early fluorocarbon experience as a model for Federal decisionmaking. The problem was not a typical one, however, and consideration of it was extremely resource intensive.

With tens of thousands of chemicals in commerce and several hundred introduced annually, the universe of possible problems is too vast to address completely in an *ad hoc* manner. Even though a small fraction of chemicals is known to have significant adverse human health and environmental effects, the chore of screening chemicals for effects is immense. The Federal government is now grappling with how to plan and coordinate the vast array of research, information, and regulatory programs relating to chemicals and their effects on man and the environment. Attention is being given to establishing priorities for regulatory actions, combining agency efforts and resources more effectively, and improving the quality of the dialogue between Federal agencies and non-Federal groups.

There are over two dozen Federal statutes relating to hazardous substances control, administered primarily by six regulatory agencies. A dozen other Federal agencies have major research or policy responsibilities in the toxic chemicals field. More than 32 Federal agencies are involved in the compilation and use of data on chemicals and their effects.

Several major interagency efforts have been operating to establish a more coherent Federal program, e.g., the Interagency Regulatory Liaison Group, the Toxic Substances Strategy Committee, the Interagency Toxic Substances Data Committee, the TSCA Interagency Testing Committee, the DHEW Committee to Coordinate Environmental and Related Programs, the National Toxicology Program, and others. It is too early to know for certain whether or not they will succeed, but there is an unusual amount of attention being given hazardous substances, at high levels of government.

All of these activities have one theme in common: they are pinning their hopes on *preventive* strategies. It is not possible to give many chemicals the attention that has been given to chlorofluorocarbons. The objective is to establish the data gathering, monitoring, and assessment capabilities needed to review new chemicals before marketing, and to control existing ones found to pose significant hazards. Once that catching up is accomplished, intensive efforts such as those expended on fluorocarbons should not be necessary except in cases where the state of the art does not permit prediction of the adverse effects prior to widespread use. (Even if a comprehensive preventive toxics strategy had been in effect in the early 1930s when fluorocarbons were first manufactured, knowledge of the stratosphere at that time was too limited for anyone to have considered possible ozone reduction effects.)

It is difficult to be very optimistic about how soon the catching up might be completed. The Interagency Toxic Substances Data Committee estimated as much as 10 years for full implementation of a Federal-wide Chemical Substances Information Network.[104] The first policies and regulations for carrying out the data gathering, testing, and regulatory requirements of the Toxic Substances Control Act have been issued recently, and others have yet to be developed. The Interagency Regulatory Liaison Group (composed of EPA, FDA, CPSC, the Occupational Safety and Health Administration of the Department of Labor and the Food Safety and Quality Service of USDA) is working on the regulation of chemicals of interest to two or more of its members. This interagency group grew out of the cooperation of three of the members on the fluorocarbon issue. Their cooperation is laudable, but very resource-intensive until more comprehensive strategies are developed for screening chemicals and setting research and regulatory priorities.

It should be possible for mankind to enjoy the immense social, economic, and public health benefits that chemicals offer while at the same time avoiding many of the risks associated with their production, use, and disposal. To do so, however, will require willingness, a dedicated effort, and unprecedented cooperation on the part of businesses and industries, labor, consumers, scientists, and government agencies. The complexities encountered during these early years of concern about stratospheric protection indicate the many challenges ahead as the risks and benefits associated with the use of other chemicals are weighed.

REFERENCES

1. Interagency Task Force on Inadvertent Modification of the Stratosphere (IMOS), Transcript of Proceedings, Thursday, 27 February 1975, Council on Environmental Quality and Federal Council for Science and Technology, Washington, D.C., 1975.
2. Johnson, K. L., Managing the Toxic Substances Control Act: An Examination of the Environmental Protection Agency's Performance in Administering TSCA, August 1978 draft report.
3. Crutzen, P. J., The influence of nitrogen oxides on the atmospheric ozone content, *Q. J. R. Meteorol. Soc.*, 96, 320, 1970.
4. Crutzen, P. J., SST's — threat to the Earth's ozone shield? *Ambio*, 1, 41, 1972.
5. Johnston, H. S., Reduction of stratospheric ozone by nitrogen oxide catalysts from SST exhaust, *Science*, 173, 517, 1971.
6. Grobecker, A. J., Coroniti, S. C., and Cannon, R. H., Jr., The Effects of Stratospheric Pollution by Aircraft, Report DOT-TST-75-50, U.S. Department of Transportation, Washington, D.C.
7. National Research Council, Climatic Impact Committee, Environmental Impact of Stratospheric Flight: Biological and Climatic Effects of Aircraft Emissions in the Stratosphere, National Academy of Sciences, Washington, D.C., 1975.
8. COMESA, The Report of the Committee on the Meteorological Effects of Stratospheric Aircraft (COMESA), 1972-1975, U. K. Meteorological Office, Bracknell, England, 1975.
9. Comité d'Etudes sur les Conséquences des Vols Stratospheriques (COVOS), Activites 1972-1976, Societe Meteorologique de France, Boulogne, 1976.
10. Lovelock, J. E., Atmospheric fluorine compounds as indicators of air movements, *Nature (London)*, 230, 379, 1971.
11. Lovelock, J. E., Atmospheric halocarbons and stratospheric ozone, *Nature (London)*, 252, 292, 1974.
12. Chemical Manufacturers Association, Summary Research Program on Effect of Chlorofluorocarbons on the Atmosphere, Original September 26, 1975, Revision #15, June 15, 1981.
13. Molina, M. J. and Rowland, F. S., Stratospheric sink for chlorofluoromethanes — chlorine atom-catalysed destruction of ozone, *Nature (London)*, 249, 810, 1974.
14. Cicerone, R. J., Stolarski, R. S., and Walter, S., Stratospheric ozone destruction by man-made chlorofluoromethanes, *Science*, 185, 1165, 1974.
15. Crutzen, P. J., Estimates of possible future ozone reductions from continued use of fluoro-chloromethanes CF_2Cl_2, $CFCl_3$, *Geophys. Res. Lett.*, 1, 205, 1974.
16. Wofsy, S. C., McElroy, M. B., and Sze, N. D., Freon consumption: implications for atmospheric ozone, *Science*, 187, 535, 1975.
17. Not with a bang, but with a pssssst! *New Times*, March 7, 1975, 27.
18. Editorial, *The New York Times*, November 12, 1974.
19. Skeen, I. L., Letter to the Editor, *Barron's*, July 21, 1975.
20. Spray can gas feared life time bomb, *The Ann Arbor News*, August 11, 1974.
21. Salisbury, D. F., Not with a bang but a psst! Do aerosol cans spell environmental doomsday? *The Christian Science Monitor*, September 24, 1974.
22. Not with a bang but a pffft? *The New York Times*, December 21, 1975.
23. Fluorocarbons — Impact on Health and Environment, Hearings Before the Subcommittee on Public Health and Environment of the Committee on Interstate and Foreign Commerce, House of Representatives, December 11 and 12, 1974, U.S. Government Printing Office, Washington, D.C., 1975.
24. Dotto, L. and Schiff, H., *The Ozone War*, Doubleday & Company, Inc., Garden City, New York, 1978, chap. 8.
25. National Research Council, press release, H. S. Gutowsky named chairman of stratospheric pollutants study, March 19, 1975.
26. Panel to assess ozone problem, *The New York Times*, February 19, 1975.
27. Lovelock, J. E., as quoted by Singer, A., Some scientists are worried about certain gases used in aerosols, *Yorkshire Post (Leeds, England)*, December 18, 1974.
28. Sullivan, W., Ozone threat seen potential weapon, *The New York Times*, February 28, 1975.
29. Hammond, A. L., Public credibility on ozone, *Science*, 187, 1182, 1975.
30. Interdepartmental Committee for Atmospheric Sciences (ICAS), Federal Council for Science and Technology, The Possible Impact of Fluorocarbons and Halocarbons on Ozone, U.S. Government Printing Office, Washington, D.C., 1975.
31. Interagency Task Force on Inadvertent Modification of the Stratosphere, Council on Environmental Quality and Federal Council for Science and Technology, Fluorocarbons and the Environment, U.S. Government Printing Office, Washington, D.C., 1975.
32. Ramanathan, V., Greenhouse effect due to chlorofluorocarbons: climatic implications, *Science*, 190, 50, 1975.

33. U.S. Environmental Protection Agency, Office of Air and Waste Management, Office of Air Quality Planning and Standards, prepared for EPA by Arthur D. Little, Inc., Preliminary Economic Impact Assessment of Possible Regulatory Action To Control Atmospheric Emissions of Selected Halocarbons, U.S. Environmental Protection Agency, Research Triangle Park, North Carolina, 1975.
34. Cosmetic industry seeks alternatives to aerosols, Lansing, Michigan *State J.*, June 24, 1975.
35. **Slom, S. H.**, Aerosol industry is facing the fallout as consumers fear hazard from sprays, *The Wall Street Journal*, May 2, 1975.
36. **Waltz, T.**, The Chlorofluorocarbon Issue: Social Decisionmaking Under Uncertainty, U.S. Department of Commerce, NOAA, Associate Administrator for Environmental Monitoring and Prediction, Program Integration Office, August 1975.
37. **Johnsen, M. A.**, Business Strategies of the Future, 10th International Aerosol Congress, London, England, September 1975. (See report in *Aerosol Age*, 20(11), 18, November 1975).
38. U.S. Department of Commerce, Bureau of Domestic Commerce, Domestic and International Business Administration, Further Information on the Fluorocarbon Industry and on Potential Economic Impacts of Restriction of Fluorocarbon Production, June 1975 (a separate insert included with Reference 31, also reprinted in Reference 43).
39. U.S. Department of Commerce, Bureau of Domestic Commerce Staff Study, Economic Significance of Fluorocarbons, U.S. Department of Commerce, Washington, D.C., 1975.
40. Public Affairs Department, untitled press release, E. I. du Pont de Nemours & Co., Wilmington, DE, June 12, 1975.
41. **Bantham, R. A.**, Look up and live. (The ozone is still there), *The New York Times*, August 28, 1975. (The newspaper identified Bantham as legal counsel to a major aerosol marketing company; the article was adapted from a speech he made to a trade association.)
42. **Salisbury, D.**, Aerosol spray debate spawns compromise, *The Christian Science Monitor*, June 26, 1975.
43. Inadvertent Modification of the Upper Atmosphere: Research and Development Relating to Halocarbons and Ozone Depletion, Hearings Before the Subcommittee on the Environment and the Atmosphere of the Committee on Science and Technology, U. S. House of Representatives; May 20, 21, 22; July 31, 1975; U.S. Government Printing Office, Washington, D.C., 1975.
44. Gambling with the ozone, *The Washington Star*, June 23, 1975.
45. **Alexander, T.**, What we know — and don't know — about the ozone shield, *Fortune*, August 1975.
46. Kaiser Aluminum & Chemical Corporation, At Issue: Fluorocarbons, Kaiser, Oakland, California, 1975.
47. **Ward, R. B., Ed.**, Science Update, Fluorocarbon/Ozone, Freon Products Laboratory, E. I. Du Pont de Nemours & Company, Wilmington, DE, January 1976 and intermittently thereafter.
48. **Alexander, G.**, Industry group launches defense of fluorocarbons, *Los Angeles Times*, July 28, 1975.
49. **Dotto, L. and Schiff, H.**, *The Ozone War*, Doubleday & Company, Inc., Garden City, New York, 1978, chap. 7.
50. **Greenhouse, S.**, Aerosol feels the ozone effect, *The New York Times*, June 22, 1975.
51. Aerosol can users convening, Michigan City (Indiana) *News Dispatch*, June 24, 1975.
52. Tiny pumps thrive on aerosol's woes, *Bus. Week*, August 25, 1975.
53. An open letter to consumers about aerosols, from Samuel C. Johnson, Chairman, Johnson Wax, Racine, Wisconsin, advertisement in *The Washington Post* (and other newspapers), June 20, 1975.
54. 40 FR 29914 (July 16, 1975).
55. U.S. Environmental Protection Agency, Registration Division, Notice to Producers, Formulators, Distributors, and Registrants of Pesticides, Pesticide Registration (PR) Notice 75-6, December 4, 1975.
56. U.S. Environmental Protection Agency, Registration Division, PR Notice 76-3, October 18, 1976.
57. 42 FR 21807, April 29, 1977; 42 FR 42780, August 24, 1977.
58. 41 FR 52071, November 26, 1976; 42 FR 22017, April 29, 1977.
59. **Hunten, D. P., Ed.**, The Stratosphere 1975-1980, Report of a Workshop May 28-30, 1975, Goddard Space Flight Center, Greenbelt, Maryland, 1975.
60. National Aeronautics and Space Administration, Office of Space Science, Executive Summary, Upper Atmospheric Research Program, December 1975.
61. National Academy of Sciences, Climatic Impact Committee, Interim Report of the Panel on Atmospheric Chemistry, National Academy of Sciences, Washington, D.C., 1975.
62. Stratospheric Ozone Depletion, Hearings Before the Subcommittee on the Upper Atmosphere of the Committee on Aeronautical and Space Sciences, U.S. Senate; Part 1 (September 8, 9, 15, and 17, 1975) and Part 2 (September 18, 19, and 23, 1975), U.S. Government Printing Office, Washington, D.C., 1975.

63. Interagency Task Force on Inadvertent Modification of the Stratosphere (IMOS), Council on Environmental Quality and Federal Council for Science and Technology, Report by the Interdepartmental Committee for Atmospheric Sciences: Potential Modifiers of the Stratosphere Other Than Fluorocarbons, Washington, D.C., 1976.
64. IMOS Subcommittee on Biological and Climatic Effects Research, Federal Council for Science and Technology, A Proposed Federal Research Program To Determine the Biological and Climatic Effects of Stratospheric Ozone Depletion, unpublished draft, February 1976. (Some parts are excerpted in Reference 68.)
65. Chlorofluorocarbon Effects and Regulations, Hearing Before the Subcommittee on the Upper Atmosphere of the Committee on Aeronautical and Space Sciences, U.S. Senate, December 15, 1976, U.S. Government Printing Office, Washington, D.C., 1977.
66. National Research Council, Assembly of Mathematical and Physical Sciences, Committee on Impacts of Stratospheric Change, Response to the Ozone Protection Sections of the Clean Air Act Amendments of 1977: An Interim Report, National Academy of Sciences, Washington, D.C., 1977.
67. **Large, A. J.**, Fluorocarbon curbs U.S. will phase in deemed insufficient to halt ozone harm, *The Wall Street Journal,* September 25, 1978.
68. Stratospheric Ozone Research and Effects, Hearings Before the Subcommittee on the Upper Atmosphere of the Committee on Aeronautical and Space Sciences, U.S. Senate, February 25 and March 1, 1976, U.S. Government Printing Office, Washington, D.C., 1976.
69. **Stoel, T. B., Jr., Compton, R. I., and Gibbons, S. M.**, International regulation of chlorofluoromethanes, *Environmental Policy and Law,* 3, 129, 1977.
70. Organization for Economic Cooperation and Development, Fluorocarbons: An Assessment of Worldwide Production, Use and Environmental Issues, First Interim Report, Paris, France, 1976.
71. Organization for Economic Cooperation and Development, Environment Directorate, Work Programme on Fluorocarbons, Paris, France, June 16, 1976.
72. Organization for Economic Cooperation and Development, Environment Committee, Chemicals Group, The Economic Impact of Restrictions on the Use of Fluorocarbons, Final Report, ENV/Chem/77.2 (Rev.), Paris, 1977.
73. Train urges international control of fluorocarbon releases to atmosphere, *Environment Reporter,* 6, 1079, 1975.
74. Department of the Environment, Central Unit on Environmental Pollution (U.K.), Pollution Paper No. 5, Chlorofluorocarbons and Their Effect on Stratospheric Ozone, Her Majesty's Stationery Office, London, 1976.
75. **Peterson, R.**, Memorandum to the President, March 2, 1976.
76. **Tannenbaum, J. A.**, Theory that aerosols deplete ozone shield is attracting support, *The Wall Street Journal,* December 3, 1975.
77. National Research Council, Assembly of Mathematical and Physical Sciences, Committee on Impacts of Stratospheric Change, Environmental Effects of Chlorofluoromethane Release, National Academy of Sciences, Washington, D.C., 1976, p. 11.
78. **Tannenbaum, J. A.**, Defenders of fluorocarbons are set back as research center concedes test error, *The Wall Street Journal,* June 30, 1976.
79. Is the 'threat' of aerosols going pfffft? *The National Observer,* May 18, 1976.
80. **Sullivan, W.**, Experts now see risk to ozone from fluorocarbons lessened, *The New York Times,* May 7, 1976.
81. **Rowland, F. S. and Molina, M.**, letter to Carroll Leslie Pegler, May 20, 1976.
82. **Shurkin, J. N.**, Aerosols hurt ozone, panel finds, *The Philadelphia Inquirer,* March 19, 1976.
83. National Academy of Sciences, Statement of the National Academy of Sciences, March 19, 1976. This was followed by a Memorandum to News Media from Howard J. Lewis, Director, Office of Information, April 5, 1976.
84. **Rowland, F. S.**, Letter to Carroll Leslie Bastian, October 15, 1976.
85. **Jesson, P.**, Testimony rebutting letter to Carroll Leslie Bastian, October 15, 1976 concerning Chlorofluorocarbon Effects and Regulations, Hearing Before the SubCommittee on the Upper Atmosphere of the Committee on Aeronautical and Space Sciences, U.S. Senate, December 15, 1976, U.S. Government Printing Office, Washington, D.C., 1977, 186.
86. **Maugh, X.**, The ozone layer: The threat from aerosol cans is real, *Science,* 194, 170, 1976.
87. National Research Council, Assembly of Mathematical and Physical Sciences, Panel on Atmospheric Chemistry, Committee on Impacts of Stratospheric Change, Halocarbons: Effects on Stratospheric Ozone, National Academy of Sciences, Washington, D.C., 1976.
88. Council on Environmental Quality press release, Industry urged to voluntarily curb aerosol use, September 17, 1976.
89. **Bastian, C. L. and Muir, W.**, memorandum to Peterson, R. and Stever, H. G., September 24, 1976.
90. EPA Administrator calls for NATO nations to phase out nonessential uses of fluorocarbons, EPA *Environmental News,* October 12, 1976.

91. Demise of aerosol can is in sight as U.S. moves to phase out fluorocarbon sprays, *The Wall Street Journal,* October 18, 1976.
92. U.S. Environmental Protection Agency, Registration Division, PR Notice 76-3, October 18, 1976.
93. **Schmidt, A. M.** (Commissioner of Food and Drugs), letter to Russell Peterson, Chairman of CEQ, October 12, 1976.
94. **Hudson, R. D., Ed.,** Chlorofluoromethanes and the Stratosphere, NASA, Goddard Space Flight Center, Reference Publication 1010, Washington, D.C., 1977.
95. NASA, Assessment Report: Effects of Chlorofluoromethanes on Stratospheric Ozone, 1977.
96. U.S. orders phaseout of almost all fluorocarbon aerosol sprays within two years, *The Washington Post,* May 12, 1977.
97. 42 FR 24535 (May 13, 1977).
98. 43 FR 11300 (March 17, 1978).
99. United Nations Environment Program, Report on the Meeting, UNEP Meeting of Experts Designated by Governments, Intergovernmental and Non-Governmental Organizations on the Ozone Layer, Washington, D.C., March 1-9, 1977, UNEP/WG.7/25/Rev. 1, 8 March 1977.
100. United Nations Environment Program, UNEP Ozone Layer Bulletin, Issue No. 1, UNEP, Nairobi, January 1978.
101. **Sullivan, W.,** Other nations give little support to the U.S. on regulating fluorocarbons to protect ozone, *The New York Times,* March 10, 1977.
102. Environmental Protection Agency, Food and Drug Administration, and Consumer Product Safety Commission, International Meeting on Regulation of Chlorofluorocarbons, Washington, D.C., April 26-28, 1977.
103. International Research and Technology Corporation, The Economic Impact of Potential Regulation of Chlorofluorocarbon-Propelled Aerosols, Prepared for the Environmental Protection Agency, Food and Drug Administration, and Consumer Product Safety Commission under EPA Contract No. 68-01-1918, 1977.
104. Council on Environmental Quality, The Feasibility of a Standard Chemical Classification System and a Standard Chemical Substances Information System, A report of the Congress prepared pursuant to Section 25(b) of the Toxic Substances Control Act of 1976 (15 U.S.C. 2601), July 1978, 32.
105. National Research Council, Assembly of Mathematical and Physical Sciences, Committee on Impacts of Stratospheric Change, Stratospheric Ozone Depletion by Halocarbons: Chemistry and Transport, National Academy of Sciences, Washington, DC, 1979.
106. National Research Council, Assembly of Mathematical and Physical Sciences, Committee on Impacts of Stratospheric Change, and Commission on Socio-technical Systems, Committee on Alternatives for the Reduction of Chlorofluorocarbon Emissions, Protection Against Depletion of Stratospheric Ozone by Chlorofluorocarbons, National Academy of Sciences, Washington, DC, 1979.
107. Department of the Environment, Central Unit on Environmental Pollution (U.K.), Pollution Paper No. 15, Chlorofluorocarbons and Their Effect on Stratospheric Ozone (Second Report), Her Majesty's Stationery Office, London, 1979.
108. United Nations Environment Program, Ozone Layer Bulletin, Issue No. 2, UNEP, Nairobi, July 1978; Ozone Layer, Volume II, No. 1, January 1979; The Ozone Layer, Volume II, No. 2, July 1979; Environmental Assessment of the Ozone Layer (Depletion and Its Impact) as of November 1979, Bulletin No. 5, January 1980; Report of Fourth Session, Bilthoven, November 11-14, 1980 (unpublished).
109. 45 FR 66726 (October 7, 1980).
110. Organization for Economic Cooperation and Development, Environment Committee, draft report on Chlorofluorocarbons, ENV (80) 32 (1st Revision) Paris, March 6, 1981.

Chapter 9

STRATOSPHERIC OZONE, THE SCIENTIFIC COMMUNITY AND PUBLIC POLICY

Harvey Brooks

TABLE OF CONTENTS

I. BACKGROUND

The period since the middle 1960s has been characterized by a rapid growth of scientific and technical issues in the content of political debate and decision-making, and the consequent growth of new roles for scientists in the political process and in interaction with politicians.[1] To a large extent this development has gone hand in hand with the growth of the environmental movement, and the pervasiveness of environmental, health, and safety issues in the agenda of politics. As this development has evolved, the sophistication and complexities of the questions that have been addressed have increased remarkably, and nowhere is this more clearly seen than in the debates and regulatory actions relating to stratospheric ozone. A subject which until the late 1960s was a rather arcane corner of the atmospheric sciences, of interest to a handful of specialists, suddenly burst upon the political scene and became the subject of voluminous Congressional hearings and government reports.

Such a development was not entirely new or unprecedented. For a brief period after World War II nuclear physicists stepped onto the political stage and played their part in creating the institutions to manage the problems engendered by the discovery and application of nuclear energy. Later an influential segment of the physics community remained active in the military-technological revolution engendered by the Cold War. But until the end of the 1960s scientific involvement in politics was largely limited to nuclear energy and national security policy,[2] and was concerned more with means than with ends.[3]

In 1963 Rachel Carson's *Silent Spring*[4] suddenly raised the public consciousness about environmental questions, and a new era of scientific involvement was launched. There arose a small group of publicly visible scientists who popularized environmental issues and turned ecology into a household word, converting a rather obscure science into what was virtually a political ideology, swimming against the euphoria about science and technology which had dominated public attitudes since the dramatic technological achievements of World War II.[5] In fact this movement gradually crystallized into a philosophy of "limits to growth" which became fashionable among literary intellectuals following the publication of the works of J. W. Forrester, D. H. Meadows et al. and the Club of Rome in 1971 to 1973.[6-8] In much of this, science and ideology were inextricably mixed, and indeed one of the contentions of the new movement was that there was no such thing as "objective" science, and that all science and technology embodied hidden presuppositions, and essentially reactionary political goals.[9,10] Technological progress, which had previously been invoked by social revolutionaries as the instrument for the liberation of man, now was denounced as the instrument of his enslavement and the destruction of the life support systems which must sustain him in the future.

Some of this new perspective, minus the more extreme ideological overtones, was embodied in legislation creating new regulatory agencies and new administrative procedures to control the impact of technology and protect the environment defined in an ever more inclusive sense. One of the most far reaching of these new laws was the National Environmental Policy Act (NEPA) of 1969. Its famous section 102, little noticed at the time, provided that each federal agency, before taking any action "significantly affecting the quality of the environment," must prepare an environmental impact statement (EIS), a public document to be submitted to extensive public criticism and review by other agencies.[11] The EIS procedure provided a vehicle for bringing scientists heavily into the decision-making process, and for attempting to weigh scientific and technical evidence in connection with a host of federal projects and industrial operations potentially subject to regulation. The Courts saw to it that "federal action" was interpreted very broadly to include the licensing or permitting of private projects.[12]

The new environmental priorities were also reflected in the research agenda of science. The environmental sciences attracted strong federal support, and everywhere "defensive research"[1,13] prospered. The term is used to denote research aimed at anticipating possible adverse effects of technology before they become highly visible, as opposed to "offensive" research whose primary motivation is to develop the basis of new products or services to meet new human needs, or old needs more cheaply and efficiently. At the same time a revolution in the sensitivity of measuring techniques made it possible to investigate ever more subtle effects and to detect ever smaller traces of pollutants. Furthermore, the development of statistical techniques made it easier to detect small adverse health effects in large populations.

As *Silent Spring* launched the environmental movement in 1963, the supersonic transport (SST) debate triggered concern about stratospheric ozone in 1967, and resulted in a new lease on life for upper atmospheric physics, hitherto one of the more "pure" aspects of the otherwise rather practical atmospheric sciences. Yet, in fact, the ozone problem emerged as an afterthought, just as the SST project was all but dead as a result of other considerations. The SST debate was carried on at two different levels. At one level were genuine problems of the sonic boom, of excessive fuel consumption, and of increasingly unfavorable economics, especially after the sonic boom concern dictated prohibition of supersonic flights over land. At another level, the SST became a highly visible symbol of complex, expensive, and government-financed high technology which was seen as serving the interests of the few at the expense of the many. The political steam for the fight against the SST seems to have derived as much from these ideological grounds as from the quite genuine technical, economic, and environmental difficulties associated with the technology. Near the close of the Senate debate, Professor Harold Johnston (University of California, Berkeley) suggested the possible effects of NO_x from SST exhausts in depleting the ozone layer,[14,15] and, as testified to by Dr. T. M. Donahue (University of Michigan), the late James MacDonald had then pointed out the possible role of ozone depletion in increasing the incidence of skin cancer in the light-skinned part of the population.[16,17] It is difficult to estimate in what degree the appearance of the dread word "cancer" in last minute testimony influenced the final vote. If it did, it seems rather ironic that, 12 years later, the effect which triggered the demise of the American SST and a major research effort on stratospheric ozone, has now been stated to be negligible, or possibly even of reverse sign.[18] Yet the research program itself has been of great benefit. The program was a major factor in the discovery of several other possible mechanisms of ozone depletion, especially the chlorofluoromethanes (CFM). It also stimulated epidemiological research on the incidence of the malignant skin cancer, melanoma, and recognition of the likelihood that recent rapid growth of the incidence of this malignancy was due to changes in lifestyle among young professional and managerial workers.[19,20] Indeed one of the recommendations of the Academy report is:[19]

> "Since ultraviolet-induced skin cancer will continue to present a serious health hazard, we need to study possible preventive medicine actions carefully, without regard to the effectiveness of CFM's in reducing ozone or decisions about their regulation."

This could turn out to be one of the most important recommendations resulting from all the studies, although it is independent of ozone depletion. Admittedly the etiology of malignant melanoma is still in a very confused state, so that its association with changes in patterns of UV exposure cannot be asserted as confidently as in the case of nonmalignant skin cancer.[21]

At the same time the efforts to understand upper atmospheric chemistry have led to new insights into geochemical cycles and a whole new line of research in this domain.[22,23]

II. THE NATURE OF LONG-TERM ENVIRONMENTAL THREATS

The stratospheric ozone problem represents one of a new class of long-term environmental threats which present unique challenges to the political process. Among the common characteristics of these threats are the following:

1. They involve difficult trade-offs between current economic or social benefits and uncertain adverse effects which are far in the future, effects which may be negligible so far as anybody currently represented in the political system is concerned. In the case of the CFM's it is calculated that half the maximum ozone depletion is not reached for more than 50 years if emissions continue at current rates; even if emissions are stopped ozone depletion is estimated to continue to build up for more than a decade and to peak at 1.5 times the current value.[24]
2. The effects are distributed worldwide, so that control ultimately requires concerted international action. In the case of the CFM's the U.S. only accounted for about 50% of the world output in 1974, and its share has now declined to one third. In other examples, such as nuclear power and combustion of fossil fuels, the U.S. accounts for about 30% of the total effect worldwide.
3. At a time when, under the most pessimistic assumptions, a decision to regulate becomes urgent, the predicted effects are subject to very large uncertainty, and hence the potential benefits of regulation are equally uncertain. For example, in the CFM case, at the time of the report of the Federal Task Force on Inadvertent Modification of the Stratosphere (IMOS report) in 1976,[25] the range of uncertainty in ozone depletion estimates was at least a factor of 10 (from 2 to 20% at 95% confidence level for continuous worldwide release at the 1973 rate). Moreover, the uncertainty becomes progressively greater the closer one gets to the consideration of human implications. Most certain is the amount of CFM in the atmosphere. Less certain is the predicted amount of ozone depletion this will lead to. Even more uncertain is the effect of a given degree of ozone depletion on health, and least certain is the effect of either ozone depletion or the greenhouse effect from infrared absorption by CFM's in the troposphere on agricultural ecosystems and the biosphere generally.
4. The effects are cumulative and depend on the total past history of emissions. On the one hand they are reversible only slowly (in the CFM case it takes from 50 to 75 years for ozone depletion to drop from peak to half value after emissions cease), but on the other hand the extra adverse effect of delaying controls is only a tiny fraction of the cumulative effect of past history (delay from 1978 to 1980 in controlling aerosols reduces estimated peak ozone reduction by only 1/18 to 1/2%, which is about 1/40 the amount of reduction which would occur if the 1973 emission rate continued indefinitely or about 1/10 the maximum reduction that would occur if emissions stopped in 1978).[26]

There are several other environmental threats that may have these same characteristics. Methyl bromide fumigants, nitrogen fertilizers, as well as SST exhausts and nitrogen fixation in fossil fuel combustion, and atmospheric nuclear explosions[27] have been suggested as possible sources of stratospheric ozone depletion, although considerable doubt has been thrown recently on the nitrogen mechanisms.[18] Other sources of chlorine compounds such as the space shuttle and large scale chlorination in connection with paper making and water treatment have also been suggested. These cases are, of course, closely analogous to the CFM case, although their magnitude now appears less significant than CFM's.

Perhaps the first threat identified with similar characteristics was the testing of nuclear weapons in the atmosphere and the resulting fall-out of radioactive debris. The distribution of radioactivity was global, and its major effects cumulative, especially with regard to long-lived fission product species such as Sr^{90}, induced long-lived activity from C^{14} and tritium, and transuranic elements of high alpha activity. The fear of irreversible long-term environmental damage was probably decisive in mobilizing public opinion against continuation of atmospheric testing and securing Senate ratification of the Partial Test Ban treaty in 1962.

Closely analogous concerns have now arisen in connection with the ultimate disposal of high level (and indeed also actinide-containing low and intermediate level) radioactive wastes from the growing nuclear power industry,[28] and similar discussions have occurred in connection with the release of long-lived radioactive gases such as Kr^{85} into the atmosphere from the reprocessing of reactor fuel for a large-scale nuclear power industry.[29,30] Again the ultimate threat is worldwide and cumulative and mainly affects future generations. Mine tailings from uranium mining and milling also present a long-term threat through the release of radon from tailings which have not been properly covered. It has recently been suggested that a similar threat from alpha-active materials could arise from the disposal of coal ash from power plants.[31]

The increase of CO_2 in the atmosphere as a result of the burning of fossil fuels, particularly coal, and of deforestation, poses very similar problems. It is worldwide, cumulative, far in the future, and virtually irreversible by the time the effect is large enough to be of concern. Indeed in one aspect the delay times involved are much longer than for the CFM case, for the possible melting of the Greenland and Antarctic ice caps and the consequent rise in ocean levels is something that takes place gradually over many hundreds of years once the CO_2 burden is in the atmosphere; the slowness of reabsorption of CO_2 from the atmosphere into the deep oceans guarantees that CO_2 emissions are a "time bomb" with a much longer fuse than the decade or so hold-up of CFM's in the troposphere.[23,32] The CO_2 situation is further complicated by the discovery that CFM's and N_2O are both infrared absorbers capable of producing a "greenhouse effect" in the Earth's heat balance which is significant though smaller than the effects of CO_2 at similar times in the future, based on extrapolation of current source activities.[33] It should be noted that even if the effects of fertilizers and of CFM release on stratospheric ozone should ultimately prove to be of little consequence, their climatic effects would remain. In particular, some case for curtailment of CFM release would remain. Due to the shape of the CFM release curve vs. time, ambient measurements of CFM's are an insensitive measure of lifetime. Interlaboratory calibration of CFM measurements in a recent evaluation was disturbingly poor.[34] The CFM's will ultimately be slowly removed by oceanic and other sinks, as well as by photochemical reaction in the stratosphere, and the rate of removal is somewhat faster than for exchange of CO_2 into the deep oceans, but the recovery time is nevertheless long enough to span generations.

The dispersion of nonbiodegradable and potentially carcinogenic or mutagenic chemical compounds in the environment from industrial activity is another environmental threat with similar characteristics. What makes it analogous to the threats just mentioned is the long induction period connected with carcinogenesis at low concentrations. Because of this, dangerous dose-commitments could accumulate before epidemiological data could reveal the threat. Indeed this has already occurred in connection with a number of substances to which manufacturing workers have been exposed, such as vinyl chloride monomer and asbestos. It is the fear of such long induction-period threats from food additives and drugs that led to the passage of the famous Delaney clause[35] publicized in connection with the recent proposed saccharin ban.

In all these examples there occurs the same issue of the cost of delay in control while more scientific information is accumulated vs. the cost of economic disruption and consumer inconvenience resulting from the immediate and rapid imposition of a ban on the suspected activity or product. Where should the burden of proof lie? Should every even slightly suspect activity be stopped or regulated until it is proved safe beyond reasonable doubt, or should the initiation of regulations be permitted only after reasonably convincing scientific evidence of damage? What is in fact reasonable scientific proof of damage, and who is to be the judge of the evidence? How should the required burden of proof be related to the maximum credible degree of damage that might be sustained through the suspect activity? What should be the relative weight attached to "worst case" estimates as compared with "most probable" estimates of adverse effects in deciding when and how much to regulate? How should the social or economic benefits of a particular activity, and the availability and timing of alternate ways of achieving the same benefits be weighed in deciding whether an activity should be controlled? How should the distribution of risks and benefits among different parts of the world's population be taken into account in establishing regulations? All of these questions have arisen at one time or another in the course of the debate about CFM's, and the scientists who participated, no matter how much they tried to "stick to the facts," found themselves unable to avoid making explicit or implicit judgments about almost every one of these essentially nonscientific value questions.

III. THE ROLES OF SCIENTISTS IN THE CFM DEBATE

One can distinguish three roles for the scientists who were involved in the development of information and policy on CFM's. These roles were:

1. As sounders of alarm, advocates of regulation, lobbyists with administrative, legislative, and judicial agencies
2. As defenders of industry, and advocates of caution in premature regulation of activities which could turn out to be harmless when all the evidence was in
3. As impartial advisers to Congress, the Executive, and the public, and as interpreters of the scientific consensus to the political process

In practice the lines between these roles proved difficult to draw. Those who tried sincerely to maintain their impartiality found themselves compelled to take a stand with respect to the burden of proof question described in the preceding section, no matter how much they tried to avoid it. In the comprehensive hearings held by Senator Bumpers in September 1975, the Senator tried to pin down each of the scientific witnesses on the question of when to regulate. For example, he questioned Dr. M. McElroy (Harvard University) in the following vein:[36]

> "Let me ask you the same question I have been asking the other witnesses, Dr. McElroy. Dr. Cicerone (University of Michigan), he suggested that we ban aerosol sprays that are not necessary immediately with a view towards perhaps a crash program to find suitable substitutes to the freons [sic] that are necessary for refrigeration. Would you agree with that essentially?"

McElroy was, perhaps, more successful than several other scientific witnesses in evading this question, and in doing so he well expressed the difficulty of the scientist's position:

> "I think, first of all, it is very difficult for those of us who are, you know, very closely and personally, intimately involved in research to be — it is very difficult to be totally objective. One has a certain involvement, emotional involvement in the work which makes it very difficult to really take a broader view."

But he went on to say "...my advice to you on that issue isn't worth any more than the advice of any informed layman on the subject..." However, he apparently felt that the National Academy panel, not having the emotional bias that he had, could make better recommendations on regulation.

McElroy's testimony reminds us of the natural affinity between a scientist emotionally involved in the establishment of a new hypothesis which he has conceived and a politician who has staked out an area of legislation or regulation that he can call uniquely his own. "There is an interesting parallel between the scientist's desire to establish priority for a discovery or invention, and a politician's search for new issues on which he can make a name for himself — issues which differentiate him from other politicians in the eyes of his constituents, just as a new finding differentiates a scientist in the eyes of his professional peers."[37] Such a politician and such a scientist become natural allies in consciously or unconsciously excluding alternative viewpoints, an increasing hazard in a time when technical questions are providing increasingly fertile ground for the politician's search for distinctive issues which he can appropriate as his own.

Other scientific witnesses in category (3) were less cautious than McElroy. Dr. J. G. Anderson, (University of Michigan), for example, tried to distinguish between his role as a scientist and as a citizen:[38]

"I should point out that the measurements that I am involved in are crucial to the question and I would like to remain neutral on such a question as you ask until I satisfy myself of the results of those measurements. That is from a scientific point of view. From the personal point of view, I feel very strongly about the issue of protecting the very delicate ozone photochemistry, and from that point of view I would urge on the basis of the data and calculations already available that action be taken."

Anderson went on to endorse the position of Cicerone, as summarized by Senator Bumpers in his question to McElroy.

Stung by criticism from industrial scientists that the CFM effect on ozone was an unproved hypothesis, the result of a purely theoretical speculation with minimal experimental support, Professor F. S. Rowland, (University of California, Irvine), hit out less cautiously:[39]

"How long should we wait for someone to find this missing factor which might serve as a basis for an alternative scientific hypothesis before we act on the conclusions which we readily derive from the only hypothesis now available?"

Asked point blank by Senator Bumpers what he would recommend to the committee, Rowland said:[40] "I think that the economic dislocations need to be given minimal weight compared to the maximum weight to the possible harm to the environment."

Cicerone, also, irritated by the constant harping on the uncertainties in the ozone theory, and the quotation of his own earlier testimony to a House committee as a justification for delay,[41] was at pains to reinterpret his statement for the record[42] and to point out that scientific uncertainty was something which cut both ways.[43] The greater the uncertainty the greater the possible underestimation of the ozone depletion effect, a contingency the industrial witnesses avoided mentioning, preferring instead to stress the possible overestimation.

The rather strong positions on regulation taken by the scientists Rowland, Cicerone, and Anderson, who attempted to be objective, is partly explained by the apparent propensity of the industrial witnesses to ignore or gloss over the experimental evidence, and to describe the ozone depletion hypothesis as purely theoretical. Anderson, for example says:[44] "Let me say that I watched very carefully the response of industrial opinion resulting from additional measurements on this question and I have not been convinced that serious scientific evidence has greatly altered their position..."

At work here was a kind of mutual polarization that drove the scientific witnesses further towards the advocacy of immediate regulation than they might have gone, if the industrial witnesses had not so continually and unanimously dismissed or ignored the experimental evidence.

In testimony on the public policy aspects of the CFM problem the author also addressed regulatory strategy, to a large extent anticipating the position later taken by the National Academy Committee:[45]

"On the whole I am sympathetic to a rather gradual approach to problems of this type...The temptation towards dramatic, draconian action should be avoided, for not only will it produce unforeseen effects, but it might result in substitute actions by industry which would create even worse problems of a different kind, as happened with the attempt to ban phosphate detergents a few years ago. Unless important new information appears on the horizon it is necessary now to plot a long-range strategy for phasing out the use of freon [sic] except in circumstances where escape to the atmosphere can be minimized. A progressive phasing out to allow time for substitution and adjustment over a period of years is much more desirable than sudden action, and attention should concentrate on those uses, such as cosmetics, which are least essential, and on those measures, such as leak tightness of new refrigeration equipment, which are most easily and cheaply implemented."

IV. THE ADVOCATES — INDUSTRY AND THE ENVIRONMENTAL LOBBY

So far our discussion has been restricted to the testimony of those scientists who were deeply involved in the research on stratospheric ozone, and who, within the limits imposed by that close involvement, tried to interpret the scientific consensus as neutrally as possible to the politicians. Now we turn our attention to the frank advocates.

The environmental advocates were represented scientifically by Dr. K. Ahmed of the Natural Resources Defense Council (NRDC). Dr. Ahmed's testimony dealt primarily with regulatory strategy, and with the presentation of evidence that the banning of aerosol propellants would be beneficial to consumers because the substitute products would be cheaper. He insisted that the "legitimate cause for concern" expressed in the IMOS report[25] should have constituted a sufficient basis for immediate regulatory action, at least against aerosol products. He said:[46]

"It is highly contradictory, we submit, that when a task force, consisting of representatives from 14 Federal agencies, express a 'serious concern,' that it further recommends delay in action to be taken by Federal agencies."

He then went on to say:[40]

"In conclusion, I would like to emphasize once more that an extraordinarily strong case has been made that the present discharges of fluorocarbon compounds will seriously affect the Earth's atmosphere, which may lead to potentially grave environmental consequences...We believe that the time to enact decisive Federal legislation has come...We urge that the subcommittee act on passage of such a bill as an item of highest priority."

Ahmed supported a bill introduced in the House which gave the administrator of the Environmental Protection Agency (EPA) "expedited regulatory authority over materials affecting stratospheric ozone." Saying that even the House bill was too limited he urged that EPA be given "unambiguous authority including clear emergency powers to curb, restrict, or ban the use of any material in any product whose discharge into the atmosphere may seriously affect the stratosphere." Clearly the burden of proof was to be shifted heavily against all such products.

By contrast the industry witnesses, while admitting in various degrees a cause for concern, uniformly advocated the delay of regulatory action or even preparations for such action. The scientific witnesses representing industry seized on any scattered pieces of evidence which might throw doubt on the validity of what they called "the ozone hypothesis." They pointed to the large range of uncertainty in the estimates of future ozone depletion but implicitly dismissed the possibility that present estimates might as well be underestimates as overestimates of the effect. Mr. R. L. Schuyler (E. I. du Pont de Nemours & Company) advocating the shifting of the burden of proof onto the regulators:[48] "as a prerequisite to regulations, the promulgating agency be required to affirmatively find a probable hazard, based on accepted scientific data."

He was rather vague on what would constitute such data, and when questioned by Senator Bumpers as to whether this meant he disagreed with the IMOS recommendation that the government initiate regulatory action by January 1978 in the absence of any evidence tending to disprove the ozone theory, he evaded the question, repeating that "I think we should go forward with numbers that are factual, not theoretical, and then make decisions on that basis." Schuyler repeatedly referred to the original scientific estimates as a "worst case" calculation "with fluorocarbons being the only source of stratospheric chlorine and the stratosphere being the only place where fluorocarbons are decomposed." In fact this was a justified caution, although subsequent evidence suggests that neither reservation actually makes much difference in the estimates.

Although the industry witnesses continually assured the Congress they were prepared to curtail CFM production if the case against the CFM's were proved, the tone of their discussion indicates that none of them truly believed what the scientists were saying would ultimately prove to be correct. This made them reluctant to contemplate publicly even advance planning for regulation or for such phasing out of CFM uses as would provide only minor inconvenience to consumers. A typical statement was that of Mr. J. W. Dickinson of the Gillette Co. Asked whether we couldn't really do without most aerosol products he said:[49]

> "Of course we could...do without a great many things in our lives if there were adequate justification, if the justification existed (as it does not) to warrant imposing a different choice on consumers and to warrant the economic disruption that would result."

Nevertheless, it seems clear that in its actual planning industry was prepared to take more action than it would admit to in its public testimony. In fact one of the justifications Dickinson gave for delaying the imposition of restrictions was that it would reduce disruption as a result of "the development and planning which companies have already initiated as prudent steps to protect their future."[50] In this context Senator Domenici suggested that industry was devoting too large a proportion of its research to disproving or discrediting the ozone theory rather than to the development of alternatives. Denying this, Dr. R. L. McCarthy of Du Pont said that Du Pont's "research effort now and our process effort is mainly centered on new compounds and not on trying to disprove the theory."[51] In fact Schuyler insisted that Du Pont had virtually stopped all efforts to find new uses for fluorocarbons and according to McCarthy their "total research effort now in products and processes" was "aimed at new fluorocarbon compounds which contain hydrogen" which are 50 to 200 times less destructive of ozone than the fluorocarbons 11 and 12.[52]

Industry witnesses made much of the cooperative research program on the ozone problem being supported by industry, but relatively little appears in the record on the concrete results of this program, in part because the industry supported program was only just getting underway at the time. Almost all the scientific evidence, important for regulation, quoted in the Senate hearings, the IMOS report, and the later reports of the NAS, came from government supported work, mostly of independent university scientists.

One aspect of the testimony is interesting in the light of subsequent scientific developments. Dr. J. P. Lodge (Consultant) in his testimony stressed the analogy between the effects of stratospheric nitrogen and the effects of fluorocarbons, implying that the nitrogen and chlorine theories would stand or fall together. As authority for this he quoted from a news article quoting Johnston, the originator of the idea of ozone depletion by NO_x exhaust from the SST. In the words of this article in *Science News:*[53]

"There are 'exact parallels' between what happens when nitrogen oxides from natural sources and jet engine exhausts destroy ozone and what happens when chlorine radicals are released from propellants. 'We have to "change the labels" in our equations,' Johnston says, 'but we don't have to redo all the basic work. The chlorine model,' he says, is 'virtually proven by analogy.' "

Lodge went on to call attention to the lack of evidence for ozone depletion resulting from Soviet high yield nuclear tests in the early 1960s, and adduces from this absence of the predicted effect that the NO_x theory is probably in error and by analogy the chlorine theory as well. This point was seized upon by several industrial witnesses to cast doubt on the CFM theory. In addition Lodge mentioned evidence from Dr. J. K. Angell and Dr. J. Korshover (National Oceanic and Atmospheric Administration), that emission of NO_x from volcanic eruptions had not resulted in observable ozone depletion as predicted.[54] He twits the IMOS committee, saying:[53]

"The paper by Angell and Korshover was made available to the Federal Task Force on Inadvertent Modification of the Stratosphere (IMOS). They apparently were so convinced of the chemistry they had been taught that they used figures from the paper without noting the conclusions."

On the first point there is now strong evidence that the NO_x theory was in fact wrong or at best highly uncertain, owing to the lack of knowledge of one important reaction,[18] but because of the same error the estimate of ozone depletion due to chlorine has more than doubled. Thus the argument from analogy proved false. Nevertheless, the confidence with which hundreds of workers viewed the nitrogen theory at the conclusion of the $50 million CIAP[55] program certainly looks embarassing vis à vis the general public in the light of the present scientific situation, and the layman might therefore be forgiven considerable skepticism regarding the conclusions about CFM. In the wisdom of hindsight the testimony of neither the industrial witnesses nor the independent scientists looks as strong as it did at the time it was given.

Furthermore, as will be indicated below, more recent work which takes into account the three dimensional dynamics of the stratosphere has thrown some doubt on the practical biological and ecological consequences of ozone depletion due to CFM's, since the largest effect in the annual average depletion occurs at high latitudes in late winter, when the UV intensity is weak, and biological exposure is minimal due to the fact that people are indoors and crops are dormant.[56,57]

The second point, the lack of effect from the Agung eruption, becomes considerably more complex in the light of recent revisions of the NO_x theory. In his testimony Lodge asserts that the argument from analogy between NO_x and chlorine is unnecessary because in fact HCl and Cl_2 are also emitted in volcanic eruptions, suggesting that the absence of ozone depletion implies no effect from chlorine as well as NO_x. He was apparently unaware of the point made by Cicerone in a submission for the record[58] that the amount of chlorine supplied volcanically is probably only between 2 and 6% the amount supplied by CFM emissions up to 1973. Hence its contribution would have been undetectable in comparison with random natural fluctuations or with the effects of NO_x. In retrospect it seems obvious that the Agung eruption probably produced a transient increase in ozone due to NO_x (which was observed, though discounted) and that this completely blanketed any effect of chlorine from the eruption. Thus in fact the volcanic evidence was irrelevant so far as the CFM problem was concerned.

V. THE NATIONAL ACADEMY OF SCIENCES

The National Research Council, through its Assembly of Mathematical and Physical Sciences, had been engaged in a variety of atmospheric studies for many years. As a consequence of concerns over the SST, the Department of Transportation (DOT) asked the National Academy of Sciences (one of the parent bodies of the National Research Council) to establish the Climatic Impact Committee (CIC) to assess "the climatic and biological impact of a projected fleet of high altitude aircraft" (including both SST's and modern wide-body subsonic jet aircraft). The CIC completed its assignment in April 1975. The committee's name was subsequently changed to the Committee on Impacts of Stratospheric Change (CISC) and in April 1975 it created a panel under Professor Herbert Gutowsky (University of Illinois), "to study the question of whether the chlorofluoromethanes would destroy ozone, and if so, what the magnitude of the effect would be and what uncertainties were associated with the prediction." Simultaneously the CISC undertook to reassess the biological and climatic effects of ozone reduction and "the appropriate policy consequences of both our present knowledge and the knowledge we are likely to have in the future." These committees had available to them the extensive research information regarding ozone compiled as a result of DOT's Climatic Impact Assessment Program (CIAP). Neither committee included any of the scientists who had been deeply involved in the research on ozone depletion, or who had testified in the extensive Congressional hearings. Nor was there any overlap with the IMOS task force. The report of CISC was made available to the President's Science Advisor, Dr. H. Guyford Stever, on September 10, 1976.

The report of the CISC was eagerly awaited to provide the definitive word on the state of knowledge on all aspects of the CFM problem. In its report of June 1975 the IMOS Task Force had said:[59]

> "If the National Academy of Sciences confirms the current task force assessment, it is recommended that Federal regulatory agencies initiate rulemaking procedures for implementing regulations to restrict fluorocarbon uses. Such restrictions could reasonably be effective by January 1978 — a date that, given the concern expressed now, should allow time for consideration of further research results and for the affected industries and consumers to initiate adjustments."

As a further basis for the recommended delay the IMOS report cited the short time the Task Force had to make its study (5 months) and the prospective availability of important new data, especially measurements of key intermediates in the stratosphere, which could be taken into account in the Academy's study.

The CISC confirmed the IMOS assessment and stated as its main conclusion that "selective regulation of CFM uses and releases is almost certain to be necessary at some time and to some degree of completeness." However, it went on to say that "neither the needed timing nor the needed severity can be reasonably specified today" and that the consequences of 2 years' delay before deciding to regulate would be negligible.[60] The report in fact addressed regulation policy more explicitly than many critics thought appropriate, and there were some apparent inconsistencies that could be read into its position. For example, it said:[61]

> "The choice of when to make decisions about regulation is a political one in the highest sense of that word. The ultimate balance — between increased impact on industry and on spray-can uses, on the one hand, and possible climatic impacts and more certain skin cancer increases, both very small for a short delay, on the

other — has inevitably to be made by those who decide for the whole of each country concerned, in the U.S. by its Congress and President."

Or again:[62]

"Having laid open the facts as best it can and stressed the fundamental importance of regulation use by use, the scientific community as represented through the National Academies of Sciences and Engineering and the National Research Council, can, we believe, properly leave decisions about timing, in this country, to the Congress of the U.S."

Here the use-by-use strategy is given a scientific justification, but the timing decision is not, and is therefore left to the political process.

Yet, having apparently "passed the buck" to the political process, CISC recommended fairly explicit timing and strategy:[63] "...in view of the small changes in ozone reductions following from a year or two delay, we wish to recommend against decision to regulate at this time."

The report also recommended regulation on a use-by-use basis, in preference to blanket controls on emissions. This strategy was given a scientific justification in that: "...in the one case where we understand in numerical detail how the release of CFM's produces a disadvantageous effect, a doubling in the release schedule is to be expected to produce roughly a doubling of the number of additional skin cancers."[64]

But it went on further to recommend that:

"No one's interest, we believe, is served by cancellations without appropriate delays. There should be adequate notice, within a procedure that recognizes the differing human values of different uses and that is intended, over a reasonable span of years, to diminish substantially the release of CFM's both in the U.S. and worldwide."

However, the committee, or at least its chairman, appears to have been enthusiastic about this strategy on more general grounds, and in his preface to the CISC report points to its recommendations as "a useful example of a regulatory attitude that, we are sure, will be frequently needed in the decades ahead."

The recommendation for delay, although at most probably no more than 9 months beyond the deadline of January 1978 set by the IMOS report, caused considerable distress among environmentalists.[65] The Academy was accused of going beyond its competence, and the scientific data, in recommending a specific regulatory strategy, especially in the light of its own disclaimers. Moreover, it made no effort to use the existing data on melanoma and nonmelanoma cancers to estimate the extra deaths and cancers that might be caused worldwide by a 2-year delay in U.S. regulation of aerosol uses; consequently there was no quantitative basis for comparing the extra deaths or illnesses with the reduced economic dislocation. Many would argue, of course, that such estimates would be so rough as to be no better than the more qualitative judgment formulated by the committee.

The CFM story and the stratospheric ozone problem continue to unfold. There is certainly no assurance that all the surprises have yet turned up, and it is difficult to guess how stable the present scientific picture may remain in the future. In its interim report to the EPA administrator in December 1977, the CISC bemoans that "to say how many more major upheavals we should expect in the future is rather like trying to foresee the unforeseeable." The Interim Report lacks the sense of confidence displayed in the September 1976 report, or in the testimony of the independent researchers in the Senate Subcommittee hearings. Instead it says:[66]

"It is unfortunately true, however, that accompanying very substantial overall progress, the recent development of our understanding of stratospheric chemistry has been dominated by major upheavals caused by the recognition of the importance of processes whose role either had not been properly appreciated (such as the addition of chlorine nitrate chemistry which delayed publication of the previous panel report) or whose rate coefficient had been grossly misjudged (such as the rate constant for the important reaction HO_2 + NO → HO + NO_2, whose recently established larger value has made chlorine catalysis of ozone much more effective than previously thought and NO_x catalysis less so)."

The Academy is now embarked on a new round of studies not only to update the information on stratospheric ozone depletion by CFM's, but also to assess the various uses of CFM's and the possibility of substitutes, design improvements to minimize emissions, and various alternate regulatory strategies and their socioeconomic impact. Thus the magnitude of the task seems to grow, and its spread of scope into more "political" areas seems to continue, even as the date of availability of definitive scientific conclusions seems to recede further into the future. The case for some regulation of CFM's no longer rests exclusively on the ozone depletion effect. This is because the accumulation of CFM's in the troposphere results in infrared absorption which adds to the "greenhouse" effect produced by CO_2 from combustion and N_2O from fertilizers. However, at comparable times the absorption attributable to CFM's would probably not exceed 10% of that due to CO_2 from combustion. However, if CFM emissions continued to increase, the Earth's climate could not tolerate their continued accumulation, even if there were no ozone depletion effect. Yet from the perspective of 1979 the cautionary and negative attitude of industry in the 1975 hearings in the face of the apparently clear experimental evidence being produced by research does not look quite as foolish as it appeared to some at the time, though it would be wrong to say that the tables had turned.

VI. EPILOGUE

Since this paper was first written much new evidence has appeared. In particular, two major comprehensive studies have been published, one by the British Department of the Environment[67] and the other by the U.S. National Academy.[68] Although these two reports utilize the same evidence and data there are substantial differences in interpretation and tone. The two reports have recently been comparatively reviewed by McElroy, Sze, and Wofsy.[69] The tone of the British report is much more cautious than that of the American report, and much more weight is placed on the many uncertainties in the models, and the discrepancies between stratospheric observations on the concentration of relevant compounds, and the concentration profiles with height of these compounds predicted in the models. There appears to be a growing concensus that the predictions of the ozone profiles are subject to considerably greater uncertainty than seemed true at the time of the 1976 reports. In addition, a new factor has appeared in the consideration of the latitudinal and seasonal distribution of the predicted ozone depletion, and its consequences for seasonally averaged UV penetration, as well as the biological significance of this UV penetration. The net effect is that the biological effects of the predicted ozone depletion could well be less than one quarter what one would infer from the decrease in average column height of the ozone layer by itself. Thus the events of the last 3 years have tended to reinforce the cautionary statements made in the three preceding paragraphs.

REFERENCES

1. Brooks, H., Knowledge and action: the dilemma of science policy in the 70's, *Daedalus,* 102, 125, 1973.
2. York, H. F., *The Advisors: Oppenheimer, Teller, and the Superbomb,* W. H. Freeman, San Francisco, 1976.
3. Nelkin, D., The political impact of technical expertise, *Soc. Studies Sci.,* 5(1), 36, 1975.
4. Carson, R. L., *Silent Spring,* Houghton Mifflin, Boston, 1962.

5. Nelkin, D., Ecologists and the public interest: establishing professional standards, *Hastings Center Report,* 10, October 1980.
6. Forrester, J. W., *World Dynamics,* Wright-Allen Press, Cambridge, MA, 1971.
7. Meadows, D. H., Meadows, D. L., Randers, J., and Behrens, W. W., III, *Limits to Growth: A Report for the Club of Rome's Project on the Predicament of Mankind,* Universe Books, New York, 1972.
8. Olsen, M. and Landsberg, H. H., Eds., *The No Growth Society,* Norton Press, New York, 1973, 119.
9. Ravetz, J., *Scientific Knowledge and its Social Problems,* Clarendon Press, Oxford, 1971.
10. Roszak, T.,The monster and the titan: science, knowledge, and gnosis, *Daedalus,* 103, 17, 1974.
11. Anderson, F. R., *NEPA in the Courts, a Legal Analysis of the National Environmental Policy Act,* Johns Hopkins University Press, Baltimore, 1973.
12. Murphy, A. W., NEPA and the licensing process: environmentalist magna carta or agency coup de grace? *Columbia Law Review,* 72(6), 963, 1972.
13. Weber, E., Teal, G. K., and Schillinger, A. G., Eds., Remarks on technological forecasting in the next decades, in *Technology Forecast for 1980,* Van Nostrand-Reinhold Co., New York, 1971, 3.
14. Statement of Johnston, H. S., in FAA Certification of the SST Concorde, Hearings, Subcommittee of the House Committee on Government Operations, 94th Congress, 1st and 2nd Sessions, 67.
15. Johnston, H. S., Reduction of stratospheric ozone by nitrogen oxide catalysts from supersonic transport exhaust, *Science,* 173, 517, 1971.
16. McDonald, J. E., Hearings before a Subcommittee of the Committee on Appropriations, House of Representatives, Civil Supersonic Aircraft Development (SST), 1971.
17. Statement of Donahue, T.M., in FAA certification of the SST Concorde, Hearings, Subcommittee of the House Committee on Government Operations, 94th Congress, 1st and 2nd Sessions, 93.
18. Tukey, J. W., Chairman, and Committee, Response to the Ozone Protection Sections of the Clean Air Act Amendments of 1977: An Interim Report, NAS, December 1977, 15.
19. Tukey, J. W., Chairman, and Committee, Halocarbons: Environmental Effects of Chlorofluoromethane Release, NAS, September 1976, App. E, Chapter 8.
20. Tukey, J. W., Chairman, and Committee, Response to the Ozone Protection Sections of the Clean Air Act Amendments of 1977: An Interim Report, NAS, December 1977, 59.
21. Cutchis, P., FAA-EQ-78-19, On the Linkage of Solar Ultraviolet Radiation to Skin Cancer, 1978.
22. McElroy, M. B., Chemical processes in the solar system: a kinetic perspective, *Int. Rev. Sci.,* 2, 9, 1976.
23. Stumm, W., Report of the Dahlem Workshop on Global Chemical Cycles and Their Alterations by Man, Berlin, November 15-19, 1976, Dahlem Konferenzen.
24. Tukey, J. W., Chairman, and Committee, Halocarbons: Environmental Effects of Chlorofluoromethan Release, NAS, September 1976, Table 3,12.
25. Fluorocarbons and the Environment, Report of Federal Task Force on Inadvertent Modification of the Stratosphere (IMOS), Council on Environmental Quality, Federal Council for Science and Technology, June 1975.
26. Tukey, J. W., Chairman, and Committee, Halocarbons: Environmental Effects of Chlorofluoromethane Release, NAS, September 1976, 6.
27. Johnston, H. S., Whitten, G., and Birks, J., The effect of nuclear explosions on stratospheric nitric oxide and ozone, *J. Geophys. Res.,* 78, 6107, 1973.
28. Report to the President by the Interagency Review Group on Nuclear Waste Management, TID-28817, USGPO, October 1978.
29. U.S. Environmental Protection Agency, Office of Radiation Programs, Technology Assessment Division, Environmental Analysis of the Uranium Fuel Cycle. II. Nuclear Power Reactors, EPA-520/9/73-003-C, November 1973.
30. U.S. Nuclear Regulatory Commission, Numerical guides for design objectives and limiting conditions for operation to meet the criterion 'as low as practicable' for radioactive material in light-water-cooled nuclear power reactor effluents, *Fed. Reg.,* 40(87), 1975.
31. Pigford, T. H., Radioactivity in stored coal ash and in nuclear power waste, *Trans. Am. Nuc. Soc.,* 30, 293, 1978.
32. Energy and Climate, Panel on Energy and Climate, Geophysics Study Committee, Geophysics Research Board, National Academy of Sciences, Washington, D. C., 1977.
33. Tukey, J. W., Chairman, and Committee, Halocarbons: Environmental Effects of Chlorofluoromethane Release, NAS, September 1976, 61.
34. Rasmussen, R. A., Interlaboratory comparison of fluorocarbon measurements, *Atmos. Environ.,* 12, 2505, 1978.
35. Brooks, H., The Impossibility of Proving Impossibilities, speech presented at Herbert Spencer Series on Technology and Society, University of Pennsylvania, Philadelphia, April 14, 1976.

36. Stratospheric Ozone Depletion, Hearings before the Subcommittee on the Upper Atmosphere of the Committee on Aeronautical and Space Sciences, U.S. Senate, 94th Congress, September 18, 19, and 23, 1975, Testimony of McElroy, 1048.
37. **Brooks, H.**, Expertise and politics — problems and tensions, *Proc. Am. Philos. Soc.,* 259, 1975.
38. Stratospheric Ozone Depletion, Hearings before the Subcommittee on the Upper Atmosphere of the Committee on Aeronautical and Space Sciences, U.S. Senate, 94th Congress, September 18, 19 and 23, 1975, Testimony of Anderson, 1022.
39. Stratospheric Ozone Depletion, Hearings before the Subcommittee on the Upper Atmosphere of the Committee on Aeronautical and Space Sciences, U.S. Senate, 94th Congress, September 18, 19 and 23, 1975, Testimony of Rowland, 796.
40. Stratospheric Ozone Depletion, Hearings before the Subcommittee on the Upper Atmosphere of the Committee on Aeronautical and Space Sciences, U.S. Senate, 94th Congress, September 18, 19 and 23, 1975, Testimony of Rowland, 939.
41. Stratospheric Ozone Depletion, Hearings before the Subcommittee on the Upper Atmosphere of the Committee on Aeronautical and Space Sciences, U.S. Senate, 94th Congress, September 18, 19 and 23, 1975, Testimony of Orfeo, 738.
42. Stratospheric Ozone Depletion, Hearings before the Subcommittee on the Upper Atmosphere of the Committee on Aeronautical and Space Sciences, U.S. Senate, 94th Congress, September 18, 19 and 23, 1975, Testimony of Cicerone, 984.
43. Stratospheric Ozone Depletion, Hearings before the Subcommittee on the Upper Atmosphere of the Committee on Aeronautical and Space Sciences, U.S. Senate, 94th Congress, September 18, 19 and 23, 1975, Testimony of Cicerone, 949.
44. Stratospheric Ozone Depletion, Hearings before the Subcommittee on the Upper Atmosphere of the Committee on Aeronautical and Space Sciences, U.S. Senate, 94th Congress, September 18, 19 and 23, 1975, Testimony of Anderson, 1023.
45. Stratospheric Ozone Depletion, Hearings before the Subcommittee on the Upper Atmosphere of the Committee on Aeronautical and Space Sciences, U.S. Senate, 94th Congress, September 18, 19 and 23, 1975, Testimony of Brooks, 782.
46. Stratospheric Ozone Depletion, Hearings before the Subcommittee on the Upper Atmosphere of the Committee on Aeronautical and Space Sciences, U.S. Senate, 94th Congress, September 18, 19 and 23, 1975, Testimony of Ahmed, 1050.
47. Stratospheric Ozone Depletion, Hearings before the Subcommittee on the Upper Atmosphere of the Committee on Aeronautical and Space Sciences, U.S. Senate, 94th Congress, September 18, 19 and 23, 1975, Testimony of Ahmed, 1051.
48. Stratospheric Ozone Depletion, Hearings before the Subcommittee on the Upper Atmosphere of the Committee on Aeronautical and Space Sciences, U.S. Senate, 94th Congress, September 18, 19 and 23, 1975, Testimony of Schuyler, 570.
49. Stratospheric Ozone Depletion, Hearings before the Subcommittee on the Upper Atmosphere of the Committee on Aeronautical and Space Sciences, U.S. Senate, 94th Congress, September 18, 19 and 23, 1975, Testimony of Dickinson, 700.
50. Stratospheric Ozone Depletion, Hearings before the Subcommittee on the Upper Atmosphere of the Committee on Aeronautical and Space Sciences, U.S. Senate, 94th Congress, September 18, 19 and 23, 1975, Testimony of Dickinson, 701.
51. Stratospheric Ozone Depletion, Hearings before the Subcommittee on the Upper Atmosphere of the Committee on Aeronautical and Space Sciences, U.S. Senate, 94th Congress, September 18, 19 and 23, 1975, Testimony of McCarthy, 615.
52. Stratospheric Ozone Depletion, Hearings before the Subcommittee on the Upper Atmosphere of the Committee on Aeronautical and Space Sciences, U.S. Senate, 94th Congress, September 18, 19 and 23, 1975, Testimony of McCarthy, 614.
53. Stratospheric Ozone Depletion, Hearings before the Subcommittee on the Upper Atmosphere of the Committee on Aeronautical and Space Sciences, U.S. Senate, 94th Congress, September 18, 19 and 23, 1975, Testimony of Lodge, 717.
54. **Angell, J. K. and Korshover, J.**, Quasi-biennial and long term fluctuations in total ozone, *Mon. Weather Rev.,* 101, 426, 1973.
55. Environmental Impact of Stratospheric Flight: Biological and Climatic Effects of Aircraft Emissions in the Stratosphere, Climatic Impact Committee, NAS, 1975.
56. **Pyle, J. A.**, A simple calculation of ozone depletion by chlorofluoromethanes using a two-dimensional model, *Nature (London),* 271, 42, 1978.
57. **Vupputuri, R. K. R.**, The structure of the natural stratosphere and the impact of chlorofluoromethanes on the ozone layer investigated in a 2-D time dependent model, *Pure Appl. Geophys.,* 117, 448, 1979.
58. Stratospheric Ozone Depletion, Hearings before the Subcommittee on the Upper Atmosphere of the Committee on Aeronautical and Space Sciences, U.S. Senate, 94th Congress, September 18, 19 and 23, 1975, Testimony of Cicerone, 974.

59. Fluorocarbons and the Environment, Report of Federal Task Force on Inadvertent Modification of the Stratosphere (IMOS), Council on Environmental Quality, Federal Council for Science and Technology, June 1975, 5.
60. Tukey, J. W., Chairman, and Committee, Halocarbons: Environmental Effects of Chlorofluoromethane Release, NAS, September 1976, 7.
61. Tukey, J. W., Chairman, and Committee, Halocarbons: Environmental Effects of Chlorofluoromethane Release, NAS, September 1976, 9.
62. Tukey, J. W., Chairman, and Committee, Halocarbons: Environmental Effects of Chlorofluoromethane Release, NAS, September 1976, 10.
63. Tukey, J. W., Chairman, and Committee, Halocarbons: Environmental Effects of Chlorofluoromethane Release, NAS, September 1976, 8.
64. Tukey, J. W., Chairman, and Committee, Halocarbons: Environmental Effects of Chlorofluoromethane Release, NAS, September 1976, 37.
65. Sullivan, W., Fluorocarbons' End Sought by Peterson, *N.Y. Times,* September 18, 1976.
66. Tukey, J. W., Chairman, and Committee, Response to the Ozone Protection Sections of the Clean Air Act Amendments of 1977: An Interim Report, NAS, December 1977, 19.
67. Department of the Environment, Chlorofluorocarbons and Their Effect on Stratospheric Ozone, Pollution Paper No. 15, Her Majesty's Stationery Office, London, 1979.
68. National Academy of Sciences, Stratospheric Ozone Depletion by Halocarbons: Chemistry and Transport, National Academy of Sciences, Washington, D.C., 1979.
69. McElroy, M. B., Sze, N. D., and Wofsy, S. C., Reports by the U.S. National Academy of Sciences and the U.K. Department of the Environment on the Impact of Chlorofluorocarbons on Ozone: Comparison and critique, unpublished memorandum, 1980.

Chapter 10

REGULATORY ACTIONS*

George F. Wirth, Perry W. Brunner, and Ferial S. Bishop

TABLE OF CONTENTS

* This chapter reflects the views of the authors and in no way is to be construed to reflect the official or unofficial view, policy, or plans of the U.S. Environmental Protection Agency.

I. THE REGULATORY ACTION IS ANNOUNCED

Regulatory action by the U.S. Government on chlorofluorocarbons (CFC's) began in October of 1976.

Unlike many environmental actions, demands for action against CFC's did not stem from highly visible and publicly observed environmental insults. The effect of CFC's on the ozone layer had been observed by no one. Awareness of the ozone problem was brought to the public by scientists with a hypothesis. In fact, concern for damage to the ozone layer by CFC's had its beginning in investigations of a completely different science and public policy issue — the impact of the supersonic transport (SST) on the stratosphere.[1]

In June of 1974, Drs. F. Sherwood Rowland and Mario J. Molina of the University of California at Irvine presented studies indicating that chlorofluoromethanes, or more generally CFC's, could harm the ozone layer.[2]

Reaction from interest groups came quickly. In November of 1974, The National Resources Defense Council (NRDC) submitted its first petition on CFC damage to the ozone layer to the Consumer Product and Safety Commission (CPSC). The CPSC denied this petition for action on the basis of insufficient information on the correctness of the hypothesis of ozone depletion by CFC's.[3]

In December of 1974, The Food and Drug Administration (FDA) requested the Council on Environmental Quality (CEQ) to designate a lead agency to coordinate preparation of an Environmental Impact Statement (EIS) should regulatory action on CFC's become necessary. No lead agency was established at this time, but the CEQ and the Federal Council on Science and Technology (FCST) formed the Interagency Task Force on the Inadvertent Modification of the Stratosphere (IMOS). The IMOS report, issued in June of 1975, stated that CFC releases are cause for concern; that it would seem necessary to restrict uses of CFC's to replacements in refrigeration, air-conditioning, and closed, recycled systems; and that regulatory action be taken if the National Academy of Sciences (NAS) confirmed the IMOS assessment in its study initiated in March of 1975.[4]

In July of 1975, the FDA published a notice requesting information on CFC's in food, drug, and cosmetic products and informed the public that serious consideration was being given to the problem.[5] Also, in July of 1975, the NRDC petitioned the FDA for restrictions on the use of CFC's. This petition was denied by the FDA on the basis that immediate action was not necessary at this time in light of the NAS's current study.[6]

On September 13, 1976, the NAS released two reports. The Committee on Impacts of Stratospheric Change released a report assessing the environmental effects of chlorofluoromethane release.[7] The Committee's Panel on Atmospheric Chemistry released a report on the effects of halocarbons on stratospheric ozone.[8] These reports became the focus of action and controversy in the regulatory process. The NAS reports will be discussed in this chapter in the context of the regulatory issues.

On October 12, 1976, then Administrator of the Environmental Protection Agency (EPA), Mr. Russell Train, announced in Brussels, Belgium, at a meeting of the Committee on the Challenges of Modern Society (an environmental group under the auspices of the North Atlantic Treaty Organization), that the U.S. was initiating the regulatory process to phase out the nonessential uses of CFC's.[9] On the same day, in a letter to then Chairman of the CEQ, Dr. Russell Peterson, the FDA indicated intent to initiate the regulatory process to phase out CFC's in nonessential uses in food, drug, and cosmetic products.[10]

On October 18, 1976, the EPA issued a request for voluntary utilization of non-CFC propellants in pesticides, a requirement for informative labels to be placed on

CFC-propelled pesticides after April 15, 1977, and a warning that the EPA would request discontinuance of the use of CFC propellants in the process of registering pesticides.[11]

On November 22, 1976, the CPSC found that CFC aerosol consumer products presented an unreasonable risk of injury to consumers by destroying the ozone layer. The CPSC directed its staff to prepare a draft notice declaring such products to be "banned hazardous products." The CPSC recognized the EPA plan to regulate CFC's and stated its intent to terminate the CPSC action if it were unnecessary after the EPA action. The CPSC made this statement in response to petitions for regulatory action again filed by the NRDC.[12]

By the end of November of 1976, the regulatory process was in full swing. Two meetings (October 27th and November 1st) had been held by the CEQ for the purpose of coordinating the CPSC, the EPA, and the FDA regulatory activities.[13] Russell Train had approved a "development plan" whereby the EPA chaired an interagency work group formed to develop the regulatory action(s).[14] The interagency work group already had met on five occasions (November 8th, 16th, 18th, 23rd, and 30th).[15]

The story of the events that led to this flurry of regulatory activity in October and November of 1976 is the essence of Chapter 8, authored by Ms. Carroll Leslie Bastian. This chapter is about the work group (Table 1) that developed the regulations — a new set of "faces" on the CFC issue. They primarily were people armed with new regulatory authority granted by Congress and a commitment from their Agency "Heads" to examine the potential for regulating the control of activity suspected of causing ozone depletion.

II. THE GROUND RULES

The interagency work group's ground rules were defined in the development plan approved by Russell Train.[14]

The development plan defined the work group membership to consist of individuals from the CPSC, the EPA, the FDA, the CEQ, the Department of Commerce (DOC), the National Oceanic and Atmospheric Administration (NOAA, part of DOC), the National Aeronautics and Space Administration (NASA), the National Science Foundation (NSF), and the Department of Transportation (DOT). The work group was chaired by Mr. George F. Wirth of the then-embryonic Office of Toxic Substances in EPA. The CPSC, the EPA, the FDA, and the DOC were the most active members and attended all meetings of the work group.

The development plan provided for regulating CFC's in a two-phase effort. The first phase would be the proposal of a regulation, by April 1977, of nonessential aerosol uses of trichlorofluoromethane (CFC-11) and dichlorodifluoromethane (CFC-12) and other CFC's with similar properties and environmental effects likely to be used as substitutes in aerosol use. The second phase would be the regulation of other uses of CFC's by June 1978. The CEQ was identified as the coordinator of regulatory responsibility among the CPSC, the EPA, and the FDA and of the preparation of the EIS.

In preparation of the development plan, the EPA designed the two-phase process in the belief that there existed nonessential uses of CFC's. The EPA believed that there were substitutes available for nonessential uses, and that a ban on nonessential uses would have far less economic impact than a ban on essential uses. Supporting this belief was an economic study of aerosols being completed by EPA at the time the work group was initiated.[16]

The April 1977 date for a proposed regulation on nonessential uses of CFC's was selected on the basis of Russell Train's announcement in Brussels, where he not only committed the U.S. to initiating the regulatory process, but also called on the major

Table 1
LIST OF WORK GROUP MEMBERS CHLOROFLUOROCARBONS WORK GROUP PHASES I AND II

(1977 through 1978)

Consumer Product Safety Commission

Dorothy Canter
Shakuntala Chaube
Colin Church
Arlen Slobodow

Council on Environmental Quality

Warren Muir

Department of Commerce

Roger Chiarodo
Peter Gabauer
Edward Shykind

Environmental Protection Agency

Ferial Bishop
David Brandwein
Perry Brunner
John Butler
Stan Coerr
Gail Cooper
Elizabeth Cotsworth
Farley Fisher
Alphouse Forziati
Kevin Gaynor
Dan Golumb
Ted Hadzi-Antich
Douglas Hale
Anne Haughton
Judy Kosovich
Pope Lawrence
Frank Letkiewicz
Larry Longanecker
Claire Matassoni
Mark Mercer
Larry O'Neill
Courtney Riordan
Jim Silverman
James Touhey
George Wirth
Herbert Wiser

Table 1 (continued)

Food and Drug Administration

Margaret Gilhooley
Buzz Hoffmann
David Klauder
Susan Reinsch
Gail Updegraff

National Aeronautics and Space Administration

Robert Seals

National Oceanographic and Atmospheric Administration

Don Hunt

National Science Foundation

Carroll Leslie Bastian

producing and consuming nations to assist in curtailing nonessential uses of CFC's. To assist in the control of nonessential uses of CFC's on an international level, Russell Train invited the regulatory heads of the major producing and consuming nations to meet for discussions in the U.S. in the spring of 1977.[9] Based on this invitation, the EPA assumed an obligation to present a proposed U.S. position, and specifically a proposed regulation, to the international community by spring of 1977.

For regulation of other uses of CFC's, an additional 18 months was believed necessary to complete economic impact studies and prepare a proposed regulation. Therefore, the summer of 1978 was selected as the target date for proposed regulation on other uses.[14]

In addition to the identification of the two-phased approach and the target dates for regulation that the work group assumed as basic guidelines to their regulatory task, there were jurisdictional guidelines that had been established in the two coordinating meetings held by CEQ in October and November of 1976. First was the identification of EPA as the lead agency and hence as the chair of the work group to be established. This leadership was delegated to EPA by consensus of the three lead agencies (CPSC, EPA, and FDA) on the basis that the EPA had the broadest regulatory authority to control CFC's in other uses and hence had the longest-term interest in CFC regulatory action. The EPA's broad regulatory authority had just been acquired by passage of the Toxic Substances Control Act (TSCA, October 11, 1976). The EPA also had the best capability to prepare the EIS because the EPA was completing the economic impact assessment of potential regulation. Two of the authors of this paper attended both CEQ meetings plus other bilateral meetings on the subject of leadership, and we believe the deciding factor on leadership was the EPA's long-term interest and regulatory authority under the TSCA.[16]

Second, in addition to the decision on the leadership from the CEQ meetings, a basic set of guidelines on the use of regulatory authorities was established. The CEQ meeting participants concluded that the public interest would not be served by separated regulations on CFC's authored by the CPSC, the EPA, and the FDA. The CEQ meeting participants believed it was possible to regulate all uses and particularly all the Toxic Substances Control Act (TSCA October 11, 1976). The EPA also had the best capability to prepare the EIS because the EPA was completing the economic impact assessment of potential regulation. Two of the authors of this paper attended nonessential aerosol uses under the TSCA and the Federal Food, Drug, and Cosmetic Act (FFDCA); and the public interest would be served if regulation were limited to not more than two published rules. A question did remain after the CEQ meetings on how or if CFC aerosol pesticides were to be controlled; but since their control was primarily under the Federal Insecticide, Fungicide, and Rodenticide Act (FIFRA) — an EPA administered authority — this problem was left to be resolved by the work group. Therefore, in summary, the work group began its deliberations with the following criteria:

1. EPA would lead an interagency work group.
2. There would be a proposed regulation or at least a position on regulation of nonessential aerosol uses of CFC's by April of 1977, and all other uses would not be proposed for regulation before the summer of 1978.
3. Regulation would be limited to no more than two actions, one under the TSCA and one under the FFDCA.
4. The work group would prepare a joint EIS.

III. THE ISSUES ARE IDENTIFIED

The work group began deliberations in November of 1976. (See Figure 1 — Chart of Key Regulatory Actions — for a schematic presentation of the chronology and interrelationship of events in the regulatory process.) Even though criteria for an assumed regulatory action was established by the CEQ meetings and the development plan, the work group assumed the responsibility to consider all options for controlling CFC releases including no control at this time.

The work group was immediately aware of the argument for postponing regulation until more scientific evidence proved the ozone depletion theory. Many work group members initially expressed a strong propensity not to regulate CFC's until a strong case for regulation could be made on ozone depletion and on resulting health and/or environmental effects. The work group was unable to resolve this issue at the onset of deliberations, but did agree to continue forward with the resolution of other issues on the assumption that the case against CFC's could be made. This issue was resolved later by deliberations on the timing of a CFC aerosol ban — a major issue of the work group discussed in Section IV.

Warning labels were considered by the work group as a partial control mechanism that would decrease CFC releases by discouragement of the market for CFC aerosol products. Although warning label regulations were eventually promulgated by the FDA, the CPSC, and informative label regulations in the case of pesticides by the EPA (discussed in Section IV), labels were rejected by the work group as a long-term effective control mechanism. The rejection was based on a number of reasons such as the impression that the market discouragement would not result in the least essential products being removed from the marketplace, the impression that significant CFC emission reduction would take a long time, the impression that such labels would create confusion in the marketplace and hence an unstable economic atmosphere among producers of aerosol products, and the impression that there was no more legal basis for labels than a ban if the ozone hypothesis was a valid basis for action of any kind.

Another approach to control of CFC emissions, suggested by some industrial firms, was to ban through regulation on a product-by-product basis after considering the availability of alternative propellants and product delivery systems for the product under threat of a ban. This approach was also rejected by the work group because of its extremely time-consuming administrative attributes, and because the work group

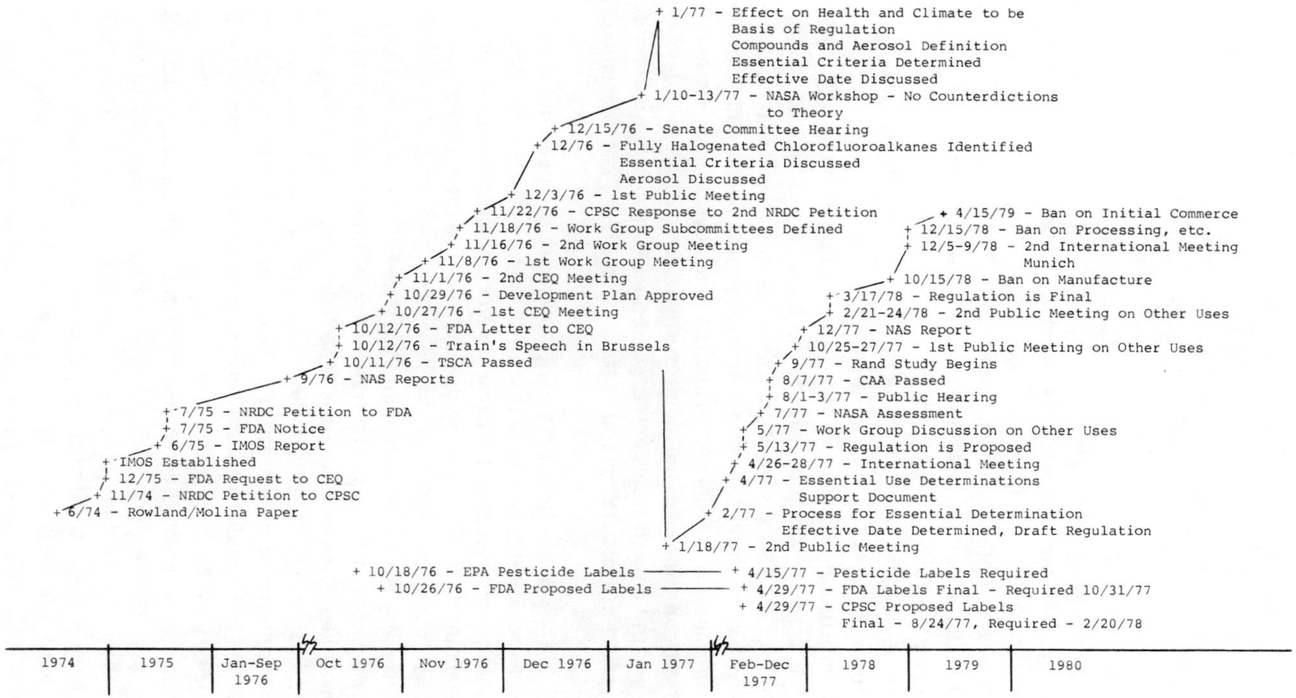

FIGURE 1. Chart of key regulatory actions — chlorofluorocarbons.

believed that a sufficient number of substitutes were available as alternatives for most of the aerosol products.

Other approaches considered were CFC production limitations, use limitations, and recovery requirements. None of these approaches seemed the least bit practical. Recovery of CFC's after release from an aerosol can is impossible from a practical standpoint. Limitations on aerosol users in private homes is not only impractical from a political and social standpoint, but also it is nearly impossible to enforce. Production limitations, although not impossible to enforce, would be costly to administer; and there appeared to be no reasoning on which to divide up some production allocation. Canada did achieve a reduction in production of CFC's for aerosols through industrial subscription to voluntary reductions. The U.S. work group did discuss such an approach only in passing and dismissed the idea based on expected political objections in the U.S. to "industry/regulatory agency deals".

From a global perspective, the work group recognized that the rest of the world discharged about the same amount of CFC's through aerosols as did the U.S. — roughly 25% of world emissions. Rather than wait until the world regulates aerosols, the work group believed that the U.S. regulatory agency heads, and particularly Russell Train, desired to lead the CFC control effort and thereby attract other countries to follow the U.S. example of worldwide environmental protection. This belief was reinforced by Russell Train's initiative in scheduling an international meeting.

The availability of CFC substitutes for aerosols was a constant question during work group deliberations. The work group was concerned about the safety of hydrocarbons, but the trade press and the market demonstrated that hydrocarbons were already then used in many products in the home, and that many aerosol products contained sufficient water to suppress flammable hydrocarbon-propelled products. Hence, flammability hazard potential for the consumer was not significantly increased.

At this time, the work group believed that good arguments for regulatory relief for necessary consumer protection against flammability hazards could be considered for each specific case.

After consideration of the various control alternatives, the work group assumed that sufficient evidence existed to reasonably conclude that CFC's depleted the ozone layer, and that this ozone depletion would eventually cause deleterious health and environmental effects. The work group then proceeded to discuss in detail the problems associated with a ban regulation.

In considering a ban, three key issues were encountered. The first key issue was how to define nonessential aerosol uses of CFC's. The framework of how nonessential aerosols should be defined was identified immediately. The framework, the work group agreed, should be two definitions. First, define the compound or compounds to be regulated and, second, define aerosol propellant use. However, the actual definition of the compound(s) to be regulated and the actual definition of aerosol propellant use was not resolved until February 1977.

The second key issue was generated by the framework of the compound and aerosol propellant definitions. It was assumed that regulation of all CFC compounds, when used as aerosol propellants, would regulate some highly specialized uses of CFC's that had no substitutes and whose product utility was essential by the judgment of reasonable people. Therefore, some essential test, criteria, or standard would have to be developed to allow exceptions to the general definitions. Much argument was to follow in the next 2 months in attempts to resolve this issue.

The third key issue was when should the regulation take effect. Although the work group had assumed a date for proposal of the regulation, no guidance was given on when the regulation should actually become effective.

The answer to the first and third key issues — what to regulate and when — were sought in the NAS report. Already by the end of November of 1976, there was argument over what the NAS said about regulation and particularly when to regulate.

To resolve these and other problems, the work group was divided into three subcommittees for the purpose of reviewing the literature and recommending solutions.[13,16] The first subcommittee, called the Problem Definition Subcommittee, was to prepare the case and educate the full work group on the overall health and environmental problem including the ozone depletion rate and the then current uncertainty surrounding the effects of the chlorine nitrate molecule on the atmospheric chemistry. This subcommittee was chaired by Mr. Courtney Riordan of the EPA Office of Research and Development who was the project manager of EPA's Biological and Climatic Effects Research (BACER) program — an EPA managed interagency research program designed to determine the effects of ozone depletion. Other EPA, NAS, and NOAA work group members sat on this subcommittee.

The second subcommittee, called the Solutions Subcommittee, was to examine and solve the problems of essential use test methods, regulation enforcement methods, the need for labeling CFC aerosol substitutes for CFC aerosol products, imports and exports, and economic impacts. This subcommittee was chaired by Mr. John Butler of the EPA Office of Planning and Evaluation who was a member of the IMOS Committee and also the Project Officer for the CFC aerosol regulation economic impact study being performed by International Research and Technology Corp. (IR&T Report).[17] Besides the EPA members, the subcommittee contained members from the FDA, the CPSC, and the DOC.

The third subcommittee, called the EIS subcommittee, was to prepare the support document for the regulation (in the form of an EIS) and resolve the issues of aerosol definition, compounds to be regulated, pesticides to be regulated, if any, and the all-important question of when the regulation should take effect. This subcommittee was chaired by Mr. Perry Brunner of the EPA's new regulations group in the Office of Toxic Substances, and contained membership from the FDA, the CPSC, and the EPA's Office of Federal Activities.

In addition to the subcommittees as a source of information gathering, two public meetings were scheduled to hear from all interested parties. These meetings were held on December 3, 1976, and January 18, 1977, both in Washington, D.C.[18,19]

IV. THE REGULATION IS DEVELOPED

By the end of December 1976, the issue of what compounds to regulate was nearing resolution. The fully halogenated chlorofluoroalkanes were identified as the compounds to be regulated. These compounds, suggested by the FDA members, were recognized as the compounds with the most active calculated impact on ozone destruction. Although the NAS study had concentrated on CFC-11 and CFC-12, the members of the Work Group with scientific backgrounds in the subject argued that there was no basis to assume that any saturated hydrocarbon molecule fully substituted with chlorine and fluorine would have any less effect than CFC-11 or CFC-12 on the destruction of ozone. The work group did recognize that CFC's which were not fully halogenated would have less effect, but this lesser effect was not well-defined and therefore could not be regulated without further research on the reaction rates. The effects on ozone of brominated compounds also were discussed. Work group members stated that the effect of bromine on ozone could be greater, on an equal weight basis, than the chlorine effect. However, after the January 18, 1977 public meeting, bromine compounds were not chosen for regulation because sufficient knowledge of their reaction rates was not yet available. The subject of brominated compounds was identified for consideration in Phase II of CFC regulation.[16,19]

The second part of the definition of what was to be regulated was a tougher issue. There was general concurrence in the Work Group that CFC's were to be regulated when used to "push things" out of a sealed can. However, the work group did not want a definition of aerosol propellant that rested on the physical act of propelling since that would involve considering vapor pressures of CFC's at various temperatures during use. Also, a definition that included a technical definition for "aerosol" was viewed as equally cumbersome. The consensus was to develop a definition that identifies the purpose of the CFC, and then regulate the CFC when used for that purpose.[16]

By the December 3, 1976 public meeting, the work group was "of a mind" to regulate CFC's when sealed in an aerosol can. However, the public meeting identified two problems with this approach. First, it was pointed out that CFC-12 is sealed in a can for use in recharging refrigeration systems — a use not to be regulated in Phase I. Secondly, a very eloquent plea from the electronics industry was received on behalf of preserving the use of CFC-12 alone in an aerosol can for electronic circuit troubleshooting. The CFC-12 is used to chill electronic components to identify intermittent component failures. Banning this very small CFC emission use would have had tremendous economic impact on repair of electronic equipment — particularly on the TV repair industry as they would have been forced to return to the repair method of "part by part replacement" until the problem is eliminated. More than anything else, the testimony at the December 3rd public hearing changed the work group's approach to aerosol definition to a definition that identified an aerosol as a product whose use or purpose is to propel another and different product from a container.[20] Such a definition was suggested by the work group at the second public meeting on January 18, 1977,[21] and basically not challenged again — or at least not until after the regulation was final when there were inquiries from manufacturers as to whether CFC's without sufficient vapor pressure to propel on their own were, in fact, regulated. The Government's answer to this question has been an emphatic "yes" if the CFC's purpose is to impart a characteristic to the propelling substance.[22]

With the definition of what to regulate developed, the work group, in early January of 1977, turned its attention to defining essential aerosol products. The start of the development of an essentiality test began with the recognition that the availability of substitutes must play some role in determining essentiality. However, at this point, two major splits in the work group developed over the essentiality question. One side of the work group appeared to be arguing that essentiality is associated with societal needs for the product (a view mostly held by some EPA staff), and the other side arguing that we should avoid making determinations on societal needs (a view held strongly by DOC staff, CPSC staff, and some EPA staff. FDA staff probably held both views at various times).

The resolution of the essentiality test was that the availability of alternatives; the economic significance of the product including effects in the marketplace; the environmental and health impacts of the product, and its substitutes; and the effects on the quality of life resulting from no longer having the product available for use would all be considered simultaneously when determining if the product was essential. Both sides got their "licks" in this criteria, but the authors believe the side not wishing to eliminate products solely on the basis that government considered them frivolous prevailed. No product was denied solely on the basis of no impact on the quality of life.[16]

The above criteria were developed into a flow diagram test (see Figure 2 — Flow Diagram for Essentiality Determination) that conformed to all the criteria simultaneously. Each product requesting an essential use exemption was passed through this test in work group sessions. Votes were taken for each test determination and the work group's opinion was written and placed in the record.[23]

In actual practice, almost all votes on each test on each product were unanimous

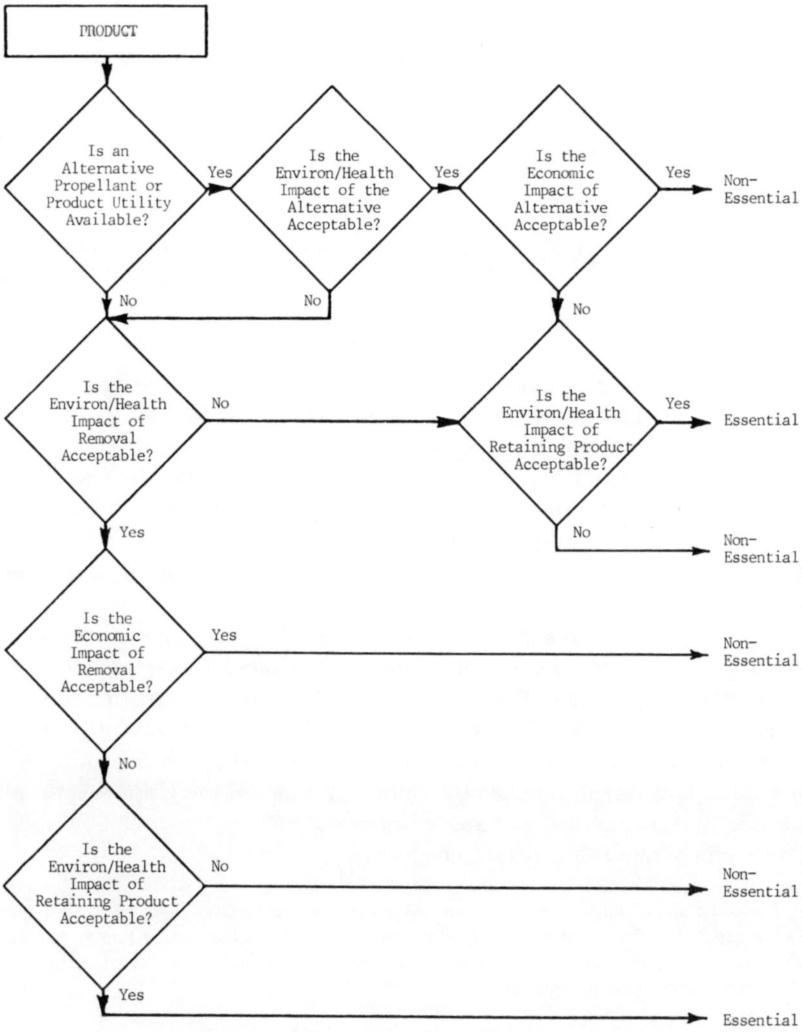

FIGURE 2. Flow diagram for essentiality determination

and the question of whether or not a product was frivolous never appeared visible or existent.

There was one very tough essentiality question not involving the criteria — that being the question of where and how the small business consideration, as required by TSCA, was to be implemented. This question remained an issue until final promulgation of the regulations and is discussed in this chapter in Section VI.

The final and most significant issue of the work group — the effective date of the regulation — involved the question whether or not there should be a regulation.

From the beginning of the regulatory process, argument was generated as to when or if the NAS said to regulate. The following two consecutive paragraphs, quoted from the Executive Summary of the NAS report,[7] "Halocarbons, Environmental Effects of Chlorofluorocarbons Release," demonstrate some of the ambiguity in the NAS position.

"CONCLUSIONS

Selective regulation of CFM uses and releases is almost certain to be necessary at some time and to some degree of completeness. Neither the needed timing nor the needed severity can be reasonably specified today. Costs of delay in decision are small, not more than a fraction of a percent change in ozone depletion for a couple of year's delay. Measurement programs now under way promise to reduce our uncertainties quite considerably in the near future.

RECOMMENDATIONS

Accordingly

1. As soon as the inadequacies in the bases of present calculations are significantly reduced, for which no more than 2 years need be allowed, and provided that ultimate ozone reductions of more than a few percent then remain a major possibility, we recommend undertaking selective regulation of the uses and releases of CFMs on the basis of ozone reduction."

As you can see in their conclusion, the NAS indicated that the timing cannot be "reasonably specified today." However, in their first recommendation, the NAS recommended selective regulation and the time of 2 years. Based on these and other statements in the NAS report, "sides lined up." One side said regulate in 2 years. Another side said wait 2 years to decide. Numerous other sides developed in between. No one in the Work Group, to the authors' recollection, including advocates for CFC's, argued not to regulate — but "when" was a key controversy during the development of the regulation.

The work group assumed the position that it would be fully in concert with the NAS recommendation if a regulation banning nonessential aerosol uses was not in effect before October of 1978. This position was further reinforced in December of 1977, by the statement of Dr. John W. Tukey, Chairman of the NAS Committee on Impacts of Stratospheric Change, before Senator Dale Bumpers at a Hearing on Chlorofluorocarbon Effects and Regulations of the Subcommittee on the Upper Atmosphere of the Committee on Aeronautical and Space Sciences.[24]

Tukey's closing paragraph in his statement was:

"In closing, I would emphasize that the committee did not recommend a fixed 2-year delay in regulation. It did indicate that, in its limited judgment, it would be wise to wait for some further knowledge, knowledge that would decrease the uncertainties of the second kind, but that it would not be wise to wait more than 2 years before making the decision to regulate."

Senator Bumpers opened his questions and the following discourse resulted:

"Senator Bumpers. Thank you, Dr. Tukey. On the last page of your testimony, you point out that the committee has "indicated that 'we wish to recommend against decision to regulate at this time.' "
Subsequent to your testimony this morning, the heads of EPA, FDA, and Consumer Product Safety Commission will testify as to the steps they are taking to regulate.
I assume that you are at least generally aware of some of the actions they have taken and some of the proposals they are making.
Do you disagree with that?
Dr. Tukey. I would call to your attention to the following quotation about half an inch further down on the last page, where the committee really disclaimed, you know, either the right or the privilege or the responsibility to decide about regulation.
Senator Bumpers. I notice where you say it is a political decision.
Dr. Tukey. Yes; right. The committee felt that from where they sat on the scientific evidence, rather than to recommend immediate regulation, they felt it was better to recommend a delay. I doubt if there were many, if any, members of the committee who would be uncomfortable with the sort of things that have been proposed by the regulatory agencies. Speaking for myself, I don't feel any great discomfort.
Now, I haven't seen details. If what is proposed by the regulatory agencies does not include a reasonable attention to escape hatches for specially important uses, then I would begin to feel uncomfortable."

Based on the above statement, the work group assumed its current activity was not in conflict with the NAS recommendation and then proceeded to argue the effective date issue with the assumption that it should not be before October 1978 if the regulation was to have the support of the NAS study.

In January of 1977, the work group problem definition subcommittee reported that they believed there existed sufficient information to establish health and environmental effects associated with ozone depletion. Members of the subcommittee presented data on a relationship between nonmelanoma skin cancer and increased UV radiation. In addition, the subcommittee also provided strong argument for basing the regulatory action on both statistically observed health effects and suspected but unquantified global climatic and nonhuman biological effects.

The case for regulation was strengthened in the eyes of the work group by a NASA workshop on CFC's on January 10th to the 13th, 1977, during which no contradictions to the ozone depletion theory were forthcoming.[25]

No more was the NAS recommendation on dates resolved in the eyes of the work group when the effective date became a minor point of controversy between the EPA and the FDA. The EPA position was that we should look to the economic impact study (the IR&T report) for guidance and select the date beyond which there are no increasing improvements in lessening the economic impact. This time period turned out, according to the economic study, to be 18 months after regulation is announced.[17] EPA argued that the regulation banning the manufacture of CFC's for aerosol products should be 18 months after the date of the proposed regulation — then scheduled for April of 1977. Such a timing would ban the manufacture of CFC's for aerosol products in October of 1978, and hence the processing of CFC's sometime after — preferably 90 days in order to use inventories existing at the time.

Since the FDA was going to ban the manufacture of food, drug, and cosmetic products, equivalent to the EPA ban on processing CFC's for aerosols, the FDA work group members were not in agreement to wait until 1979 for their regulation to go into effect — arguing that the FDA was not so constrained by economic considerations under its authority as the EPA was under TSCA.

This issue was "hammered away" in work group and bilateral meetings well into the nights of February 1977. The two sides may have walked away from each other in previous such disagreements, but in this case, work group members of the respective agencies felt some compelling reason to stay at the negotiating table as long as necessary in order to preserve the goal of promulgating a joint regulation. Where the idea of a joint regulation was created in the work group is not known by these authors. The authors' recollection is that it was "always" assumed we would publish regulations together, and that the regulations would be in complete agreement with each other.

Finally, the date was resolved by selecting October 15, 1978, as the date for the ban on the manufacture of CFC's for nonessential aerosol uses and December 15, 1978, as the date for the ban on the processing of CFC's for nonessential aerosol uses and the manufacture of nonessential CFC aerosol food, drug, and cosmetic products.

With this issue resolved, the regulation was drafted into a proposal and the EIS and other support documents were prepared in a "crash effort" in order to be reviewed by the Agency Heads and hence be available by the April 26th international meeting then drawing near.

We have not highlighted the science — the ozone depletion theory and resultant health and environmental effects — as a major issue in work group proceedings. This omission is not to imply that the theory was accepted without challenge by the work group. However, there were no major arguments among work group members that took issue with the degree of risk involved. The science was discussed at almost every work group meeting. Presentations on particular aspects of the ozone depletion theory or the health and environmental effects developed by Riordan and his Problem Definition Subcommittee were made at almost every work group meeting. On this task, Dr. Herbert Wiser of EPA, and Dr. D. S. Nachtwey of NASA, provided the work group with extensive reporting on the scientific information then being developed.

The work group's consensus after hearing all arguments on the ozone depletion theory and on the cancer incidence increase associated with UV radiation was that the risk of not controlling CFC emissions was too great in view of the generally acceptable economic impact associated with the conversion of CFC aerosols to substitutes. This finding was believed to be particularly true at the time for the risk associated with the least determinable effect hypothesized — climatic change.[16]

The aerosol-propelled pesticide issue was resolved by the work group attorneys. They concluded that CFC manufacture for nonfood, nondrug, or noncosmetic aerosols in general could be regulated under the TSCA because in manufacture no CFC can be identified as a pesticide.[26] However, an essential use exemption was granted for flying insect pesticides for use in nonresidential food handling areas except when applied by total release or metered valve devices and for space spraying of aircraft. The exclusion of the metered valve device as an essential use was made in the belief that automatic release tended to overuse pesticides by not tailoring the pesticide release amount and timing to the specific pest problem.[16] A suit on this issue was filed after the final regulation was promulgated, and the EPA granted the essential use exemption to metered valve devices. An argument for granting essential use status to metered valve devices was that such devices with dose and timing controls were found available in the marketplace. Also at issue was the effectiveness of a pesticide delivery system that the EPA believed was more properly resolved in proceedings under the FIFRA.[27]

Labels for CFC aerosols were considered a control mechanism for CFC emissions from the beginning of the regulatory action. The EPA pesticide program announced CFC content labeling requirements on October 18, 1976, and the FDA proposed a requirement for CFC warning labels on November 26, 1976. The FDA proposal was made final on April 29, 1977.[28]

On April 29, 1977, the CPSC proposed a labeling and reporting requirement under the Consumer Product Safety Act requiring manufacturers of such products to notify CPSC of the consumer products containing chlorofluorocarbon propellants and to notify consumers at the point of purchase by a warning label identical to the label proposed by the FDA. The CPSC proposal was made final on August 24, 1977.[29]

V. THE REGULATION IS PROPOSED

The international meeting of environmental regulators was held on April 26, 27, and 28, 1977, in Washington, D.C., with the CPSC, the EPA, the FDA, and the Department of State acting as co-hosts. The scientific and economic situation of CFC's and a draft of the proposed U.S. regulations were discussed. Canada made a presentation on its voluntary agreement with the Canadian aerosol industry. The response to the U.S. and Canadian positions varied from country to country. Non-CFC-producing countries indicated they might be prepared to follow either the U.S. or the Canadian approach. Most CFC-producing countries indicated that they would wait until more research confirmed the existence of the problem before making a decision on regulatory action. The main point of "the wait for more research" countries was that their legislative frameworks allowed them to proceed more rapidly to control CFC emissions than did the authorities granted the United States regulatory agencies. Therefore, if a "wait for more research" country decided to regulate aerosol uses in the future, the effective date of their regulation might be very close to the U.S. effective dates.[30]

Since that meeting, the Netherlands has a labeling requirement for CFC-propelled aerosol products. Sweden enacted legislation to ban, beginning July 1, 1979, most nonessential CFC aerosol products. Norway enacted similar legislation to be effective in 1980.[16]

On May 13, 1977, the EPA and the FDA published their proposed regulations prohibiting nonessential aerosol uses of CFC's. CPSC published a notice stating that the Commission had reviewed the proposed regulations and had determined that a specific ban by CPSC was unnecessary.[31]

The announcement of the joint regulations along with the international meeting provided early opportunity for the newly appointed Administrator of the EPA (Mr. Douglas Costle) and Commissioner of the FDA (Mr. Donald Kennedy) to meet each other and to meet the Chairman of the CPSC (Mr. John Byington). Probably before, but certainly at the press conference announcing the proposed regulations, the Agency Heads talked in private about continuance of joint and coordinated regulatory actions.[16] This discussion gave birth to the Interagency Regulatory Liaison Group (IRLG) created by the three Agency Heads along with Ms. Eula Bingham of the Occupational Safety and Health Administration (OSHA). The IRLG is in existence today and consists of the four Agency Heads along with various working committees to plan and review actions of mutual interest.

The publishing of the proposed regulations signaled the start of argument for and against the regulations. However, before the EPA public hearing on the proposed regulation, NASA added to the case for regulation by release of its report, "Chlorofluoromethanes and the Stratosphere". The NASA assessment represented a consensus of the scientists working on prediction of ozone depletion. The report generally reaffirmed the findings of the NAS through a workshop held in January of 1977 and then increased the ozone depletion estimate by a factor of two (10.8% to 16.5%) through a July 1977 workshop. The increase was based on calculations using the existing models and a new reaction rate for $HO_2 + NO \rightarrow OH + NO_2$.[32]

VI. THE REGULATION IS ARGUED

The public hearing was held on August 1st, 2nd, and 3rd, 1977, in Washington, D.C., to solicit comments on the proposed regulation. According to the rules of an informal hearing under TSCA, members of the work group constituted the hearing panel — including the FDA, CPSC, and DOC members.[33]

There were numerous presentations for additional essential use exemptions, but these arguments were limited to technical facts concerning the aerosol use of CFC's and substitutes.

A new argument was heard for the case of postponing the regulation until ozone depletion was confirmed. From the industrial interest, a mathematical analysis technique was presented that was claimed to be able to identify ozone depletion in 5 years when used to do trend analysis of observations of ozone levels. A postponement of the regulation was recommended until ozone depletion was observed.[34]

The work group's posthearing review of the postponement recommendation was not favorable. Against the recommendation was the fact that in the U.S., regulatory action under all existing legal authorities required procedures or arguments that would not allow instant action on CFC aerosols. In addition, quick drastic action would have far greater economic impacts than longer planned actions. Finally, outside of the U.S., control over the non-U.S. emissions could not be instantly obtained or predicted. Therefore, the work group decided to recommend the regulation dates basically as proposed.[16]

The final regulation was not promulgated until March 17, 1978, because of two items. The first item was the decision to wait for the NAS report to be published in December of 1977. Since the NAS had played a crucial role in the CFC issue, their latest findings were believed essential to the strength of the argument for regulation. The NAS report was being prepared in response to a Congressional mandate in Section 153(d) of the Clean Air Act Amendments of 1977 (CAA). The new NAS report, although stating a need for research before definite answers could be given, supported the argument for regulation by making the following points.

1. The estimated seriousness of chlorofluorocarbon emissions in terms of ozone reduction was roughly double that predicted in September 1976.
2. The relationship between increased UV radiation and harmful human health effects was strengthened.[35]

With the NAS report, the work group was ready to recommend final promulgation of the regulation. However, one item remained a tough issue and was not resolved until late January of 1978. The issue was what special, if any, consideration should be given to businesses with claims for essentiality where the businesses' primary claims are based on their small size. The TSCA directs EPA to consider impacts on small business in administering the law.

This issue emanated from one essential use exemption request from a diamond grit spray concern operated by one person on a part-time basis. Work group sympathy favored the exemption of one such tiny business, but the justification of the exemption generated opinions too numerous to count. Difficulty in making the decision was partly caused by an expectation of numerous requests for exemptions based only on small size of the business after the rule was final. The work group believed it was not directed by TSCA to grant exemptions only on the basis of small business.

The exemption was finally granted although an analysis of "why" does not yet appear describable — probably because of some meeting on the criteria, some consideration of the smallness of the business, the considerable perseverance of the owner of the business, and some fear that such small entrepreneurship in the U.S. cannot be "stepped-on" by big government.[16]

VII. THE REGULATION IS FINAL

The final CFC aerosol regulations were promulgated on March 17, 1978, by the EPA and the FDA with a statement of concurrence by the CPSC.[36]

The final EPA regulation prohibited:

1. The manufacture of all fully halogenated chlorofluoroalkanes for aerosol propellant use subject to the TSCA on October 15, 1978
2. The processing and distribution of fully halogenated chlorofluoroalkanes for aerosol propellant use subject to the TSCA on December 15, 1978
3. The processing for export of fully halogenated chlorofluoroalkanes for aerosol propellant use subject to the TSCA on December 15, 1978
4. The importation of fully halogenated chlorofluoroalkanes as a chemical substance or as part of an article for any aerosol propellant use subject to the TSCA on December 15, 1978

The final FDA regulation prohibited:

1. The manufacture or packaging of food, drug, device, or cosmetic products containing fully halogenated chlorofluoroalkanes as an aerosol propellant subject to the FFDCA on December 15, 1978

2. The initial introduction into interstate commerce of finished products subject to the FFDCA on April 15, 1979.

The EPA and the FDA regulations both exempted certain essential aerosol products. The EPA exempted Department of Defense products, electronic cleaners, aircraft maintenance products, mine warning devices, plastic mold release agent products, some pesticide products, and diamond grit spray. The FDA exempted certain metered dose drug products used for oral inhalation and contraceptive foam products.

With promulgation of the CFC aerosol regulations, the first phase of chlorofluorocarbon regulations was complete.

VIII. BIGGER PROBLEMS

The first discussion by the work group of other nonaerosol uses of CFC's took place in May of 1977 immediately following the completion of the proposed regulation on aerosols.[13] Since there was no economic impact study underway for other uses, the first order of business was to plan and initiate a contract for study of controls on other uses of CFC's.

The Rand Corporation of Santa Monica, California, was selected as the contractor to study the economic impact of regulating other uses of CFC's. The study, supported by the EPA, the FDA, and the CPSC, was designed for two steps. First, Rand was to collect data and provide the most definitive report to date of the production, uses, emissions, and quantities associated with each activity on other uses of CFC's. Second, and with input from the work group, Rand was to analyze the economic impact of various control alternatives for reducing emissions from other uses of CFC's. The word "control" was used at this time as opposed to the word "regulation" because it was recognized that conventional government regulation may not be the best form of reducing CFC emissions from other uses.

Particularly, the Planning and Evaluation Office staff in EPA was interested in examining the viability of economic incentive-type controls such as marketable permits — a permit sold by auction for permission to produce CFC's. Marketable permits hopefully would limit CFC use to the highest value use and eliminate CFC use in areas where economics favored the use of other substances. Rand began work on their contract in September 1977.[16]

On August 7, 1977, the President signed into law the Clean Air Amendments of 1977 (CAA) and hence changed the existing division of legal authorities over CFC's.[37]

Part B of the CAA directs the Administrator of the EPA to understand the effects of human actions on the stratosphere by directing and supporting research and monitoring studies undertaken throughout the Government, the universities, and the private sector. The Administrator of EPA is required to report to Congress on specific dates on the results of research findings, the progress of international cooperation on protecting the stratosphere, and recommendations for additional research.

Part B of CAA also authorizes the Administrator of EPA to regulate substances, practices, processes, and activities to prevent or abate any risk to the stratosphere. The Administrator is required to report to Congress, on specific dates, all regulatory activity by EPA and all other Federal agencies.

The CAA specifically stated that these new regulatory provisions were not to affect any proposed rule published under the TSCA prior to the enactment of the CAA nor any State law or regulation pertaining to control of halocarbons used as propellants in aerosol sprays. However, the CAA strongly implies that any future action to protect the stratosphere should be taken under the provisions of the CAA, and State laws and regulations may not contain provisions that are not identical to Federal regulations.

In short, although not wholly preemptive, future control beyond CFC aerosol spray regulations are to be accomplished under the authority granted in the CAA.

In enacting the CAA provisions, Congress did follow closely the progress of the CFC aerosol regulatory action and did preserve the progress and future of the actions being taken by the three agencies.[16]

The CAA does appear to grant broader regulatory authority for control of stratospheric pollutants to the EPA than does the TSCA. The Administrator for EPA shall promulgate regulations under the CAA when in his "...judgment, any substance, practice, process, or activity may reasonably be anticipated to endanger public health or welfare,..." This regulatory criteria appears less restrictive than the "unreasonable risk of injury to health or the environment" criteria of Sections 6(a) and 6(c) of the TSCA. The Administrator of EPA is required to notify Congress when he promulgates rules under CAA for protection of the stratosphere.

With the CAA enacted, the work group faced the problem of how broad is its regulatory assignment. Does the work group continue to examine regulatory needs for CFC's only or does it expand its deliberations to assume the regulatory responsibilities of the CAA and consider all chemicals harmful to the stratosphere? Since EPA chaired the work group, and EPA has the responsibility for administering the CAA, the logical choice was for the CFC work group to expand its study of needed regulatory actions to cover all chemicals harmful to the stratosphere. However, there was an organizational problem to this logical choice. The EPA staff of the work group were from the resources of the TSCA. Several meetings were required in EPA before agreement was reached that the CFC work group, and more particularly, the TSCA staff of the work group, should continue to act as the group to initiate, recommend, and report regulatory action authorized by the CAA. In addition, a policy on air pollution control regulation recommending conversion to the use of certain solvents for photochemical smog control versus future possible regulation of the recommended solvent for ozone depletion protection had to be resolved in order to avoid conflicting policies from the two programs. Agreement to these problems posed by passage of the Clean Air Act Amendments was finally resolved by the spring of 1978.[16]

The work group continues today with the broad charge to recommend actions as directed by the CAA. Full participation of the non-EPA members continues. No suggestion nor any issue has been forthcoming concerning the participation of non-EPA agencies in an activity where all future regulatory authority for control of stratospheric pollutants probably rests within EPA.

In October of 1977, the first public meeting to consider control of other uses of CFC's was held in Washington, D.C. The three-day public meeting produced a surprising 129 witnesses and a large audience. Almost all of the witnesses represented certain other uses of CFC's and all told of difficulties and impossibilities in attempting to convert to substitutes. The work group received "the message" that other uses of CFC's involve nearly a dozen completely different businesses and uses of CFC's, and that most of these uses were without viable substitutes for essential products in our society — or at least essential enough that the loss of the products to producers and consumers would result in a significant change in lifestyle.[38]

From testimony and submittals from the October 1977 public meeting, the work group was able to piece together its own crude analysis of the CFC other use categories, annual production rates, and annual emission rates. A crude estimate of the emission reduction possible if all conceivable technologic and economic controls were applied was also calculated. This analysis, along with "guessed" cost estimates of the controls were compared to the worldwide rough estimates of future production and hence emissions. This comparative analysis demonstrated to the work group that the reduction of aerosol emissions outside the U.S. would probably be far less costly to the global economy than control of other U.S. uses of CFC's. In short, in light of little control on CFC emissions in the rest of the world, regulation of other uses of CFC's in the U.S. could not be made into a viable argument by the summer of 1978 — the original development plan target date for some control on other uses of CFC's.[16]

Consequently, at the second public meeting on other uses of CFC's, held in February of 1978, the work group announced that the Agencies had decided that no regulation of other uses of CFC's would be forthcoming before the spring of 1979. This date corresponded with the completion of the Rand work which was seen as necessary before any detailed determination of the viability of any control action could be made.[39]

Regulatory action on CFC's ends here as of this writing (December 1978). At every* opportunity possible, the U.S. attempts to convince the representatives of the CFC-producing and consuming countries that reduction and elimination of CFC's in aerosols is a prudent and economically viable step to take for insurance against depletion of the ozone layer. This was the U.S. theme at the meeting in December of 1978 in Munich, Germany, where the international group of regulators, convened first in the U.S. in April of 1977, again gathered to discuss CFC's and ozone depletion.[40]

IX. PERSPECTIVE

Recognizing the biases of the authors, the regulation of CFC's produced a new type of regulatory action and some unusual public administration action. Two of these actions and why they came about are worth discussing.

The first action, and the most significant, was the joint development of regulations by a work group consisting of representatives from three regulatory agencies each with different legislative authorities and a fourth agency, DOC, basically with no authority over the problem. DOC played a strong role in developing the regulations and adroitly represented its constituent group. In the end, the joint coordinated government regulation was well-received by all interests and is believed to be a plus in public administration. How and why did it come about?

There is no one particular ground rule, authority, or person that can be identified with the cooperative spirit and result that emanated from the CFC regulatory effort. However, observations of other than normal conditions can be made and one can speculate that these conditions contributed to the result.

One condition is that the main part of the work group's activities took place between November of 1976 and February of 1977 — a period of transition in the Government from one political party to another. There were strong personalities that defined the ground rules at the CEQ meetings on initiation of the regulatory effort, but most of them had left their respective agencies by the time the regulation was proposed. The work group appears to have been nearly without political direction during most of its deliberations. However, the work group did have the direction of the senior scientific institution in the U.S. — the NAS. The work group believed the NAS was saying that it is time to start on regulatory action. We speculate that the NAS position was responsible for commitment to the action.

The explanation of the cooperative activity among the Federal Agencies could be explained by the membership of the Agency leaders on the work group. All were seasoned bureaucrats, roughly of the same age, years of experience in their particular agencies, and in total — highly sensitized to current public resentment against complex, costly, and particularly contradictory regulations from various agencies with overlapping authority over the same things. With their long-term career work under criticism,

* See Section X of this chapter.

the chief representatives of each individual agency sought a counteraction; and we believe this desire most often caused the cooperation among them on most issues. The arguments in the work group were mostly generated by differences of opinion within the same agency.

Also, short-term selfishness may have driven all to cooperation. During the main part of the CFC deliberations, there were no Agency Heads to impress because they were yet to be appointed. Under this condition, maybe each career civil servant saw the impression of himself being made to the new Agency Head only in the result achieved — that being a cooperatively developed coherent regulation. And, how could anyone criticize that!

What about commitment to the belief in the ozone depletion theory? As stated in the beginning, most work group members consisted of regulatory people, not the scientists that devoted years of their life to this problem.

The second action worth discussing is the fact that a health and environmentally protective regulation was promulgated on the basis of a scientific hypothesis. How was such relatively strong action taken short of full scientific proof? Work group cooperation could not have "sold" this regulation.

We believe three factors moved this regulation into being. First was the science. Albeit a theory, the ozone depletion theory is plausible, and over the 2 years of deliberations on the subject to February 1979, the theory has been reinforced by numerous checks on the data and additional research. In short, the theory stands up thus far, and industry, to its credit, basically admits it — and the NAS blessed it.

Second, and more importantly, we believe that a "grass roots consumer revolt" against the aerosol product was the biggest driving force. We believed that many consumers came to the belief by 1976 that 4 ounces of a product in a 12-ounce aerosol can could not be as "good" as 12 ounces of product in a pump or pour-out container. Of course, such a belief may ignore the effectiveness of the delivery system. However, when the aerosol was accused of damaging the environment on a global basis, the impression of "not getting the most for my money", combined with an accusation of "environmental damage", caused many consumers to "damn the aerosol" in the marketplace. In short, regulatory action only hastened what may have already taken place in the market to a lesser degree.

Finally, the third part of the scenario was the economics. We believe that there is good reason for industry and consumers to expect higher profits and/or lower prices in the long-term. The substitutes are cheaper. Therefore, after the industry got used to the idea and read a market prediction, it found its time was better spent converting than fighting.

All speculation, but some speculation on "why", is important for a regulation that impacted a greater than half-billion dollar industry, that was conceived on a hypothesis, and that was fought with less vigor than most other environmental control actions.

X. APPENDIX

Since the preparation of this chapter, the authors careers have led them away from this issue. Mr. Wirth, presently on a leave of absence from EPA, and the Editors concluded that an Appendix by the Editors would be appropriate.

The principal developments during 1979 and 1980 included the issuance of four reports in the U.S. and two in Europe. In the U.S. in 1979, the NAS Panel on Stratospheric Chemistry and Transport updated previous reports from 1976[7,8] and 1977[35] and reported calculations that long-term ozone depletion, if emissions continued at 1977 rates, would reach 16.5% with an uncertainty range of 5 to 28%.[41] The NAS

Committee on Impacts of Stratospheric Change reported at the same time on the effects on ozone depletion if it were to occur.[42] NAS stated that human skin cancer would increase, crop yields from several species would likely be reduced, larval forms of several important seafood species would suffer appreciable killing, and an average warming of the surface of the Earth by a few tenths of a degree Celsius would occur by the middle of the 21st century. These conclusions were stated in the key findings, with the relevant uncertainties discussed more fully in the body of the reports and in technical appendices.

Also the NAS Committee on Alternatives for the Reduction of Chlorofluorocarbon Emissions,[43] and separately the Rand Corporation,[44] reviewed the economic significance of CFC's in nonaerosol uses as well as the availability and limitation of alternative chemicals and processes. The reviews also compared the efficacy of regulation based on traditional mandatory controls (or technology-based emissions standards) with the efficacy of novel regulation based on economic incentives.

In 1979 and 1980 two European reports were issued. The United Kingdom Department of the Environment[45] updated its previous position paper and the Commission of the European Communities reported to the Council of Ministers.[46] The descriptions of the theory and the results of model calculations were similar to the NAS report. However, the key difference between the European reports and the U.S. reports hinged on the perceived validity of the calculated ozone depletion estimates and the perceived present need for regulation. The U.K. report concluded that the uncertainties were large and unquantifiable, and both European reports concluded that further regulation was not immediately necessary beyond actions already underway in the European Economic Community.

EPA, citing the NAS findings on calculated depletion and the effects that may result if such depletion occurred, issued an Advance Notice of Proposed Rulemaking (ANPR)[47] in October 1980. In the ANPR EPA reviewed the 1979 NAS assessments of the ozone depletion theory[41] and the effects of such depletion.[42] In discussing the need for regulation, EPA reviewed the international and domestic implications of three strategies: (1) wait-and-see, (2) no-growth, and (3) substantial emissions reductions. EPA concluded that a "no-growth" strategy was advisable in the short-term, but that the only acceptable long-term strategy would be "substantial emissions reductions." The ANPR then compared approaches to implementing a "no-growth" strategy, describing six options:

Mandatory Controls Approach	
Option I:	Technology-based control
Option II:	Selective product bans
Option III:	Use ceiling combined with a significant new use rule (SNUR)
Economic Incentives Approach	
Option IV:	Direct allocation [of CFC production rights] to manufacturers
Option V:	Direct allocation of CFC [production or use] permits to users
Option VI:	Government auction of [CFC production or use] permits

The ANPR stated that EPA "believes the [economic incentives] approach to be a more efficient method of reducing the environmental and human health risk."

EPA concluded the ANPR by soliciting comments on the validity of the ozone depletion theory, on the risks of continued world emissions of CFC's on the strategies for CFC control, on international aspects of regulation, and on approaches to achieving a production ceiling domestically.

A total of over 2000 written comments were received by EPA in response to the ANPR, mostly from affected industries. Many of the comments contained in-depth analyses and updates on the science, world use and emissions, and CFC alternatives, as well as detailed discussion of the various regulatory options.

At the same time it was becoming apparent that CFC use was not growing worldwide at its historic (pre-1975) rate and the ozone depletion calculations, updated for new scientific information, were being revised downward. A change in Administration was also occurring.

The result of these collective influences was that, in the spring of 1981, EPA postponed the issuance of any proposed CFC regulation until at least December 1981.[48]

REFERENCES

1. Climatic Impact Committee, Environmental Impact of Stratospheric Flight: Biological and Climatic Effects of Aircraft Emissions in the Stratosphere, National Academy of Science, Washington, D.C., 1975.
2. **Molina, M. J. and Rowland, F. S.**, Stratospheric sink for chlorofluoromethanes: chlorine atom catalyzed destruction of ozone, *Nature (London)*, 249, 810, 1974.
3. **Butts, S. D.**, Letter from the former Assistant Secretary for Consumer Product Safety Commission to Thomas B. Stoel, Jr., Natural Resources Defense Council, Washington, D.C., August 27, 1976.
4. Report of Federal Task Force on Inadvertent Modification of the Stratosphere (IMOS), Fluorocarbons and the Environment, Council on Environmental Quality and Federal Council for Science and Technology, G.P.O., June 1975.
5. Federal Register Notice; 40 F.R. 29914; July 16, 1975; Food, Drugs, and Cosmetics Containing Fluorocarbon-11, Fluorocarbon-12, and other Fluorocarbon Propellants.
6. **Schmidt, A. M.**, Letter from the Commissioner of Food and Drug Administration to Thomas B. Stoel, Jr., Natural Resources Defense Council, Washington, D.C., September 8, 1975.
7. **Committee on Impacts of Stratospheric Change, Halocarbons: Environmental Effects of Chlorofluoromethane Release, National Academy of Sciences, National Research Council, Assembly of Mathematical and Physical Sciences, Washington, D.C., 1976.**
8. Panel on Atmospheric Chemistry, Halocarbons: Effects on Stratospheric Ozone, National Academy of Sciences, National Research Council, Assembly of Mathematical and Physical Sciences, Washington, D.C., 1976.
9. **Train, R. E.**, former Administrator, Environmental Protection Agency, Fluorocarbon Depletion of the Ozone, Speech before the NATO/CCMS Fall Plenary, Brussels, Belgium, October 12, 1976.
10. **Schmidt, A. M.**, Letter from the Commissioner of the Food and Drug Administration to Dr. Russell W. Peterson, Chairman, Council on Environmental Quality, October 12, 1976.
11. Federal Register Notice: 42 F.R. 9205; February 15, 1977; PR Notice 76-3, Chlorofluorocarbons used as Aerosol Propellants; PR Notice 77-1, Permissible Labeling Statement Pertaining to Chlorofluorocarbons in Aerosol Pesticide Products.
12. Consumer Product Safety Commission, Executive Session of the Commission; John Byington, Chairman; November 22, 1976.
13. **Wirth, G. F.**, Personal notes of Chairman of interagency work group on CFCs.
14. **Breidenbach, A. W.**, "Request for Approval of a Development Plan to Initiate the Regulatory Process to Phase Out the Nonessential Use of F-11 and F-12 Chlorofluorocarbons — DECISION MEMORANDUM and CHLOROFLUOROCARBONS DEVELOPMENT PLAN, U.S. EPA, Washington, D.C. 20460, October 29, 1976.
15. Official Record of Rulemaking, Fully Halogenated Chlorofluoroalkanes, CFC Work Group Meeting Material, Meeting Minutes — November 8th, 16th, 23rd and 30th, 1976, U.S. EPA.
16. **Brunner, W. and Wirth, G. F.**, Personal recollections of members of interagency work group on CFCs.
17. International Research and Technology Corporation, The Economic Impact of Potential Regulation of Chlorofluorocarbon-Propelled Aerosols, Final Report, IR&T, McLean, VA., April 1977.

18. Federal Register Notice; 41 F.R. 49514; November 9, 1976; Chlorofluorocarbons, Formation of Chlorofluorocarbon Working Group; Public Meeting; Solicitation of Comments.
19. Federal Register Notice; 43 F.R. 1066; January 5, 1977; Chlorofluorocarbons; Notice of Public Meeting; Solicitation of Comments.
20. Official Record of Rulemaking, Fully Halogenated Chlorofluorocarbons, Chlorofluorocarbon Work Group, Public Session, Stenographic Transcript of Hearings, December 3, 1976, U.S. EPA.
21. Official Record of Rulemaking, Fully Halogenated Chlorofluoroalkanes, Public Participation Meeting on Chlorofluorocarbons, Stenographic Transcript of Hearings, January 18, 1977, U.S. EPA.
22. Federal Register Notice; 43 F.R. 55241; November 27, 1978; Fully Halogenated Chlorofluorocarbons — Clarification of Final Rule.
23. Environmental Protection Agency, Essential Use Determinations — Revised Support Document to Final Regulation on Fully Halogenated Chlorofluoroalkanes, March 17, 1978.
24. Hearing before the Subcommittee on the Upper Atmosphere of the Committee on Aeronautical and Space Sciences, United States Senate; Chlorofluorocarbon Effects and Regulations, Second Session, Ninety-fourth Congress; December 15, 1976.
25. NASA, Goddard Space Flight Center Laboratory for Planetary Atmospheres, Chlorofluoromethane Assessment Workshop Report, March 1977.
26. Official Record of Rulemaking, Fully Halogenated Chlorofluoroalkanes, Denny, R. J., Memorandum from the Associate General Counsel to the Administrator, EPA; October 8, 1976.
27. Federal Register Notice; 43 F.R. 59500 December 21, 1978; Fully Halogenated Chlorofluoroalkanes; Final Rule amending pesticide essential use exemption.
28. Federal Register Notice; 42 F.R. 22017; April 29, 1977; Certain Fluorocarbon (Chorofluorocarbons) Propellants in Self-Pressurized Containers.
29. Federal Register Notice: 42 F.R. 42780: August 24, 1977; Self-Pressurized Consumer Products Containing Chlorofluorocarbons Propellants-Labeling and Data Submission Requirements.
30. Official Record of Rulemaking, Fully Halogenated Chlorofluorocarbons, Communications, Foreign Governments and United Nations, International Meeting on Regulation of Chlorofluorocarbons, Report of the Meeting, Washington, D.C., April 26-28, 1977.
31. Federal Register Notice; 42 F.R. 24536; May 13, 1977; Fully Halogenated Chlorofluoroalkanes; Proposed Prohibitions and Notice of CPSC Action.
32. NASA, Chlorofluoromethanes and the Stratosphere, NASA Reference Publication 1010, Greenbelt, Maryland, August 1977.
33. Federal Register Notice; 42 F.R. 61259; December 2, 1977; "Procedures for Rulemaking Under Section 6 of the Toxic Substances Control Act."
34. Official Record of Rulemaking, Fully Halogenated Chlorofluorocarbons, EPA Panel on Fully Halogenated Chlorofluoroalkanes, on proposed regulations banning nonessential uses of chlorofluorocarbons as aerosol propellants, Stenographic Transcript of Hearings, August 1-3, 1977, U.S. EPA.
35. Committee on the Impacts of Stratospheric Change National Academy of Sciences; Response to the Ozone Protection Sections of the Clean Air Act Amendments of 1977: an Interim Report; National Academy of Sciences, Washington, D.C., December 15, 1977.
36. Federal Register Notice; 43 F.R. 11318; March 17, 1978. Fully Halogenated Chlorofluoroalkanes.
37. Public Law 95-95. The Clean Air Act, as Amended; August 7, 1977.
38. Official Record of Rulemaking, Fully Halogenated Chlorofluorocarbons, Fully Halogenated Halocarbons, Non-Aerosol Propellant Uses — Public Meeting, Stenographic Transcript of Hearings, October 25-27, 1977, U.S. EPA.
39. Official Record of Rulemaking, Fully Halogenated Chlorofluorocarbons, Public Meeting on the Non-Aerosol Uses of Fully Halogenated Chlorofluoroalkanes (Chlorofluorocarbons), Stenographic Transcript of Hearings, February 21-24, 1978, U.S. EPA.
40. Blum, B., Deputy Administrator, EPA; Speech before the International Conference on Chlorofluoromethanes; Munich, Germany; December 6-8, 1978.
41. Panel on Stratospheric Chemistry and Transport, Stratospheric Ozone Depletion by Halocarbons: Chemistry and Transport, National Academy of Sciences, National Research Council, Assembly of Mathematical and Physical Sciences, Washington, D.C., 1979.
42. Committee on Impacts of Stratospheric Change, in Protection Against Depletion of Stratospheric Ozone by Chlorofluorocarbons, National Academy of Sciences, National Research Council, Assembly of Mathematical and Physical Sciences, Washington, D.C., 1979.
43. Committee on Alternatives for the Reduction of Chlorofluorocarbon Emissions, in Protection Against Depletion of Stratospheric Ozone by Chlorofluorocarbons, National Academy of Sciences, National Research Council, Commission on Sociotechnical Systems, Washington, D.C., 1979.
44. Palmer, A. R., Mooz, W. E., Quinn, T. E., and Wolf, K. A., Economic Implications of Regulating, Chlorofluorocarbon Emissions for Nonaerosol Applications, Rand Corporation, Santa Monica, Ca, June 1980. (Published by EPA as EPA-560/12-80-001, October 1980).

45. Department of the Environment, Chlorofluorocarbons and Their Effect on Stratospheric Ozone (Second Report), Pollution Paper 15, Her Majesty's Stationery Office, London, England, 1979.
46. Commission of the European Communities, Chlorofluorocarbons on the Environment, Presentation to the Council, Brussels, Belgium, 1980.
47. Federal Register Notice; 45 F.R. 66726; October 7, 1980; Ozone-Depleting Chlorofluorocarbons; Proposed Regulatory Restriction.
48. Federal Register Notice; 46 F.R. 23629; April 27, 1981; [EPA] Agenda of Regulations, Docket No. OPTS-62009, Proposed Production Restriction for Chlorofluorocarbons.

EPILOGUE

James E. Lovelock

Mankind stands accused of polluting the stratosphere and of causing damage to the ozone layer there. This is a serious charge, for ozone in the stratosphere is said to be crucial for the very existence of life on Earth.

You have heard the case for the prosecution, delivered with vigor and with forensic skill. You have also heard what seems, by contrast, a feeble and muted defense. Now we have to decide whether or not this charge is proven beyond reasonable doubt. The decision is not as easy as the lawyers would like to believe. The evidence, although drawn from scientific observations, is not of the quality usually expected from such a source, and there are signs of bias by selection. Ordinarily, an epilogue would draw the threads of agreement so as to reach a consensus, but this affair has been so one-sided and so adversary as to require not a sort of summary but a judicial summing up.

Let us go back to the beginning again, to the time when Berkner and Marshall[1] introduced their famous theory about the evolution of oxygen in the world atmosphere. Crucial to this theory was the assumption that before ozone was present in the air, the flux of lethal UV radiation prevented life from colonizing the surface — indeed, that life must have originated in water deep enough to shield against this harmful radiation. It was only after oxygen appeared and so allowed an ozone shield to develop that life grew upon the surface of the land and sea. Only then could it be abundant, and only then could oxygen increase in the atmosphere to its present level of 21%.

We owe an immense debt to Berkner and Marshall who were the first to change the conventional wisdom of science away from that cold and sterile view of the Earth as a mere chemical entity whose atmosphere and surface composition were a simple consequence of equilibrium chemical reactions, and whose environment was in passive radiation balance with the input of energy from the sun. In such a world, life played no active part but merely accepted an environment which just happened to be right for it. Berkner and Marshall, like Hutchinson before them, presented a world in which life was an active participant, but they were the first to establish this new concept in the wisdom of science.

Some details of their hypothesis we now know to be wrong, such as that oxygen was at times more abundant than it is now. But this is no discredit; the information needed to test such notions was not then available. On the other hand, the relatively minor postulate of the hypothesis, namely, that stratospheric ozone as a UV shield was a necessary component of the atmosphere for life to emerge on the surface, seems to have become established as if it were a proven fact. Whereas more probably it is one of those giant myths whose larvae feed on that gray fabric which forms part of the barrier between the disciplines of science — a myth like that of that so-called "normal" distribution of statistics. This had a long life with biologists believing it to be a proven theorem of mathematics and mathematicians believing it to be a well-established fact of observation in biology. Since neither spoke the arcane recondite language of the other, it was many years before sufficient information had been exchanged to reveal that the "normal" distribution was, in fact, abnormal, and rarely ever found in the natural world.

What, then, is our new evidence about the function of the ozone layer as a life-protecting UV shield? First of all, it is almost certain that during at least the first billion years of life's existence on Earth, the atmosphere was oxygen free; consequently, there could have been no ozone to shield the surface life from the UV radiation of the sun. Yet there is strong evidence from the fossil record that photosynthetic microorganisms grew and flourished at the surface in spite of their exposure to this radiation. Furthermore, Margulis[2] and colleagues have shown that present-day photosynthetic algae in their natural mat-like habitats are tolerant to UV intensities much greater than those to be expected if there were no ozone layer at all.

More direct evidence comes from a consideration of the contemporary Earth where life is exposed during the year to a daily flux ranging from 0.05 Wm^{-2} in the northern temperate regions to 0.2 Wm^{-2} in the tropics. While it is true that certain varieties of the human species (fair-skinned Celts, living in such places as Queensland, Australia) are more liable to skin damage than are the rest of us, there is no indication whatever that the natural ecosystems of such highly irradiated regions are in any way limited by UV radiation. Even with the sensitive among the humankind, the hazard is not so great as to deter their voluntary migration there, nor is it equivalent to that of driving a car.

There are many regions on Earth where life is limited by adverse natural environments. Ecosystems do not flourish in the cold, nor in deserts, nor in salt lakes. But in spite of the large range of UV fluxes encountered in the different parts of the Earth, there is nowhere anything corresponding to an UV desert. To be sure, numerous experiments made under artificial laboratory environments, using both selected microorganisms as well as animal and plant tissue, find UV-B to be harmful. But there are no experiments where natural ecosystems are exposed to the daily and seasonal rhythms of UV-B change as might take place with an ozone depletion. To extrapolate such laboratory findings to the real environment is almost as illogical as concluding that water is poisonous on the grounds that a man immersed in it does not long survive. Or, if you prefer a closer analogy, many microorganisms, especially in laboratory cultures, are killed by visible light. It would be absurd to argue from this that life was endangered by the visible light from the sun.

Maybe in the tropics of an ozone-free Earth there could be UV deserts. It seems, however, very unlikely that these would persist for long. It is the nature of life to be a highly inventive and adaptive system. It would find that protection against so weakly penetrating a radiation was easy to evolve — much easier, it would seem than the problems of coping with drought, cold, or salt.

None of the foregoing is intended as a justification for the passive acceptance of the erosion of the ozone layer. No one yet knows the consequences of a change, say 20%, whether an increase or a decrease, of ozone which persisted for many years. Its purpose is to act as a counter-irritant to the chronic inflammation over allegations of impending doom. These have been all too prominent in the case made by the prosecution. Berkner and Marshall set the scene, but it was Harold Johnston who first drew public attention to the ozone problem in his spirited attempt to warn of the dangers of nitrogen oxide pollution of the stratosphere by supersonic transports. Although this action was successful only in slowing the development of the supersonic transport in the U.S. and later was recognized to be inappropriate at least for some forms of supersonic transport, it did serve to sensitize public opinion to the possible dangers of stratospheric pollution.

Strangely, in public affairs, repeated exposures to false alarms do not always go as did the fable of the boy who cried wolf. No single group can have cried wolf so often over imaginary rather than real dangers than have the environmentalists. Even so, there seems always to be a ready public hearing for anyone who cries doom, especially if the ultimate threat is somehow connected with cancer. It is normal in biological systems for the repeated exposure to a foreign protein to cause it to be accepted as a natural part of the environment. The less common abnormal response of allergy and hypersensitivity does occur, and so it seems parallels the public response to the problem of ozone depletion.

The principal participants in the case of man and ozone have been citizens of the U.S. and it is not surprising, therefore, that the hypersensitivity seems greatest in that country. You may wish to reject this notion on the grounds that it is only in the U.S. that there is an active forward view of the environment and its problems. By contrast, in Europe and elsewhere there is that alleged well-known cynical disregard of pollution and insensitivity towards global problems! Even were this so, it needs be recalled that sense and sensibility tend to be relative rather than absolute. Most people in Northern Europe, for example, regard a summer day with a maximum temperature of 69°F as hot. Also, in the U.K. and France, whether justified or not, many believe that the environmental objections to supersonic transport were bogus, inspired by chauvinism in the U.S., and were specifically anti-Concorde. Consequently, instead of a sensitization towards the idea of ozone depletion, the very reverse has occurred in these countries.

Perhaps the most well-developed example of this hypersensitivity was among the scientific community, and where that other source of fear and dread, nuclear warfare, acted as the reinforcer. In 1975, the National Academy of Sciences[3] published a report on the global consequences of nuclear warfare involving 10,000 Mton of nuclear explosive. It was concluded that the dangers from radioactivity to life, and to the natural and the manmade ecosystems of the world outside the participating countries, would be small. It was also said that these dangers would not persist for long after the war was over.

The report drew much criticism, mainly on the grounds that it was an irresponsible encouragement to the bomb-happy among the military to let fly. Now none of this is our present concern — what is, is the remarkable conclusion of the report that probably the greatest environmental danger would come from the depletion of stratospheric ozone by the nitrogen oxides injected into the stratosphere by the nuclear explosions. When all of the unimagined and incalculable possible horrors are added to the expected devastation of a major nuclear war, it surely indicates the existence of a morbid hypersensitivity for a majority among 56 scientists to conclude that ozone depletion was the greatest of them. With hindsight, we now know that nitrogen oxides are less destructive of ozone than believed then, and that an examination of the records of ozone observations shows no significant depletion following nuclear test explosions. But the preference for prejudice over evidence persists. In such a public environment where stratospheric ozone is a sacred thing, the chlorofluorocarbons never stood a chance; the moment they were suspected, and later in theory, at least, proven to be ozone destroyers, their fate was certain. To an alien observer, it seems as if in the U.S. even holy water would be banned if it were suspected of depleting ozone.

The conduct of the issue about man and stratospheric ozone as a trial rather than a debate has made the courtroom, rather than the laboratory or lecture hall, the venue. Anyone who attended the gathering at Logan, Utah, in 1976 in anticipation of a scientific meeting will recall how the lecture halls were co-opted as extra-mural courtrooms.

The scientific and forensic approaches to the use of evidence have developed in their divergent ways and are continually evolving to meet contemporary needs. The leisurely process of model building followed by destructive testing by experiment would be inappropriate in human affairs. In the same way, the need in law to reach a just settlement in disputes or to decide between innocence and guilt requires a selection and weighting of evidence in a way utterly wrong in science. To those lawyers and legislators who found scientists at Logan to be timid or vacillating, I would say that at least one scientist there found the legislators and their executioners to be like men crossing a frozen lake with no inkling of the thinness of the ice.

At Logan, the prosecution was in full cry for execution. In one encounter, Dr. Kaufman, who represented the panel of the National Academy of Sciences concerned with halocarbons and ozone, was attacked by Commissioner Pittle for failing to provide a simple yes or no answer to the complex question raised. In certain ways, it recalled that other encounter long ago between Galileo and the authority of his time. In such a court, it was easy to see why the defense kept so low a profile, the best to be hoped for was a partial execution, even if the client was innocent.

There are many other reasonable doubts concerning the consequences of chlorofluorocarbon accumulation which might have led, in more reasonable times, to the consideration of their regulation rather than anything so drastic as their banning. Perhaps, to bring things in perspective, consider what might have happened if the Molina/Rowland hypothesis had been introduced in 1965 instead of 1975. Would not the same arguments have been almost as valid then as they are now? What might we have lost in the 15 years of delay, if instead of banning them we had allowed their emissions to go uncontrolled until today? Ozone still, if anything, is rising in density in the stratosphere, not falling, and there are no signs of any adverse effects attributable to chlorofluorocarbons in the atmosphere. One thing that might well have happened was the replacement of chlorofluorocarbons as aerosol propellants by vinyl chloride. It was not then known to be a carcinogen and was regarded as safe and suitable. Indeed, many aerosol cans with vinyl chloride as the propellant gas were manufactured and were widely used in several countries, including the U.S.

The Achilles' heel of the chlorofluorocarbons is not their capacity to convey chlorine to the stratosphere nor to absorb infrared radiation. It is their great chemical and physical stability which prevents their rapid removal from the air by natural scavenging processes. Although we are still uncertain about the possible existence of some weaker scavenging process, let us for the moment assume that they do accumulate and that the only process for their removal is by UV photolysis in the stratosphere. In such circumstances, it was inevitable that if exponential growth continued the chlorofluorocarbon release would lead to an atmospheric concentration at sometime where the cost of the damage of their presence exceeded the impressive benefits of their present use. Indeed, to follow this argument further, the ever-increasing continued use of even a totally inert gas would eventually become objectionable when its concentration was so great as to dilute the oxygen of the air. The question seemed to be: when do we start to regulate and how severely.

It is easy to see why the public sales of items such as plutonium, botulinus toxin, or even antibiotics are undesirable and need be banned by law. It seems almost an 'Alice in Wonderland' overkill to apply the same final abrupt 'off with his head' control to substances which present only a slight conceivable harm yet many positive benefits, as do the chlorofluorocarbons. Such arbitrary action might eventually bring the law into disrepute. For inevitably it will be compared with the inability to ban truly harmful processes even such as that of cigarette smoking.

It is a fact that as soon as the hypothetical dangers of the continued release of chlorofluorocarbons were known, their total production ceased to grow exponentially and has, since 1974, remained more or less constant. Society and industry in current times do seem to be responsive and give evidence in this incident at least of a capacity for self-regulation. The environmentalists argue that any release of an ozone destroyer will be followed at some time by further death from skin cancer. Such a statement is not simply naive — it is all too often made willfully to persuade and influence the gullible. No one knows in fact whether or not the continued release of chlorofluorocarbons at their present level can do harm, but we do know for certain that already the substitution of hydrocarbons for fluorocarbons as a propellant gas, as a result of environmental pressure, has killed and maimed by burning. Of all the industrial chemicals taken into the home, none are safer nor more benign than are the chlorofluorocarbons. This fact alone speaks for a more flexible and sensitive method of control than outright banning. Because of the emotional pressures, and even more because of the false sense of urgency that they generated, the scientists involved in scientific decision-making have tended to be less objective than they might otherwise have been. Results of experiments which best fitted models were selected against those which did not. Let us consider some of the evidence in the chlorofluorocarbon affair that may be of doubtful quality.

One of the key postulates of the Molina/Rowland hypothesis is that there are no tropospheric sinks for the chlorofluorocarbons, and that these substances release all of their chlorine in the stratosphere. Reasonable though this assumption may be for a working hypothesis, it will not do as part of a fully-fledged theory. We need experimental proof that there are no processes in the troposphere which remove a significant quantity of chlorine from the chlorofluorocarbons before they reach the stratosphere. It was noticed as long ago as 1971 that carbon tetrachloride in the atmosphere was less abundant in the air coming from the Sahara Desert, and this led to the suggestion that maybe there was some process involving adsorption on desert dust particles which led to its removal from the atmosphere. Subsequently, laboratory experiments independently made by Ausloos and by Korte and his colleagues both found that chlorofluorocarbons adsorbed on silicate dust particles are vulnerable to photolysis by solar radiation, even in the troposhere. Later, Alyea[4] and his colleagues showed by model calculations that such a process of adsorption and photolysis could form a significant sink for chlorofluorocarbons in the global atmosphere if it took place in the major desert areas of the world. In the real world, only two observations have been made — the one mentioned earlier and another over the Sahara in 1977 which found no depletion of the halocarbons. Single expeditions such as these are almost futile if we are seeking to confirm or deny the presence of long-term processes for chlorofluorocarbon removal. They are no substitute for patient long-term monitoring both in such regions and in the world at large. Such observations are now in course but it may be several years before the questions they ask are answered.

Another approach is to compare the quantity of chlorofluorocarbon known to have been emitted with that which is found in the atmosphere. There is a fair agreement between what is expected to be present and what is found, if it is assumed that all that happened was a dilution in the volume of the atmosphere. It was just such a "back of the envelope" calculation using the measurements taken on the RV Shackleton in 1971 which provided Molina and Rowland with the basic evidence for their hypothesis. But if a sink of 2% per year for chlorofluorocarbons existed, we could not have known it then and we still cannot know it now. We are limited by the accuracy of our analytical methods; yet if it did exist, it would halve the expected depletion of ozone in the case of F-11 and reduce it by three in the case of F-12.

So the thief may have taken only half the ozone, says the lawyer, but he is still guilty. Yes says the scientist, but this is only one small part of the whole argument. Be patient and listen to more.

Consider the chlorine budget of the whole atmosphere. Regardless of the nature of the compounds in which they reside, the sum total of the ratio of chlorine atoms to other atmospheric components should be constant at all levels up through the atmosphere. We find, in fact, that the proportion sometimes found in the stratosphere is about 10 ppb, whereas in the upper troposphere it is between 2 and 3 ppb, of which less than half comes from the chlorofluorocarbons. If these and other discrepancies concerning stratospheric chemistry are confirmed, they cast serious doubts on our present capacity to understand the stratosphere.

Then there are the natural cycles of chlorine in the atmosphere. It is only a few years since methyl chloride was first discovered to be present, and no long-term monitoring of this compound has yet been made or is even contemplated. It is very difficult to analyze it accurately, but casual measurements have indicated concentrations varying from 30 ppb to 0.6 ppb — a range of 50 to 1. Even at the lowest concentration, it is the greatest chlorine carrier of the atmosphere, and it is certainly no product of industry. It comes to some extent from natural sources such as the sea and from forest fires of natural origin, but mostly it comes from agricultural burning practices, particularly those associated with the third world agriculture. We cannot just dismiss the presence of methyl chloride as unimportant because it is known to be rapidly destroyed by reactions in the troposphere. We know that sometimes there are high concentrations of methyl chloride near the surface and also of chlorine in the upper stratosphere. Perhaps there are unusual but effective processes for the upward transfer of surface air in the tropics.

Even if it should in the end turn out that the Molina/Rowland hypothesis was accurately predictive, it would still not follow that the continued release of chlorofluorocarbons would deplete ozone. A recent and very much more developed model of the atmosphere, which includes the effects of CO_2, of atmospheric transport and of radiation, has been introduced by Tuck and his colleagues.[5] This model suggests that a possible direct consequence of the growth of carbon dioxide concentration in the world atmosphere is an increase in stratospheric ozone, and that this increase will be greatest in the summer months and in those regions of the Earth most exposed to direct sunshine. It is conceivable that this effect could diminish or even reverse the trend of ozone depletion to be expected from the chlorine cycle. This model is the first put forward so far which offers an explanation of the upward trend in ozone density in spite of the ever-increasing burden of depleters. It deserves more attention than it has received.

So farewell, then, F-11 and F-12. We are sad to see you go, but your trial has been a great and spectacular show, and rightly so, for you are otherwise the most innocent of all chemicals which enter our homes, pure and free of noxious qualities. So were those virgins once sacrificed to other gods said to live in other upper atmospheric spheres. Your birth also was lowly and industrial; your parents, like theirs, seem to bear your fate with patience and with dignity.

Now that we can look back upon it, the banning of the CFM propellants on the basis of the ozone depletion theory in four short years of its introduction is a momentous event in man's history. It is the first time that action has been taken to restrict the freedom of our species on a global scale on the basis of a theory of science. Yet, it is important to recognize that if the ozone layer is indeed threatened, then the real culprits are over-population and the over-indulgent use of energy by mankind. If there were only 500 million people living at the energy use levels of the 1920s, no significant global threats would exist from chlorofluorocarbons, fertilizers, or other real or imaginary threats to ozone in the stratosphere.

The need to sustain our current numbers with their ever-growing expectations of a better life raises much more serious possible threats to mankind and to life on Earth than does ozone depletion. We hardly need to be reminded that the four horsemen still wait to ride, and there is the possibility that the environmental changes consequent upon the ever-increasing atmospheric burden of CO_2 will at least at some time render fertile and populous sections of our planet a desert.

Consider, for example, the drought of the Sahel region of Africa. Whatever its cause may be, there is no question that a very minor climatic change on a global scale can be a scourge for those who bear its local brunt. Then there is what many consider to be the most serious of all threats — that of the systematic destruction of ecosystems, the obsessive deletion of all species on Earth other than livestock, crops, and people in a vain attempt to farm the Earth and Sea to support our vast and still increasing numbers. Consider that unusual view of the Earth from space, described and shown by Croft[6] *Scientific American.* In it the Earth at night is seen to be blazing with light not only from the cities of northern temperate regions, but equally from the fires of burning forests on the land of almost all of the tropical regions of the world. The flames of these will have consumed so much that the bulk of our forest will have been eliminated to form farm land or desert by the year 2000.

This is not to suggest that we should have turned aside from the ozone depletion issue merely because other potentially more serious problems are foreseen. A man who discovers that he is due to die of cancer in a year does not neglect his toothache on that account. Indeed our very impotence in the face of the greater dangers may be why so intense an effort has been expended on the comparatively minor problem of ozone depletion. Compared with the potentially mortal environmental problems I have listed, it is trivial. The reason why industry, governmental agencies, and academic scientists have for once worked in such a harmonious way is not an accident of history, but represents a commonly-felt human need to achieve positively, even though the end is comparatively unimportant in itself.

Maybe also it represents part of a learning process. For we need to know a great deal more before the much more difficult problems consequent upon such as the carbon dioxide increase can be tackled. The ozone depletion affair has been discussed almost exclusively among Western middle class professionals with their common cultural background. With the well-known competitiveness of their society and the conflicting interests involved, it was remarkable that so much agreement took place. But can we conceive a similar consensus reached among the numerous other political, tribal, and religious communities of the world who need so much more urgently to agree on what should be done about the depletion of the forests and the living creatures within them. Indeed, the depletion of the Biosphere itself.

REFERENCES

1. **Berkner, L. V. and Marshall, L. C.,** On the origin and rise of oxygen concentration in the earth's atmosphere, *J. Atmos. Sci.,* 22, 225, 1965.
2. **Margulis, L., Walker, J. C. G., and Rambler, M.,** Reassessment of roles of oxygen and ultraviolet light in precambrian evolution, *Nature (London),* 264, 620, 1976.
3. National Academy of Sciences Report, Long Term World-Wide Effects of Multiple Nuclear Weapons Detonation, Chairman Nier, A.O.C., N.A.S. NRC, Washington, D.C., 1975.
4. **Alyea, F. N., Cunnold, D. M. and Prinn, R. G.,** Meteorological constraints on tropospheric halocarbon and nitrous oxide destruction by siliceous land surfaces, *Atmos. Environ.,* 12, 1009, 1978.
5. **Groves, K. S., Mattingly, S. R. and Tuck, A. F.,** Increased carbon dioxide and stratospheric ozone, *Nature (London),* 273, 711, 1978.
6. **Croft, T. A.,** Nightime images of the Earth from space, *Sci. Am.,* 239, 68, 1978.

ACRONYMS

Used in this and related fields. Excluded are chemical and mathematical symbols and customary abbreviations, for instance, of countries and states.

1-D, 2-D, 3-D	- One, Two, and Three-Dimensional. Used to describe computer models of the atmosphere. 1-D Vertical (and temporal); 2-D Vertical, N-S (and temporal); and 3-D Vertical, N-S, E-W (and temporal)
AAS	- Australian Academy of Science
AE	- Atmospheric Explorer (satellite, identified by final letter, as AE-C)
ARO	- Army Research Office
ARPA	- Advanced Research Projects Agency (Department of Defense)
AURA	- Association of Universities for Research in Astronomy, Inc.
BACER	- Biological and Climatic Effects Research Program (Multi-Agency, EPA lead)
BOD	- Biological Oxygen Demand — measure of biologically degradable material present in waste water
BUV	- Backscattered Ultraviolet (Radiation) — method for atmospheric measurement of species
CAA (CAAA)	- Clean Air Act (Amendments)
CAB	- Civil Aeronautics Board
CAO	- Committee on Atmosphere and Oceans (subunit of FCCSET)
CCMS	- Committee on Challenges of Modern Society (North Atlantic Treaty Organization)
CCOL	- Coordinating Committee on the Ozone Layer (subunit of UNEP)
CEQ	- (Federal) Council on Environmental Quality
CFC	- Chlorofluorocarbon (includes chlorofluoromethanes, q.v., and also F-113, F-114, F-115 and the hydrogen-containing F-22)
CFM	- Chlorofluoromethane (F-11 and F-12 in practice)
CIAP	- Climatic Impact Assessment Program (related to the supersonic transport development)
CIC	- Climatic Impact Committee (of the National Academy of Sciences)
CIEA	- Committee on International Economic Affairs (State Department)
CISC	- Committee on Impacts of Stratospheric Change (of the National Academy of Sciences)
CMA	- Chemical Manufacturers Association (recently changed from MCA — Manufacturing Chemists Association)
COAS	- (Industry) Council on Atmospheric Sciences
COMESA	- Committee on Meteorological Effects of Stratospheric Aviation (U.K.)
COVOS	- Committee on the Consequences of Stratospheric Flight (France, title translated)
CPSC	- Consumer Product Safety Commission
CSIRO	- (Australian) Commonwealth Scientific and Industrial Research Organization
DMSP	- Defense Meteorological Satellite Program (Department of Defense)
DNA	- Defense Nuclear Agency (Department of Defense) also Deoxyribonucleic Acid

DOC	- Department of Commerce
DOD	- Department of Defense
DOE	- Department of Energy, *also* (U.K.) Department of the Environment
DOT	- Department of Transportation
DU	- Dobson Unit (ozone column in milliatmosphere - cm or cm $\times 10^{-3}$ at STP)
ECC	- Electrochemical concentration cell (ozone measurement technique)
EDS	- Environmental Data Service (NOAA)
EEC	- European Economic Community
EIS	- Environmental Impact Statement
EPA	- Environmental Protection Agency
ERDA	- Energy Research and Development Administration
ERL	- Environmental Research Laboratories (NOAA)
EROS	- Earth-Resources Observation System
ESR	- Electron Spin Resonance (Spectroscopy) (analytical technique used to measure radicals)
F-	- Numerical system to identify chlorofluorocarbons (see FC-)
FAA	- Federal Aviation Administration (Department of Transportation)
FAO	- Food and Agricultural Organization (United Nations)
FC-	- Numerical system to identify fluorocarbons

FC-11 CCl_3F
FC-12 CCl_2F_2
FC-21 $CHCl_2F$
FC-22 $CHClF_2$
FC-113 $CCl_2F \cdot CClF_2$
FC-114 $(CClF_2)_2$
FC-115 $CClF_2 \cdot CF_3$

FCCSET	- Federal Coordinating Council for Science, Engineering and Technology
FCST	- Federal Council for Science and Technology
FDA	- Food and Drug Administration
FFDCA	- Federal Food, Drug, and Cosmetic Act
FGGE	- First GARP Global Experiment (NSF)
FIFRA	- Federal Insecticide, Fungicide, and Rodenticide Act
FPP	- Fluorocarbon Program Panel of the CMA (see TPFR)
FTC	- Federal Trade Commission
FY	- Fiscal Year (in the U.S. the fiscal year ends September 30)
GAMETAG	- Global Atmospheric Measurements Experiment on Tropospheric Aerosols and Gases (NSF)
GARP	- Global Atmospheric Research Program (NSF)
GCM	- General Circulation Model (of the atmosphere)
GCS	- Methodology developed to calculate Solar UV-B Irradiance, from the initials of the workers involved
GFDL	- Geophysical Fluid Dynamics Laboratory (Princeton, NJ)
GMCC	- Global Monitoring for Climatic Change (NOAA)
GOES	- Geostationary Operational Environmental Satellite
GSS	- See GCS
HALOE	- Halogen Occultation Experiment
HAPP	- High Altitude Pollution Program (Federal Aviation Administration)
HIRS	- High Resolution Infrared Sounder (satellite instrumentation)
IAG	- Interagency Agreement (e.g., between EPA and NSF)

ICAS	- Interdepartmental Committee on Atmospheric Sciences
ICSOP	- Interagency Committee for Stratospheric Ozone Protection
ICSU/SCOPE	- International Council of Scientific Unions/Scientific Committee on Problems of the Environment
IGY	- International Geophysical Year
IMOS	- (Federal Task Force on) Inadvertent Modification of the Stratosphere (or report of)
IMS	- International Magnetospheric Study (NASA)
IRIS	- Infrared interferometer spectrometer (satellite instrumentation)
IRLG	- Interagency Regulatory Liaison Group
IR&T	- International Research and Technology Corporation (Consultants)
ISOS	- International Southern Oceans Studies
JPL	- Jet Propulsion Laboratory
KPNO	- Kitt Peak National Observatory (NSF)
LIDAR	- Laser Infrared Detection and Ranging
LLAMB	- Long-Lived Atmospheric Monitoring Balloons (NSF)
MCA	- Manufacturing Chemists Association (recently changed to CMA — Chemical Manufacturers Association)
MSU	- Microwave Sounding Unit (satellite instrumentation)
NAS	- National Academy of Sciences
NASA	- National Aeronautics and Space Administration
NATO	- North Atlantic Treaty Organization
NBS	- National Bureau of Standards
NCAR	- National Center for Atmospheric Research
NCI	- National Cancer Institute
NEMS	- NIMBUS E Microwave Spectrometer (satellite instrumentation)
NEPA	- National Environmental Policy Act (of 1969)
NESS	- National Environmental Satellite Service (NOAA)
NIMBUS	- Not an acronym. Name of series of satellites. Satellites are labeled with a letter (e.g., TIROS-A) while being built, launched and checked-out in orbit, i.e., while under NASA control. Strictly research and development satellites keep the letter designation, remaining under NASA control. Satellite designations change to a numeral (e.g., TIROS-1) when the satellite is operational and when control is passed to NESS, National Environmental Satellite Service
NMC	- National Meterological Center (NOAA)
NOAA	- National Oceanic and Atmospheric Administration
NOS	- National Ocean Survey (NOAA)
NRDC	- Natural Resources Defense Council
NRL	- Naval Research Laboratory
NSBF	- National Scientific Balloon Facility (NSF)
NSF	- National Science Foundation
NWS	- National Weather Service (NOAA)
OECD	- Organization for Economic Cooperation and Development
OMB	- (Federal) Office of Management and Budget
OSHA	- Occupational Safety and Health Administration
OSTP	- Office of Science and Technology Policy
PMR	- Pressure Modulated Radiometer (satellite instrumentation)
POLYMODE	- MODE stands for Mid-Ocean Dynamics Experiment. When the study was made a joint study with the USSR, it was renamed POLYMODE

PPM, PPB, PPT	- Mixing ratio, normally by volume. Parts per million, ppm, 10^{-6}; parts per billion, ppb, 10^{-9}; and parts per trillion, ppt, 10^{-12}
RAF	- Radiation Amplification Factor relating the change in UV-B(BE) to decrease in ozone
R-B (meter)	- Robertson-Berger meter (spectroradiometric instrument)
RC	- Radiative-Convective, also 1-D RC, one dimensional RC. Describes a form of model to simulate the thermal characteristics of the atmosphere
RLD	- Replacement Limiting Dose — maximum UV-B dose which permits a population to survive
RNA	- Ribonucleic Acid
ROCSOZ	- Rocket Sounding for Ozone
SAGE	- Stratospheric Aerosol and Gas Experiment (satellite experiment)
SCAMS	- Scanning Microwave Spectrometer (satellite instrumentation)
SCEP	- Study of Critical Environmental Problems. (A study made in 1970 in connection with development of the Boeing SST)
SCR	- Selective Chopper Radiometer (satellite instrumentation)
SIRA	- Stratospheric Impact Research and Assessment (Multi-Agency — EPA lead, see also BACER)
SME	- Solar Mesosphere Explorer (satellite)
SMS	- Synchronous Meteorological Satellite (same as GOES, Geostationary Operational Environmental satellite
SNAFU	- Situation normal, all [fouled] up. A military expression describing recurrent chaos and confusion
SPO	- Sacramento Peak Observatory (NSF)
SST	- Supersonic Transport (aircraft)
SSU	- Stratospheric Sounding Unit (satellite instrumentation)
STPO	- Science and Technology Policy Office
STRAC	- (U.K.) Stratospheric Research Advisory Committee
SU	- Sunburn Unit, equivalent to 400 counts on an R-B Meter
TDRSS	- Tracking and Data Relay Satellite System
TIROS	- Television Infrared Observing Satellite (See also NIMBUS)
TOMS	- Total Ozone Mapping Spectrometer (satellite instrumentation)
TOVS	- TIROS Operational Vertical Sounder (satellite instrumentation)
TPFR	- Technical Panel on Fluorocarbon Research (of the MCA). Since June 1979 the Fluorocarbon Project Panel and later Fluorocarbon Panel (FPP) of the Chemical Manufacturers Association (CMA)
TSCA	- Toxic Substances Control Act
TSU	- Tropospheric Sounding Unit (satellite instrumentation)
UARS	- Upper Atmosphere Research Satellite
UNEP	- United Nations Environment Programme
UNESCA	- United Nations Department of Economic and Social Affairs
UNESCO	- United Nations Educational, Social and Cultural Organization
USDA	- United States Department of Agriculture
UV	- Ultraviolet radiation
UV-A, UV-B, UV-C	- Ultraviolet radiation Bands A, B and C. UV-A is UV of wavelength longer than 320 nm, i.e., longer than UV-B. UV-B is the portion of the UV spectrum which penetrates to the ground with wavelength <320 nm. Definitions vary between 280 to 320 nm and 290 to 320 nm for the band width. See Chapter 4 Volume II for detailed discussion, including terms such as UV-B (BE). UV-C is UV radiation of wavelength shorter than $\sim$ 290 nm. Such radiation does not penetrate through the earth's atmosphere

VTPR	- Vertical Temperature Profile Radiometer (satellite instrumentation)
WHO	- World Health Organization (United Nations)
WMO	- World Meterological Organization
WSU	- Washington State University

Index

INDEX

A

B

C

G

H

I

O

V

W

Z